Developments in Mathematics

Volume 67

Aims and Scope

The Developments in Mathematics (DEVM) book series is devoted to publishing well-written monographs within the broad spectrum of pure and applied mathematics. Ideally, each book should be self-contained and fairly comprehensive in treating a particular subject. Topics in the forefront of mathematical research that present new results and/or a unique and engaging approach with a potential relationship to other fields are most welcome. High-quality edited volumes conveying current state-of-the-art research will occasionally also be considered for publication. The DEVM series appeals to a variety of audiences including researchers, postdocs, and advanced graduate students.

More information about this series at http://www.springer.com/series/5834

Sergio Macías

Set Function $\mathcal{T}$

An Account on F. B. Jones' Contributions to Topology

Sergio Macías
Instituto de Matemáticas
National Autonomous University of Mexico
Ciudad de México, Mexico

ISSN 1389-2177 ISSN 2197-795X (electronic)
Developments in Mathematics
ISBN 978-3-030-65080-3 ISBN 978-3-030-65081-0 (eBook)
https://doi.org/10.1007/978-3-030-65081-0

Mathematics Subject Classification: 54B20, 54C60, 54F15

This Springer imprint is published by the registered company Springer Nature Switzerland AG
The registered company address is: Gewerbestrasse 11, 6330 Cham, Switzerland

Elsa:

¿Cómo poder expresarte
Todo lo que hay dentro de mí?
Tu forma de ser me indica
Que todavía hay gente hermosa.
Tu presencia me motiva a seguir
Luchando por lo que creo.
En este momento me gustaría,
además de ser lo que soy,
Ser el más grande poeta
Y así poder escribirte
Los versos más bellos
Que nunca nadie
En la tierra escribiera
Y decirte, de esa manera
Muy hermosa,
Todo lo que por ti siento…
Te doy gracias porque eres
Mi compañera.

S. M.

In Memoriam

Alfredo Macías

María Alvarez

James T. Rogers, Jr.

Sam B. Nadler, Jr.

Preface

This book is timely. The literature on this subject has become so extensive that it is an overwhelming task for a new researcher to become familiar with what is known. It is nice to see a book organized so elegantly, which will make the subject more readily accessible. There are also a lot of opportunity for further work on the subject, so this book is much more than a complete history of the topic. Such a volume as that could not be written now, as there are too many gaps in our knowledge.

I am not including any definitions in this Preface, as they can be found in the text itself. The notion of aposyndesis, introduced by F. B. Jones, dates back more than three quarters of a century, to the early 1940's. This concept is a useful tool for the study of continua, especially those of intermediate complexity, that is to say, neither locally connected nor indecomposable. The set-valued set function $\mathcal{T}$ arises from the analysis of fine detail of when and how a continuum may be, or fail to be, aposyndetic. Professor Macías has provided this book just when it is needed. There is a substantial amount of information available in the literature on the set function $\mathcal{T}$ now, and this book provides a coherent and organized presentation of it, along with the most comprehensive bibliography I have seen on the subject. Both current researchers and students will find it to be a helpful reference book. It is also written in such a way that it can be used as a basis for a seminar, or possibly even a single individual, working alone, wishing to learn about the subject from scratch.

Through a quirk of history, the set function $\mathcal{T}$, in its earliest form, was called $\mathcal{L}$ and was also introduced by Professor Jones. The notation was mostly changed from $\mathcal{L}$ to $\mathcal{T}$ in the 1950's and early 1960's. Jones also introduced a companion set function $\mathcal{K}$; the name of $\mathcal{K}$ has not changed. In conversation, at least, $\mathcal{T}$ and $\mathcal{K}$ have been referred to as dual functions or adjoint functions; I have encountered both terms, but never in print, I think. The function $\mathcal{T}$ has been the primary focus of research over the years; $\mathcal{K}$ typically has been studied only when its use sheds light on properties of $\mathcal{T}$. The present volume continues that convention; I hope that it may lead to more study of $\mathcal{K}$ in its own right.

The function $\mathcal{T}$ has been both a subject of research in its own right and, increasingly, a powerful tool for investigating problems that, at first glance, seem to have nothing to do with $\mathcal{T}$.

The first chapter provides general background in topology and specifically the theory of continua. The discussion of the function $\mathcal{T}$ begins in earnest in the second chapter. I am very glad to have had an opportunity to peruse this volume. I expect that others will also find it interesting and useful.

Professor Emeritus of Mathematical Sciences David P. Bellamy
The University of Delaware
Newark, Delaware, USA

Introduction

The main purpose of this book is to present Professor F. Burton Jones' set function $\mathcal{T}$. This topic was treated in [92, Chapters 3 and 5] for metric continua. Here, we present most of those results for Hausdorff continua throughout the book. Whenever we could not find the way to remove the metric hypothesis, the result is presented in its original form. We include new topics not contained in [92, Chapters 3 and 5], as we mention below. The set function $\mathcal{T}$ has been used by many authors, for example as a tool to prove results about the semigroup structure of continua [71, 72] and [62], about the existence of a metric continuum that cannot be mapped onto its cone [10] or to characterize spheres [34]. In fact, it seems that [72] is the first paper to use the notation $\mathcal{T}$ in print. Even though [71] appeared before [72], the latter was written before than the former.

The book has eight chapters. In Chapter 1, we present the basic knowledge needed for the rest of the book on continuous decompositions, topological products and inverse limits (including generalized inverse limits with a single bonding function), uniformities, (Hausdorff) continua, uniformly completely regular maps and hyperspaces. The part on decomposition, continua and hyperspaces is brought from [92, Sections 1.2, 1.7 and 1.8] with the appropriate changes to Hausdorff spaces.

Chapter 2 contains the main properties of the set function $\mathcal{T}$. We consider its symmetry, additivity, idempotency and finite powers of it. In Chapter 3, we consider decomposition theorems using $\mathcal{T}$. We present three general decomposition theorems. Also, we give a weak version of Professor F. Burton Jones' Aposyndetic Decomposition Theorem for homogeneous Hausdorff continua using only the property of Kelley and then present the full version for such continua using the uniform property of Effros. We use the set function $\mathcal{T}$ to prove Professor Janusz R. Prajs' Mutual Aposyndetic Decomposition Theorem for Hausdorff homogeneous continua using the uniform property of Effros and give some relationships between this and Jones' theorem. We include the statements of most of the results of [92, Section 5.3] to prove Professor James T. Rogers, Jr.'s Terminal Decomposition Theorem for metric homogeneous continua.

Chapter 4 is about $\mathcal{T}$-closed sets, a topic studied by several authors, for instance [46] and [122]. We present the main properties of this class of sets, a couple of characterizations of $\mathcal{T}$-closed sets: one for Hausdorff continua and the other one for metric continua. A necessary condition to be a $\mathcal{T}$-closed set is given, and we prove that this condition is also sufficient for the class of metric continua with the property of Kelley. We study the class of minimal $\mathcal{T}$-closed sets and the set function $\mathcal{T}^{\infty}$. We end with the concept of $\mathcal{T}$-growth bound and give an upper bound of the $\mathcal{T}$-growth bound for the classes of homogeneous Hausdorff continua with the property of Kelley and type λ continua. We present a relationship of the $\mathcal{T}$-growth bound of continua and monotone and open monotone maps.

Chapter 5 deals with the continuity of the set function $\mathcal{T}$. The main difference between the results presented in Section 5.1 and the ones in [92, Section 3.3] is that new results appeared after the publication of [92], and these allow us to remove the hypothesis of *point $\mathcal{T}$-symmetry* of the continuum in many of the theorems. We present classes of metric continua for which $\mathcal{T}$ is continuous. In addition, we include results about the continuity of $\mathcal{T}$ on continua. In particular, we show the equivalence of the continuity of $\mathcal{T}$ and the continuity of its restriction to continua for the product of two continua. We end this chapter proving that the continuity of the set function $\mathcal{T}$ implies the continuity of Professor Jones' set function $\mathcal{K}$.

In Chapter 6, we study the images under the set function $\mathcal{T}$ of the hyperspace of closed subsets, 2^X, and the hyperspace of singletons, known as the first symmetric product, $\mathcal{F}_1(X)$, of a metric continuum X. We show that $\mathcal{T}(2^X)$ is an analytic set. We consider finite and countable images of $\mathcal{T}$. We also study connected and compact images of $\mathcal{T}$.

Chapter 7 contains applications of the set function $\mathcal{T}$. We study continuously irreducible continua and their hyperspace of subcontinua. Also, we characterize continuously type A' metric θ-continua. We present sufficient conditions for continua to be not contractible. We prove that arc-smooth metric continua are strict point $\mathcal{T}$-asymmetric and the reverse implication is true for fans. We consider R-subcontinua in dendroids. We present two characterizations of local connectedness. We give sufficient conditions for the image under $\mathcal{T}$ of a nonempty closed subset of a metric continuum to be a shore set. We consider generalized inverse limits of nonaposyndetic homogeneous metric continua X using the set function $\mathcal{T}|_{\mathcal{F}_1(X)}$ as a bonding function. We end the chapter giving more relationships between the set functions $\mathcal{T}$ and $\mathcal{K}$.

Chapter 8 contains the questions from [92, Section 9.2]. We give an update on the advances in answering those questions and include new open ones.

I thank Javier Camargo for letting me use some of the pictures he produced.

I thank Professor David P. Bellamy for the valuable suggestions he made to improve the book. I really appreciate the time and effort he spent reading the whole book in order to write a very nice Preface.

I thank my wife, Elsa, for all her love, help and support during the preparation of this manuscript.

I thank the people at Springer, especially Jan Holland, Robinson dos Santos, Elizabeth Loew, Christoph Baumann, Thomas Hempfling, Saveetha Balasundaram, Gomathi Mohanarangan and the Series Editors: Krishnaswami Alladi, Pham Huu Tiep, and Loring W. Tu, for all their help.

Ciudad de México, Mexico Sergio Macías

Contents

Chapter 1
Preliminaries

We gather some of the results of topology of Hausdorff spaces which are useful for the rest of the book. We assume the reader is familiar with the notion of topological space and its elementary properties. We present the proofs of most of the results; we give an appropriate reference otherwise.

The topics reviewed in this chapter are: Continuous decompositions, product topology, inverse limits, generalized inverse limits, uniformities, compacta, continua, generalized metric continua and hyperspaces.

1.1 Continuous Decompositions

We present a method to construct "new" spaces from "old" ones by "shrinking" certain subsets to points.

1.1.1 Definition. A *compactum* is a compact Hausdorff space. A *map* is a continuous function.

1.1.2 Definition. A *decomposition* of a topological Hausdorff X is a collection of nonempty, pairwise disjoint sets whose union is X. The decomposition is said to be *closed* if each of its element is a closed subset of X.

1.1.3 Definition. Let $\mathcal{G}$ be a decomposition of a Hausdorff topological space X. We define $X/\mathcal{G}$ as the set whose elements are the elements of the decomposition $\mathcal{G}$. $X/\mathcal{G}$ is called the *quotient space*. The function $q\colon X \twoheadrightarrow X/\mathcal{G}$, which sends each point x of X to the unique element G of $\mathcal{G}$ such that $x \in G$, is called the *quotient map*.

1.1.4 Remark. Given a decomposition of a Hausdorff topological space X, note that $q(x) = q(y)$ if and only if x and y belong to the same element of $\mathcal{G}$. We give a topology to $X/\mathcal{G}$ in such a way that the function q is continuous and it is the biggest with this property.

S. Macías, *Set Function $\mathcal{T}$*, Developments in Mathematics 67,
https://doi.org/10.1007/978-3-030-65081-0_1

1.1.5 Definition. Let X be a Hausdorff topological space, let $\mathcal{G}$ be a decomposition of X and let $q\colon X \twoheadrightarrow X/\mathcal{G}$ be the quotient map. Then the topology

$$\mathfrak{U} = \{\Gamma \subset X/\mathcal{G} \mid q^{-1}(\Gamma) \text{ is open in } X\}$$

is called the *quotient topology for* $X/\mathcal{G}$.

1.1.6 Remark. Let $\mathcal{G}$ be a decomposition of a Hausdorff topological space X, and let $q\colon X \twoheadrightarrow X/\mathcal{G}$ be the quotient map. Then a subset Γ of $X/\mathcal{G}$ is open (closed, respectively) if and only if $q^{-1}(\Gamma)$ is an open (closed, respectively) subset of X.

1.1.7 Definition. Let $f\colon X \twoheadrightarrow Y$ be a surjective map between Hausdorff topological spaces. Since f is a function, $\mathcal{G}_f = \{f^{-1}(y) \mid y \in Y\}$ is a decomposition of X. The function $\varphi_f\colon X/\mathcal{G}_f \twoheadrightarrow Y$ given by $\varphi_f(q(x)) = f(x)$ is of special interest. Note that φ_f is well defined; in fact, it is a bijection, see the diagram on Figure 1.1.

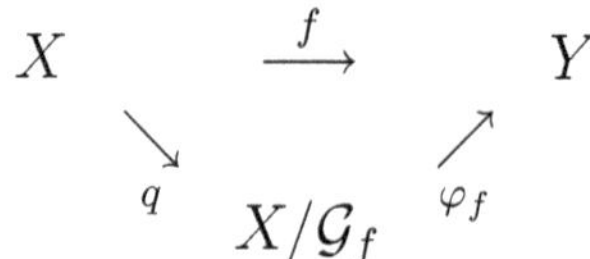

Fig. 1.1 The function φ_f

The next lemma is a special case of the Transgression Theorem [36, 3.2, p. 123].

1.1.8 Lemma. *Let $f\colon X \twoheadrightarrow Y$ be a surjective map between Hausdorff topological spaces. If $X/\mathcal{G}_f$ has the quotient topology, then the function φ_f is continuous.*

Proof. If U is an open subset of Y, then $\varphi_f^{-1}(U) = qf^{-1}(U)$. Since $q^{-1}\varphi_f^{-1}(U) = q^{-1}qf^{-1}(U) = f^{-1}(U)$ and $f^{-1}(U)$ is an open subset of X, we have, by the definition of quotient topology, that $\varphi_f^{-1}(U)$ is an open subset of $X/\mathcal{G}_f$. Therefore, φ_f is continuous. □

1.1.9 Example. Let $X = [0, 2\pi)$ and let $f\colon X \twoheadrightarrow \mathcal{S}^1$, where $\mathcal{S}^1$ is the unit circle, be given by $f(t) = \exp(t) = e^{it}$. Then f is a continuous bijection. Since $\mathcal{G}_f$ is, "essentially," X, it follows that $X/\mathcal{G}_f$ is homeomorphic to X. On the other hand, X is not homeomorphic to $\mathcal{S}^1$, since X is not compact and $\mathcal{S}^1$ is. Therefore, φ_f is not a homeomorphism.

1.1.10 Definition. A map $f\colon X \to Y$ between Hausdorff topological spaces is said to be *open* (*closed*) provided that for each open (closed) subset K of X, $f(K)$ is open (closed) in Y.

The following theorem gives sufficient conditions to ensure that φ_f is a homeomorphism:

1.1.11 Theorem. *Let $f\colon X \twoheadrightarrow Y$ be a surjective map between Hausdorff topological spaces. If f is open or closed, then the map $\varphi_f\colon X/\mathcal{G}_f \twoheadrightarrow Y$ is a homeomorphism.*

Proof. Suppose f is an open map. Since φ_f is a bijective map, it is enough to show that φ_f is open. Let Γ be an open subset of $X/\mathcal{G}_f$. Since $\varphi_f(\Gamma) = fq^{-1}(\Gamma)$, $\varphi_f(\Gamma)$ is an open subset of Y. Therefore, φ_f is an open map. The proof of the case when f is closed is similar. □

Decompositions are also used to construct the cone and suspension over a given space.

1.1.12 Definition. Let X be a Hausdorff topological space and let $\mathcal{G} = \{\{(x,t)\} \mid x \in X \text{ and } t \in [0,1)\} \cup \{(X \times \{1\})\}$. Then $\mathcal{G}$ is a decomposition of $X \times [0,1]$. The *cone over* X, denoted by $K(X)$, is the quotient space $(X \times [0,1])/\mathcal{G}$. The element $\{X \times \{1\}\}$ of $(X \times [0,1])/\mathcal{G}$ is called the *vertex* of the cone and it is denoted by ν_X (Figure 1.2).

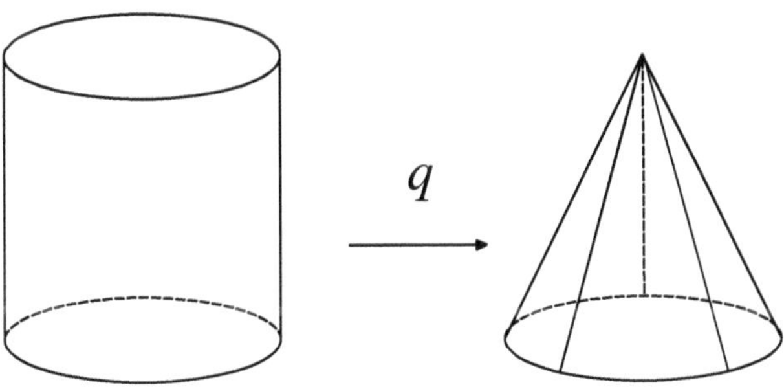

Fig. 1.2 Cone over a space

A proof of the following proposition may be found in [36, 5.2, p. 127].

1.1.13 Proposition. *Let $f\colon X \to Y$ be a map between Hausdorff topological spaces. Then f induces a map $K(f)\colon K(X) \to K(Y)$ given by*

$$K(f)(\omega) = \begin{cases} \nu_Y, & \text{if } \omega = \nu_X \in K(X); \\ (f(x),t), & \text{if } \omega = (x,t) \in K(X) \setminus \{\nu_X\}. \end{cases}$$

1.1.14 Definition. Let X be a Hausdorff topological space and let $\mathcal{G} = \{\{(x,t)\} \mid x \in X \text{ and } t \in (0,1)\} \cup \{(X \times \{0\}), (X \times \{1\})\}$. Then $\mathcal{G}$ is a decomposition of $X \times [0,1]$. The *suspension over* X, denoted by $\Sigma(X)$, is the quotient space $(X \times [0,1])/\mathcal{G}$. The elements $\{X \times \{0\}\}$ and $\{X \times \{1\}\}$ of $(X \times [0,1])/\mathcal{G}$ are called the *vertexes* of the suspension and are denoted by ν_X^- and ν_X^+, respectively (Figure 1.3).

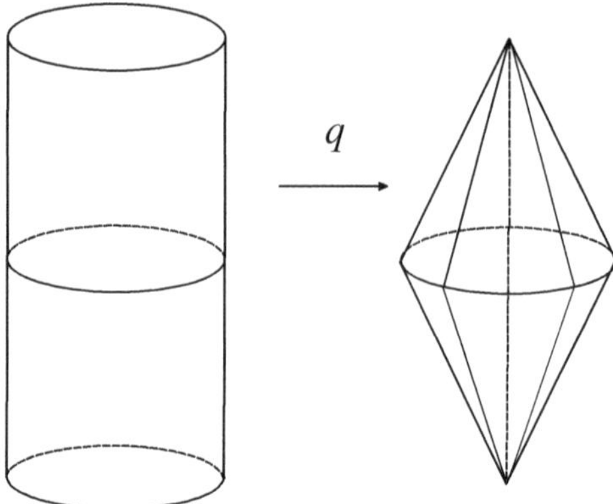

Fig. 1.3 Suspension over a space

1.1.15 Definition. Let X be a Hausdorff topological space and let $\mathcal{G}$ be a decomposition of X. We say that $\mathcal{G}$ is *upper semicontinuous* if for each $G \in \mathcal{G}$ and each open subset U of X such that $G \subset U$, there exists an open subset V of X such that $G \subset V$ and such that if $G' \in \mathcal{G}$ and $G' \cap V \neq \emptyset$, then $G' \subset U$. We say that $\mathcal{G}$ is *lower semicontinuous* provided that for each $G \in \mathcal{G}$ any two points x and y of G and each open subset U of X such that $x \in U$, there exists an open subset V of X such that $y \in V$ and such that if $G' \in \mathcal{G}$ and $G' \cap V \neq \emptyset$, then $G \cap U \neq \emptyset$. Finally, we say that $\mathcal{G}$ is *continuous* if $\mathcal{G}$ is both upper and lower semicontinuous.

1.1.16 Example. Let $X = ([-1,1] \times [0,1]) \cup (\{0\} \times [0,2])$. For each $t \in [-1,1] \setminus \{0\}$, let $G_t = \{t\} \times [0,1]$, and for $t = 0$, let $G_0 = \{0\} \times [0,2]$. Let $\mathcal{G} = \{G_t \mid t \in [0,1]\}$. Then $\mathcal{G}$ is an upper semicontinuous decomposition of X (Figure 1.4).

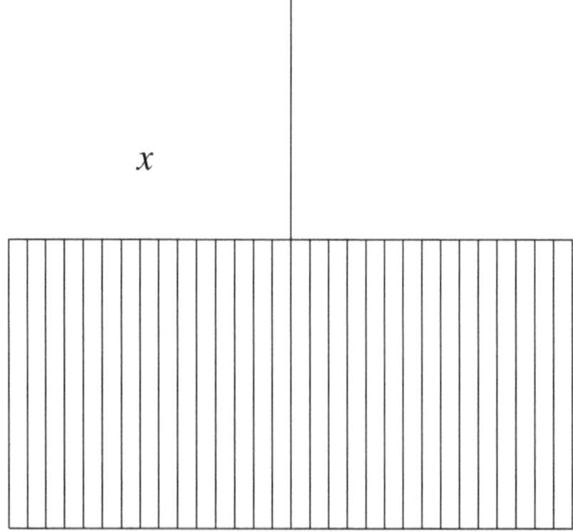

Fig. 1.4 Upper semicontinuous decomposition

1.1.17 Example. Let

$$X = ([0,1] \times [0,1]) .$$

For each $t \in [0,1)$, let $G_t = \{t\} \times [0,1]$, let $G_1 = \{1\} \times \left[0, \frac{1}{3}\right]$, let $G_1' = \{1\} \times \left[\frac{2}{3}, 1\right]$, and for each $t \in \left(\frac{1}{3}, \frac{2}{3}\right)$, let $G_{1,t} = \{(1,t)\}$. Let $\mathcal{G} = \{G_t \mid t \in [0,1]\} \cup \{G_{1,t} \mid t \in \left(\frac{1}{3}, \frac{2}{3}\right)\} \cup \{G_1, G_1'\}$. Then $\mathcal{G}$ is a lower semicontinuous decomposition of X (Figure 1.5).

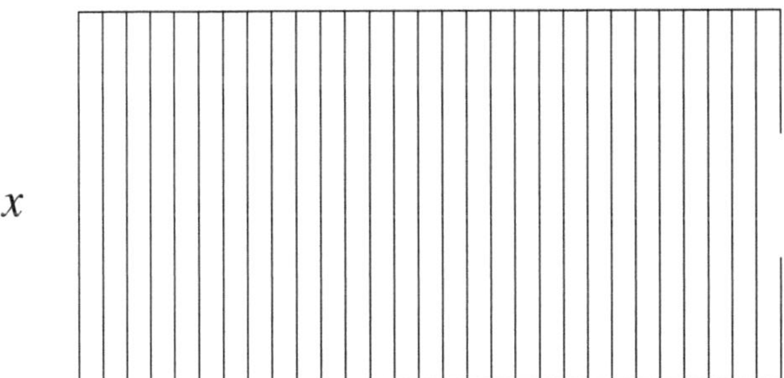

Fig. 1.5 Lower semicontinuous decomposition

The following theorem gives us a way to obtain upper semicontinuous decompositions of compacta.

1.1.18 Theorem. *Let $f\colon X \twoheadrightarrow Y$ be a surjective map between compacta. If $\mathcal{G}_f = \{f^{-1}(y) \mid y \in Y\}$, then $\mathcal{G}_f$ is an upper semicontinuous decomposition of X.*

Proof. Let U be an open subset of X such that $f^{-1}(y) \subset U$. Note that $X \setminus U$ is a closed subset of X. Hence, $X \setminus U$ is compact in X. Then $f(X \setminus U)$ is a compact subset of Y. Thus, $f(X \setminus U)$ is closed in Y and $y \notin f(X \setminus U)$. Hence, $Y \setminus f(X \setminus U)$ is an open subset of Y containing y.

If $V = f^{-1}(Y \setminus f(X \setminus U))$, then V is an open subset of X such that $f^{-1}(y) \subset V \subset U$. Since $V = \bigcup\{f^{-1}(y) \mid y \in Y \setminus f(X \setminus U)\}$, V satisfies the required property of the definition of upper semicontinuous decomposition. □

1.1.19 Remark. Let us note that Theorem 1.1.18 is not true without the compactness of X. Let X be the Euclidean plane $\mathbb{R}^2$ and let $\pi\colon \mathbb{R}^2 \twoheadrightarrow \mathbb{R}$ be given by $\pi((x,y)) = x$. Then $\mathcal{G}_\pi$ is a decomposition of X which is not upper semicontinuous. To see this, let $U = \left\{(x,y) \in X \;\middle|\; x \neq 0 \text{ and } y < \frac{1}{x}\right\} \cup \{0\} \times \mathbb{R}$. Then U is an open set of X such that $\pi^{-1}(0) \subset U$, whose boundary is asymptotic to $\pi^{-1}(0)$. Hence, for each $t \in \mathbb{R} \setminus \{0\}$, $\pi^{-1}(t) \cap (X \setminus U) \neq \emptyset$.

The next theorem gives three other ways to think about upper semicontinuous decompositions.

1.1.20 Theorem. *If X is a Hausdorff topological space and $\mathcal{G}$ is a decomposition of X, then the following conditions are equivalent:*

(a) *$\mathcal{G}$ is an upper semicontinuous decomposition;*
(b) *the quotient map $q: X \twoheadrightarrow X/\mathcal{G}$ is closed;*
(c) *if U is an open subset of X, then $W_U = \bigcup\{G \in \mathcal{G} \mid G \subset U\}$ is an open subset of X; W_U is a saturated open set;*
(d) *if D is a closed subset of X, then $K_D = \bigcup\{G \in \mathcal{G} \mid G \cap D \neq \emptyset\}$ is a closed subset of X.*

Proof. Suppose $\mathcal{G}$ is an upper semicontinuous decomposition. Let D be a closed subset of X. By Remark 1.1.6, we have that $q(D)$ is closed in $X/\mathcal{G}$ if and only if $q^{-1}q(D)$ is closed in X. We show that $X \setminus q^{-1}q(D)$ is open in X. Let $x \in X \setminus q^{-1}q(D)$. Then $q(x) \in X/\mathcal{G} \setminus q(D)$. This implies that $q^{-1}q(x) \subset X \setminus D$. Therefore, since $X \setminus D$ is open, by Definition 1.1.15, there exists an open set V of X such that $q^{-1}q(x) \subset V$ and for each $y \in V$, $q^{-1}q(y) \subset X \setminus D$. Note that $x \in V$ and $q(V) \subset X/\mathcal{G} \setminus q(D)$. Thus, $V \subset X \setminus q^{-1}q(D)$. Therefore, $X \setminus q^{-1}q(D)$ is open, since $x \in V \subset X \setminus q^{-1}q(D)$.

Now, suppose q is a closed map. Let U be an open subset of X. Since q is a closed map, we have that $q^{-1}(X/\mathcal{G} \setminus q(X \setminus U))$ is an open subset of X such that $q^{-1}(X/\mathcal{G} \setminus q(X \setminus U)) = W_U$. (If $x \in q^{-1}(X/\mathcal{G} \setminus q(X \setminus U))$, then $q(x) \in X/\mathcal{G} \setminus q(X \setminus U)$. Hence, $q^{-1}q(x) \subset X \setminus q^{-1}(q(X \setminus U)) \subset X \setminus (X \setminus U)) = U$. Thus, $x \in W_U$. The other inclusion is clear.)

Next, suppose W_U is open for each open subset U of X. Let D be a closed subset of X. Then $X \setminus D$ is open in X. Hence, $W_{X \setminus D}$ is open in X. Since, clearly, $K_D = X \setminus W_{X \setminus D}$, we have that K_D is closed.

Finally, suppose K_D is closed for each closed subset D of X. To see $\mathcal{G}$ is upper semicontinuous, let $G \in \mathcal{G}$ and let U be an open subset of X such that $G \subset U$. Note that $X \setminus U$ is a closed subset of X. Hence, $K_{X \setminus U}$ is a closed subset of X. Let $V = X \setminus K_{X \setminus U}$. Then V is open, $G \subset V \subset U$ and if $G' \in \mathcal{G}$ and $G' \cap V \neq \emptyset$, then $G' \subset V$. Therefore, $\mathcal{G}$ is upper semicontinuous. □

1.1.21 Corollary. *Let X be a Hausdorff topological space. If $\mathcal{G}$ is an upper semicontinuous decomposition of X, then the elements of $\mathcal{G}$ are closed.*

Proof. Let $G \in \mathcal{G}$. Let $x \in G$ and let $q: X \twoheadrightarrow X/\mathcal{G}$ be the quotient map. Since X is a Hausdorff topological space, $\{x\}$ is closed in X. By Theorem 1.1.20, $q(\{x\})$ is closed in $X/\mathcal{G}$. Since q is continuous and $q^{-1}q(\{x\}) = G$, G is a closed subset of X. □

1.1.22 Theorem. *If X is a compactum and $\mathcal{G}$ is an upper semicontinuous decomposition of X, then $X/\mathcal{G}$ is a Hausdorff space.*

Proof. To show that $X/\mathcal{G}$ is Hausdorff, let $q: X \twoheadrightarrow X/\mathcal{G}$ be the quotient map and let χ_1 and χ_2 be two distinct points of $X/\mathcal{G}$. Then $q^{-1}(\chi_1)$ and $q^{-1}(\chi_2)$ are two disjoint closed subsets of X. Since X is normal, there exist two disjoint open subsets U_1 and U_2 of X such that $q^{-1}(\chi_1) \subset U_1$ and $q^{-1}(\chi_2) \subset U_2$. Note that, since $\mathcal{G}$ is an upper semicontinuous decomposition, there exist two saturated open subsets W_{U_1} and W_{U_2} of X (Theorem 1.1.20) such that $q^{-1}(\chi_1) \subset W_{U_1} \subset U_1$ and $q^{-1}(\chi_2) \subset W_{U_2} \subset U_2$. Then $q(W_{U_1})$ and $q(W_{U_2})$ are two disjoint open subsets of $X/\mathcal{G}$ such that $\chi_1 \in q(W_{U_1})$ and $\chi_2 \in q(W_{U_2})$. Therefore, $X/\mathcal{G}$ is a Hausdorff space. □

The next theorem gives a characterization of lower semicontinuous decompositions.

1.1.23 Theorem. *Let X be a Hausdorff topological space and let $\mathcal{G}$ be a decomposition of X. Then $\mathcal{G}$ is lower semicontinuous if and only if the quotient map $q\colon X \twoheadrightarrow X/\mathcal{G}$ is open.*

Proof. Suppose $\mathcal{G}$ is lower semicontinuous. Let U be an open subset of X. We show $q(U)$ is an open subset of $X/\mathcal{G}$. To this end, by Remark 1.1.6, we only need to prove that $q^{-1}q(U)$ is an open subset of X.

Let $y \in q^{-1}q(U)$. Then $q(y) \in q(U)$, and there exists a point x in U such that $q(x) = q(y)$. Since $\mathcal{G}$ is a lower semicontinuous decomposition, there exists an open subset V of X containing y such that if $G \in \mathcal{G}$ and $G \cap V \neq \emptyset$, then $G \cap U \neq \emptyset$. Hence, $V \subset q^{-1}q(U)$. Therefore, q is open.

Now, suppose q is open. Let $G \in \mathcal{G}$, let x and y be two elements G and let U be an open subset of X such that $x \in U$. Since q is open, $V = q^{-1}q(U)$ is an open subset of X such that $G \subset V$. In particular, $y \in V$. Let $G' \in \mathcal{G}$ be such that $G' \cap V \neq \emptyset$. Then $G' \subset V$. Thus, $q(G') \in q(U)$. Hence, there exists a point $u \in U$ such that $q(u) = q(G')$. Since $q^{-1}q(G') = G'$, $u \in G'$. Thus, $G' \cap U \neq \emptyset$. Therefore, $\mathcal{G}$ is lower semicontinuous. □

The following corollary is a consequence of Theorems 1.1.20 and 1.1.23:

1.1.24 Corollary. *Let X be a Hausdorff topological space and let $\mathcal{G}$ be a decomposition of X. Then $\mathcal{G}$ is continuous if and only if the quotient map is both open and closed.*

The next theorem gives us a necessary and sufficient condition on a map $f\colon X \twoheadrightarrow Y$ between compacta, to have that $\mathcal{G}_f = \{f^{-1}(y) \mid y \in Y\}$ is a continuous decomposition.

1.1.25 Theorem. *Let X and Y be compacta and let $f\colon X \twoheadrightarrow Y$ be a surjective map. Then $\mathcal{G}_f = \{f^{-1}(y) \mid y \in Y\}$ is continuous if and only if f is open.*

Proof. If $\mathcal{G}_f$ is a continuous decomposition of X, by Theorem 1.1.23, the quotient map $q\colon X \twoheadrightarrow X/\mathcal{G}_f$ is open. By Theorem 1.1.11, $\varphi_f\colon X/\mathcal{G}_f \twoheadrightarrow Y$ is a homeomorphism. Hence, $f = \varphi_f \circ q$ is an open map.

Now, suppose f is open. By Theorem 1.1.18, $\mathcal{G}_f$ is upper semicontinuous. Since $q = \varphi_f^{-1} \circ f$ and f is open, q is open. By Theorem 1.1.23, $\mathcal{G}_f$ is a lower semicontinuous decomposition. Therefore, $\mathcal{G}_f$ is continuous. □

In the following definition a notion of convergence of sets is introduced.

1.1.26 Notation. The symbol $\mathbb{N}$ denotes the set of positive integers.

1.1.27 Definition. Let $\{X_n\}_{n=1}^{\infty}$ be a sequence of subsets of the Hausdorff topological space X. Then:

(1) the *limit inferior* of the sequence $\{X_n\}_{n=1}^{\infty}$ is defined as follows:

$$\liminf X_n = \{x \in X \mid \text{for each open subset } U \text{ of } X \text{ such that}$$

$$x \in U, U \cap X_n \neq \emptyset \text{ for each } n \in \mathbb{N}, \text{ save, possibly, finitely many}\}.$$

(2) the *limit superior* of the sequence $\{X_n\}_{n=1}^{\infty}$ is defined as follows:

$$\limsup X_n = \{x \in X \mid \text{for every open subset } U \text{ of } X \text{ such that}$$

$$x \in U, U \cap X_n \neq \emptyset \text{ for infinitely many indexes } n \in \mathbb{N}\}.$$

Clearly, $\liminf X_n \subset \limsup X_n$. If $\liminf X_n = \limsup X_n = L$, then we say that the sequence $\{X_n\}_{n=1}^{\infty}$ is a *convergent sequence* with limit $L = \lim_{n\to\infty} X_n$ (Figure 1.6).

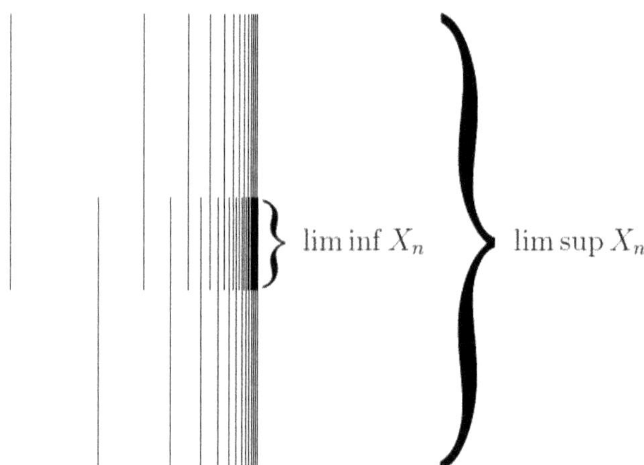

Fig. 1.6 lim inf and lim sup

1.1.28 Lemma. *Let $\{X_n\}_{n=1}^{\infty}$ be a sequence of subsets of the Hausdorff topological space X. Then* $\liminf X_n$ *and* $\limsup X_n$ *are both closed subsets of X.*

Proof. Let $x \in \mathrm{Cl}(\liminf X_n)$. Let U be an open subset of X such that $x \in U$. Since $x \in \mathrm{Cl}(\liminf X_n) \cap U$, we have that $\liminf X_n \cap U \neq \emptyset$. Hence, $U \cap X_n \neq \emptyset$ for each $n \in \mathbb{N}$, save, possibly, finitely many. Therefore, $x \in \liminf X_n$. The proof for lim sup is similar. □

The next theorem tells us that second countable Hausdorff spaces behave like sequentially compact spaces using the notion of convergence just introduced.

1.1.29 Theorem. *Each sequence $\{X_n\}_{n=1}^{\infty}$ of closed subsets of a second countable Hausdorff space X has a convergent subsequence.*

Proof. Let $\{U_m\}_{m=1}^{\infty}$ be a countable basis for X. Let $\{X_n^1\}_{n=1}^{\infty} = \{X_n\}_{n=1}^{\infty}$. Suppose, inductively, that we have defined the sequence $\{X_n^m\}_{n=1}^{\infty}$. We define the sequence $\{X_n^{m+1}\}_{n=1}^{\infty}$ as follows:

(1) If $\{X_n^m\}_{n=1}^{\infty}$ has a subsequence $\{X_{n_k}^m\}_{k=1}^{\infty}$ such that $\limsup X_{n_k}^m \cap U_m = \emptyset$, then let $\{X_n^{m+1}\}_{n=1}^{\infty}$ be such subsequence of $\{X_n^m\}_{n=1}^{\infty}$.
(2) If for each subsequence $\{X_{n_k}^m\}_{k=1}^{\infty}$ of $\{X_n^m\}_{n=1}^{\infty}$, we have that $\limsup X_{n_k}^m \cap U_m \neq \emptyset$, we define $\{X_n^{m+1}\}_{n=1}^{\infty}$ as $\{X_n^m\}_{n=1}^{\infty}$.

Since we have the subsequences $\{X_n^m\}_{n=1}^{\infty}$, let us consider the "diagonal subsequence" $\{X_m^m\}_{m=1}^{\infty}$. By construction, $\{X_m^m\}_{m=1}^{\infty}$ is a subsequence of $\{X_n\}_{n=1}^{\infty}$. We see that $\{X_m^m\}_{m=1}^{\infty}$ converges.

Let us assume that $\{X_m^m\}_{m=1}^{\infty}$ does not converge. Hence, there exists $p \in \limsup X_m^m \setminus \liminf X_m^m$. Let U_k be a basic open set such that $p \in U_k$ and $U_k \cap X_{m_\ell}^{m_\ell} = \emptyset$ for some subsequence $\{X_{m_\ell}^{m_\ell}\}_{\ell=1}^{\infty}$ of $\{X_m^m\}_{m=1}^{\infty}$ ($\liminf X_n$ is a closed subset of X by Lemma 1.1.28). Clearly, $\{X_{m_\ell}^{m_\ell}\}_{\ell=k}^{\infty}$ is a subsequence of $\{X_n^k\}_{n=1}^{\infty}$. Thus, $\{X_n^k\}_{n=1}^{\infty}$ satisfies condition (1), with k in place of m. Hence, $\limsup X_n^{k+1} \cap U_k = \emptyset$. Since $\{X_m^m\}_{m=k+1}^{\infty}$ is a subsequence of $\{X_n^{k+1}\}_{n=1}^{\infty}$ and $\limsup X_m^m \subset \limsup X_n^{k+1}$, it follows that $\limsup X_m^m \cap U_k = \emptyset$. Now, recall that $p \in \limsup X_m^m \cap U_k$. Thus, we obtain a contradiction. Therefore, $\{X_m^m\}_{m=1}^{\infty}$ converges. □

1.1.30 Theorem. *Let X be a sequentially compact Hausdorff space. If $\{X_n\}_{n=1}^{\infty}$ is a sequence of connected subsets of X and $\liminf X_n \neq \emptyset$, then $\limsup X_n$ is connected.*

Proof. Suppose, to the contrary, that $\limsup X_n$ is not connected. Since $\limsup X_n$ is closed, Lemma 1.1.28, we assume, without loss of generality, that there exist two disjoint closed subsets A and B of X such that $\limsup X_n = A \cup B$. Since X is a normal space, there exist two disjoint open subsets U and V of X such that $A \subset U$ and $B \subset V$. Thus, there exists $N' \in \mathbb{N}$ such that if $n \geq N'$, then $X_n \subset U \cup V$. To show this, suppose it is not true. Then for each $n \in \mathbb{N}$, there exists $m_n > n$ such that $X_{m_n} \setminus (U \cup V) \neq \emptyset$. Let $x_{m_n} \in X_{m_n} \setminus (U \cup V)$ for each $n \in \mathbb{N}$. The set $\{x_{m_n}\}_{n=1}^{\infty}$ is either finite or has a limit point. Hence, there exists a point x in X such that every open subset of X containing x contains infinitely many elements of $\{x_{m_n}\}_{n=1}^{\infty}$. Note that $x \in X \setminus (U \cup V)$ and, by construction, $x \in \limsup X_n$, a contradiction. Therefore, there exists $N' \in \mathbb{N}$ such that if $n \geq N'$, then $X_n \subset U \cup V$.

Since $\liminf X_n \neq \emptyset$ and $\liminf X_n \subset \limsup X_n$, we assume, without loss of generality, that $\liminf X_n \cap U \neq \emptyset$. Then there exists $N'' \in \mathbb{N}$ such that if $n \geq N''$, $U \cap X_n \neq \emptyset$. Let $N = \max\{N', N''\}$. Hence, if $n \geq N$, then $X_n \subset U \cup V$ and

$U \cap X_n \neq \emptyset$. Since X_n is connected for every $n \in \mathbb{N}$, $X_n \cap V = \emptyset$ for each $n \geq N$, a contradiction. Therefore, $\limsup X_n$ is connected. □

1.1.31 Notation. If (X, d) is a metric space, p is a point of X and $\varepsilon > 0$, then $\mathcal{V}_\varepsilon^d(p) = \{x \in X \mid d(p, x) < \varepsilon\}$.

1.1.32 Lemma. *Let X be a compact metric space, with metric d, and let $\{A_n\}_{n=1}^\infty$ be a sequence of connected subsets of X such that $\limsup A_n$ is not connected. Then, for each component L of $\limsup A_n$, there exists a subsequence $\{A_{n_k}\}_{k=1}^\infty$ of $\{A_n\}_{n=1}^\infty$ such that $\limsup A_{n_k} = L$.*

Proof. Let L be a component of $\limsup A_n$ and let $P = \{p_k\}_{k=1}^\infty$ be a countable dense subset of L. For each $p_k \in P$, consider $\mathcal{V}_{\frac{1}{k}}^d(p_k)$ and let $A_{n_k} \in \{A_n\}_{n=1}^\infty$ be such that $A_{n_k} \cap \mathcal{V}_{\frac{1}{k}}(p_k) \neq \emptyset$. Observe that $P \subset \limsup A_{n_k}$. Since $\limsup A_{n_k}$ is closed (Lemma 1.1.28), we have that $L \subset \limsup A_{n_k}$.

Let $q \in \limsup A_{n_k}$ and let U be an open subset of X containing q. Thus, there exists a subsequence $\{A_{n_{k_l}}\}_{l=1}^\infty$ of $\{A_{n_k}\}_{k=1}^\infty$ such that $q \in \liminf A_{n_{k_l}}$. It follows, from the definition of $\{A_{n_k}\}_{k=1}^\infty$, that there exists a point $p_{k_l} \in P$ such that $A_{n_{k_l}} \cap \mathcal{V}_{\frac{1}{k_l}}^d(p_{k_l}) \neq \emptyset$. Since X is a compact metric space, without loss of generality, we may assume that the sequence $\{p_{k_l}\}_{l=1}^\infty$ converges to a point a of X. This implies that $a \in \limsup A_{n_{k_l}}$. By Theorem 1.1.30, $\limsup A_{n_{k_l}}$ is connected. Since a and q both belong to $\limsup A_{n_{k_l}}$, we have that $\limsup A_{n_{k_l}} \subset L$. Thus, $q \in L$. Therefore, $\limsup A_{n_{k_l}} = L$. □

The next theorem gives us a characterization of an upper semicontinuous decomposition of a compact metric space in terms of limits inferior and superior.

1.1.33 Theorem. *Let X be a compact metric space, with metric d. Then a decomposition $\mathcal{G}$ of X is upper semicontinuous if and only if $\mathcal{G}$ is a closed decomposition and for each sequence $\{X_n\}_{n=1}^\infty$ of elements of $\mathcal{G}$ and each element Y of $\mathcal{G}$ such that $\liminf X_n \cap Y \neq \emptyset$, then $\limsup X_n \subset Y$.*

Proof. Suppose $\mathcal{G}$ is an upper semicontinuous decomposition of X. By Corollary 1.1.21, $\mathcal{G}$ is a closed decomposition. Let $\{X_n\}_{n=1}^\infty$ be a sequence of elements of $\mathcal{G}$ and let Y be an element of $\mathcal{G}$ such that $\liminf X_n \cap Y \neq \emptyset$.

Suppose there exists $p \in \limsup X_n \setminus Y$. Since Y is closed and p is not an element of Y, there exists an open set W of X such that $p \in W$ and $\mathrm{Cl}(W) \cap Y = \emptyset$. Let $U = X \setminus \mathrm{Cl}(W)$. Since $\mathcal{G}$ is upper semicontinuous, there exists an open set V of X such that $Y \subset V$ and if $G \in \mathcal{G}$ such that $G \cap V \neq \emptyset$, $G \subset U$.

Let $q \in \liminf X_n \cap Y$. Then $q \in \liminf X_n \cap V$. Hence, $V \cap X_n \neq \emptyset$ for each $n \in \mathbb{N}$, save, possibly, finitely many. Thus, $W \cap X_n = \emptyset$ for each $n \in \mathbb{N}$, save, possibly, finitely many. This contradicts the fact that $p \in W \cap \limsup X_n$. Therefore, $\limsup X_n \subset Y$.

Now, suppose $\mathcal{G}$ is a closed decomposition and let Y be an element of $\mathcal{G}$. Suppose that if $\{X_n\}_{n=1}^\infty$ is a sequence of elements of $\mathcal{G}$ such that $\liminf X_n \cap Y \neq \emptyset$, then $\limsup X_n \subset Y$.

To see $\mathcal{G}$ is upper semicontinuous, let U be an open subset of X such that $Y \subset U$. For each $n \in \mathbb{N}$, let $V_n = \mathcal{V}^d_{\frac{1}{n}}(Y)$. Suppose that for each $n \in \mathbb{N}$, there exists an element X_n of $\mathcal{G}$ such that $X_n \cap V_n \neq \emptyset$ and $X_n \not\subset U$. For each $n \in \mathbb{N}$, let $p_n \in X_n \cap V_n$. Since X is a compact metric space, $\{p_n\}_{n=1}^{\infty}$ has a convergent subsequence $\{p_{n_k}\}_{k=1}^{\infty}$. Let p be the point of convergence of $\{p_{n_k}\}_{k=1}^{\infty}$. Note that $p \in \liminf X_{n_k} \cap Y$. Hence, $\limsup X_{n_k} \subset Y$.

For each $k \in \mathbb{N}$, let $q_k \in X_{n_k} \setminus U$. Since X is a compact metric space, the sequence $\{q_{n_k}\}_{k=1}^{\infty}$ has a convergent subsequence $\{q_{n_{k_\ell}}\}_{\ell=1}^{\infty}$. Let q be the point of convergence of $\{q_{n_{k_\ell}}\}_{\ell=1}^{\infty}$. Note that $q \notin Y$ and $q \in \limsup X_{n_{k_\ell}} \subset \limsup X_{n_k}$, a contradiction. Therefore, $\mathcal{G}$ is upper semicontinuous. □

The following theorem gives us a characterization of a continuous decomposition of a compact metric space in terms of limits inferior and superior.

1.1.34 Theorem. *Let X be a compact metric space, with metric d. Then a decomposition $\mathcal{G}$ of X is continuous if and only if $\mathcal{G}$ is a closed decomposition and for each sequence $\{X_n\}_{n=1}^{\infty}$ of elements of $\mathcal{G}$ and each element Y of $\mathcal{G}$ such that* $\liminf X_n \cap Y \neq \emptyset$, *then* $\limsup X_n = Y$.

Proof. Suppose $\mathcal{G}$ is a continuous decomposition of X. By Corollary 1.1.21, $\mathcal{G}$ is a closed decomposition. Let $\{X_n\}_{n=1}^{\infty}$ be a sequence of elements of $\mathcal{G}$ and let Y be an element of $\mathcal{G}$ such that $\liminf X_n \cap Y \neq \emptyset$. By Theorem 1.1.33, $\limsup X_n \subset Y$. Suppose there exists $p \in Y \setminus \limsup X_n$. Let U be an open subset of X such that $p \in U$ and $U \cap \limsup X_n = \emptyset$ (by Lemma 1.1.28, $\limsup X_n$ is closed). Let $q \in \limsup X_n \subset Y$. Since $\mathcal{G}$ is a lower semicontinuous decomposition, there exists an open subset V of X such that $q \in V$ and if $G \in \mathcal{G}$ and $G \cap V \neq \emptyset$, then $G \cap U \neq \emptyset$. Since $q \in \limsup X_n$ and V is an open set containing q, $V \cap X_n \neq \emptyset$ for infinitely many indexes $n \in \mathbb{N}$. Hence, $U \cap X_n \neq \emptyset$ for infinitely many indexes $n \in \mathbb{N}$. Then $U \cap \limsup X_n \neq \emptyset$, a contradiction. Therefore, $\limsup X_n = Y$.

Now, suppose that $\mathcal{G}$ is a closed decomposition of X such that for each sequence $\{X_n\}_{n=1}^{\infty}$ of elements of $\mathcal{G}$ and each element Y of $\mathcal{G}$ such that $\liminf X_n \cap Y \neq \emptyset$, then $\limsup X_n = Y$. By Theorem 1.1.33, $\mathcal{G}$ is upper semicontinuous.

Let Y be an element of $\mathcal{G}$. Let p and q be two points of Y, and let U be an open subset of X such that $p \in U$. For each $n \in \mathbb{N}$, let $V_n = \mathcal{V}^d_{\frac{1}{n}}(q)$. Suppose that for each $n \in \mathbb{N}$, there exists an element X_n of $\mathcal{G}$ such that $X_n \cap V_n \neq \emptyset$ and $X_n \cap U = \emptyset$. Hence, $\limsup X_n \cap U = \emptyset$. For each $n \in \mathbb{N}$, let $q_n \in X_n \cap V_n$. Clearly, $\{q_n\}_{n=1}^{\infty}$ converges to q. Thus, $q \in \liminf X_n \cap Y$. By hypothesis, $Y = \limsup X_n$. Hence, $Y \cap U \neq \emptyset$, a contradiction. Therefore, $\mathcal{G}$ is a continuous decomposition. □

1.1.35 Theorem. *Let X be a compactum and let $\mathcal{G}$ be an upper semicontinuous decomposition of X. If $\mathcal{D} = \{D \mid D$ is a component of G, for some $G \in \mathcal{G}\}$, then $\mathcal{D}$ is an upper semicontinuous decomposition of X.*

Proof. Note that $\mathcal{D}$ is a decomposition of X, and its elements are closed in X. Let $D \in \mathcal{D}$ and let $G \in \mathcal{G}$ be such that D is a component of G. Let U be an open subset of X containing D. Since X is normal, there exists an open subset U' of X such that $U' \subset U$ and $(\mathrm{Cl}(U') \setminus U') \cap G = \emptyset$. Let $U^\star = U' \cup (X \setminus \mathrm{Cl}(U'))$ and let V

be an open subset of X such that $G \subset V$ and if $G' \cap V \neq \emptyset$, then $G' \subset U^\star$. Let $V' = V \cap U'$. If a component D' of an element G' of $\mathcal{G}$ intersects V', then it lies in $U^\star$ because $G' \cap V \neq \emptyset$. Since U' and $X \setminus \mathrm{Cl}(U')$ are disjoint open subsets and D' is contained in their union, $D' \subset U'$. Hence, $\mathcal{D}$ is upper semicontinuous. □

1.2 Topological Product and Inverse Limits

We present basic properties of products of topological spaces, inverse limits of compacta and generalized inverse limits of compacta.

1.2.1 Definition. Let $\{X_\lambda\}_{\lambda\in\Lambda}$ be a family of nonempty subsets. Then its *product* is the set:

$$\prod_{\lambda\in\Lambda} X_\lambda = \left\{ f\colon \Lambda \to \bigcup_{\lambda\in\Lambda} X_\lambda \;\middle|\; f(\lambda) \in X_\lambda \text{ for all } \lambda \in \Lambda \right\}.$$

If $\lambda_0 \in \Lambda$, then $\pi_{\lambda_0}\colon \prod_{\lambda\in\Lambda} X_\lambda \twoheadrightarrow X_{\lambda_0}$ given by $\pi_{\lambda_0}(f) = f(\lambda_0)$ is called the λ_0^{th} *projection map*

1.2.2 Remark. Given a family $\{X_\lambda\}_{\lambda\in\Lambda}$ of nonempty subsets, the Axiom of Choice guarantees that $\prod_{\lambda\in\Lambda} X_\lambda \neq \emptyset$ if and only if $X_\lambda \neq \emptyset$ for each $\lambda \in \Lambda$.

1.2.3 Notation. If $\{X_\lambda\}_{\lambda\in\Lambda}$ is a family of nonempty subsets, then an element $f \in \prod_{\lambda\in\Lambda} X_\lambda$ is denoted by $\{x_\lambda\}_{\lambda\in\Lambda}$, where $x_\lambda = f(\lambda)$.

1.2.4 Definition. Let $\{X_\lambda\}_{\lambda\in\Lambda}$ be a family of nonempty topological spaces. The *product topology* for $\prod_{\lambda\in\Lambda} X_\lambda$ has as a base

$$\mathcal{B} = \left\{ \bigcap_{j=1}^{n} \pi_{\lambda_j}^{-1}(U_{\lambda_j}) \;\middle|\; U_{\lambda_j} \text{ is open in } X_{\lambda_j}, j \in \{1, \ldots, n\} \text{ and } n \in \mathbb{N} \right\}.$$

1.2.5 Theorem. *Let Z be a topological space and let $\{X_\lambda\}_{\lambda\in\Lambda}$ be a family of nonempty topological spaces. Then a function $f\colon Z \to \prod_{\lambda\in\Lambda} X_\lambda$ is continuous if and only if $\pi_\lambda \circ f$ is continuous for every $\lambda \in \Lambda$.*

Proof. Suppose $\pi_\lambda \circ f$ is continuous for each $\lambda \in \Lambda$. Let $\bigcap_{j=1}^{n} \pi_{\lambda_j}^{-1}(U_{\lambda_j})$ be a basic open subset of $\prod_{\lambda\in\Lambda} X_\lambda$. Since

$$f^{-1}\left(\bigcap_{j=1}^{n} \pi_{\lambda_j}^{-1}(U_{\lambda_j}) \right) = \bigcap_{j=1}^{n} f^{-1}(\pi_{\lambda_j}^{-1}(U_{\lambda_j}))$$

$$= \bigcap_{j=1}^{n} (\pi_{\lambda_j} \circ f)^{-1}(U_{\lambda_j}),$$

we obtain that $f^{-1}\left(\bigcap_{j=1}^{n} \pi_{\lambda_j}^{-1}(U_{\lambda_j}\right)$ is an open subset of Z. Therefore, f is continuous. The other implication is clear. □

1.2.6 Theorem. *Let $\{X_\lambda\}_{\lambda\in\Lambda}$ and $\{Y_\lambda\}_{\lambda\in\Lambda}$ be families of nonempty topological spaces. Suppose that for each $\lambda \in \Lambda$, there exists a map $f_\lambda\colon X_\lambda \to Y_\lambda$. Then the function $\prod_{\lambda\in\Lambda} f_\lambda\colon \prod_{\lambda\in\Lambda} X_\lambda \to \prod_{\lambda\in\Lambda} Y_\lambda$ given by $\prod_{\lambda\in\Lambda} f_\lambda(\{x_\lambda\}_{\lambda\in\Lambda}) = \{f_\lambda(x_\lambda)\}_{\lambda\in\Lambda}$ is continuous.*

Proof. For each $\lambda_0 \in \Lambda$, let $\pi_{\lambda_0}\colon \prod_{\lambda\in\Lambda} X_\lambda \to X_{\lambda_0}$ and $\pi'_{\lambda_0}\colon \prod_{\lambda\in\Lambda} Y_\lambda \to Y_{\lambda_0}$ be the projection maps. Let $\{x_\lambda\}_{\lambda\in\Lambda}$ be an element of $\prod_{\lambda\in\Lambda} X_\lambda$ and let $\lambda_0 \in \Lambda$. Then $\pi'_{\lambda_0} \circ \prod_{\lambda\in\Lambda} f_\lambda(\{x_\lambda\}_{\lambda\in\Lambda}) = \pi'_{\lambda_0}(\{f_\lambda(x_\lambda)\}_{\lambda\in\Lambda}) = f_{\lambda_0}(x_{\lambda_0}) = f_{\lambda_0} \circ \pi_{\lambda_0}(\{x_\lambda\}_{\lambda\in\Lambda})$. Therefore, by Theorem 1.2.5, $\prod_{\lambda\in\Lambda} f_\lambda$ is continuous. □

The following is *Tychonoff's Theorem*, a proof of it may be found in [36, Theorem 1.4, p. 224].

1.2.7 Theorem. *Let $\{X_\lambda\}_{\lambda\in\Lambda}$ be a family of nonempty spaces. Then $\prod_{\lambda\in\Lambda} X_\lambda$ is compact if and only if X_λ is compact for each $\lambda \in \Lambda$.*

1.2.8 Definition. Let Λ be a nonempty set with an order relation "$\leq$" such that

(1) $\lambda \leq \lambda$ for all $\lambda \in \Lambda$,
(2) If λ_1, λ_2 and λ_3 belong to Λ and $\lambda_1 \leq \lambda_2$ and $\lambda_2 \leq \lambda_3$, then $\lambda_1 \leq \lambda_3$,
(3) If λ_1 and λ_2 belong to Λ, then there exists $\lambda_3 \in \Lambda$ such that $\lambda_1 \leq \lambda_3$ and $\lambda_2 \leq \lambda_3$.

Then Λ is called a *directed set*. A subset Λ_0 of Λ is a *cofinal* subset of Λ provided that for each $\lambda \in \Lambda$, there exists $\lambda_0 \in \Lambda_0$ such that $\lambda \leq \lambda_0$.

1.2.9 Definition. Let Λ be a directed set and let $\{X_\lambda\}_{\lambda\in\Lambda}$ be a family of nonempty topological spaces with Λ as its indexing set. If for each λ and α in Λ, with $\lambda \leq \alpha$, there exists a map $f_\lambda^\alpha\colon X_\alpha \to X_\lambda$ such that:

(1) $f_\lambda^\lambda = 1_{X_\lambda}$, where 1_{X_λ} is the identity map on X_λ,

and

(2) whenever α, β and λ belong to Λ and $\lambda \leq \beta \leq \alpha$, we have $f_\lambda^\beta \circ f_\beta^\alpha = f_\lambda^\alpha$.

Then $\{X_\lambda, f_\lambda^\alpha, \Lambda\}$ is an *inverse system*.

1.2.10 Definition. Let $\{X_\lambda, f_\lambda^\alpha, \Lambda\}$ be an inverse system. The *inverse limit* of the inverse system $\{X_\lambda, f_\lambda^\alpha, \Lambda\}$, denoted by $\mathcal{X}_\infty$ or by $\varprojlim\{X_\lambda, f_\lambda^\alpha, \Lambda\}$, is the set

$$\left\{ \{x_\lambda\}_{\lambda\in\Lambda} \in \prod_{\lambda\in\Lambda} X_\lambda \;\middle|\; f_\lambda^\alpha(x_\alpha) = x_\lambda, \text{ whenever } \lambda \leq \alpha \right\}.$$

The elements of the inverse limit are called *threads* and the maps f_λ^α are called *bonding maps*. We give $\mathcal{X}_\infty$ the relative topology from the product topology.

1.2.11 Notation. Let $\{X_\lambda, f_\lambda^\alpha, \Lambda\}$ be an inverse system whose inverse limit is $\mathcal{X}_\infty$. If $\lambda \in \Lambda$, then the restriction of the projection map π_λ to $\mathcal{X}_\infty$ is denoted by f_λ.

1.2.12 Lemma. *Let $\{X_\lambda, f_\lambda^\alpha, \Lambda\}$ be an inverse system and let λ_0 and α be elements of Λ such that $\lambda_0 \leq \alpha$. Then $f_{\lambda_0} = f_{\lambda_0}^\alpha \circ f_\alpha$.*

Proof. Let $\{x_\lambda\}_{\lambda\in\Lambda}$ be an element of X_∞. Then $f_{\lambda_0}^\alpha \circ f_\alpha(\{x_\lambda\}_{\lambda\in\Lambda}) = f_{\lambda_0}^\alpha(x_\alpha) = x_{\lambda_0} = f_{\lambda_0}(\{x_\lambda\}_{\lambda\in\Lambda})$. □

1.2.13 Theorem. *Let $\{X_\lambda, f_\lambda^\alpha, \Lambda\}$ be an inverse system whose inverse limit is $\mathcal{X}_\infty$. Then*

$$\mathfrak{B} = \{f_\lambda^{-1}(U_\lambda) \mid \lambda \in \Lambda \text{ and } U_\lambda \text{ is open in } X_\lambda\}$$

is a base for the topology of $\mathcal{X}_\infty$.

Proof. Clearly, $\mathfrak{B}$ is a subbase for $\mathcal{X}_\infty$. This implies that the family of finite intersections of elements of $\mathfrak{B}$ forms a base for $\mathcal{X}_\infty$. Let $\mathcal{U} = \bigcap_{j=1}^n f_{\lambda_j}^{-1}(U_{\lambda_j})$ be a basic open set for $\mathcal{X}_\infty$. We show that $\mathcal{U} \in \mathfrak{B}$.

Since Λ is a directed set, there exists $\lambda \in \Lambda$ such that $\lambda_j \leq \lambda$ for every $j \in \{1, \ldots, n\}$. Let

$$U_\lambda = \bigcap_{j=1}^n \left(f_{\lambda_j}^\lambda\right)^{-1}(U_{\lambda_j}).$$

Observe that

$$f_\lambda^{-1}(U_\lambda) = f_\lambda^{-1}\left(\bigcap_{j=1}^n f_{\lambda_j}^{-1}(U_{\lambda_j})\right) = \bigcap_{j=1}^n f_\lambda^{-1}\left(\left(f_{\lambda_j}^\lambda\right)^{-1}(U_{\lambda_j})\right) =$$

$$\bigcap_{j=1}^n (f_{\lambda_j}^\lambda \circ f_\lambda)^{-1}(U_{\lambda_j}) = \bigcap_{j=1}^n f_{\lambda_j}^{-1}(U_{\lambda_j}) = \mathcal{U}.$$

Therefore, $\mathcal{U} \in \mathfrak{B}$. □

1.2.14 Lemma. *Let $\{X_\lambda, f_\lambda^\alpha, \Lambda\}$ be an inverse system of compacta whose inverse limit is $\mathcal{X}_\infty$. Given two elements λ_0 and α of Λ such that $\lambda_0 < \alpha$, let $S_{\lambda_0}^\alpha = \{\{x_\lambda\}_{\lambda\in\Lambda} \mid x_{\lambda_0} = f_{\lambda_0}^\alpha(x_\alpha)\}$. Then $S_{\lambda_0}^\alpha$ is closed in $\prod_{\lambda\in\Lambda} X_\lambda$ and*

$$\mathcal{X}_\infty = \bigcap\{S_{\lambda_0}^\alpha \mid \lambda_0 \text{ and } \alpha \text{ belong to } \Lambda \text{ and } \lambda_0 < \alpha\}.$$

Proof. Let $\{x_\lambda\}_{\lambda\in\Lambda} \in \prod_{\lambda\in\Lambda} X_\lambda \setminus \mathcal{S}^{\alpha}_{\lambda_0}$. This implies that $x_{\lambda_0} \neq f^{\alpha}_{\lambda_0}(x_\alpha)$. Since X_{λ_0} is a Hausdorff space, there exist two disjoint open subsets U_{λ_0} and W_{λ_0} of X_{λ_0} such that $f^{\alpha}_{\lambda_0}(x_\alpha) \in U_{\lambda_0}$ and $x_{\lambda_0} \in W_{\lambda_0}$. Since $f^{\alpha}_{\lambda_0}$ is continuous, there exists an open subset V_α of X_α such that $x_\alpha \in V_\alpha$ and $f^{\alpha}_{\lambda_0}(V_\alpha) \subset U_{\lambda_0}$. Let $\mathcal{U} = \pi^{-1}_{\lambda_0}(W_{\lambda_0}) \cap \pi^{-1}_{\alpha}(V_\alpha)$. Then $\mathcal{U}$ is an open subset of $\prod_{\lambda\in\Lambda} X_\lambda$, $\{x_\lambda\}_{\lambda\in\Lambda} \in \mathcal{U}$ and $\mathcal{U} \cap \mathcal{S}_{\lambda_0} = \emptyset$. Hence, $\mathcal{S}_{\lambda_0}$ is closed in $\prod_{\lambda\in\Lambda} X_\lambda$. Clearly,

$$\mathcal{X}_\infty = \bigcap\{\mathcal{S}^{\alpha}_{\lambda_0} \mid \lambda_0 \text{ and } \alpha \text{ belong to } \Lambda \text{ and } \lambda_0 < \alpha\}.$$

□

1.2.15 Theorem. *If $\{X_\lambda, f^{\alpha}_{\lambda}, \Lambda\}$ is an inverse system of compacta whose inverse limit is $\mathcal{X}_\infty$, then $\mathcal{X}_\infty$ is a nonempty compactum.*

Proof. By Theorem 1.2.7, $\prod_{\lambda\in\Lambda} X_\lambda$ is a compact space. Let

$$\mathfrak{S} = \{\mathcal{S}^{\alpha}_{\lambda_0} \mid \lambda_0 \text{ and } \alpha \text{ belong to } \Lambda \text{ and } \lambda_0 < \alpha\}.$$

Then $\mathfrak{S}$ is a collection of closed subsets (Lemma 1.2.14) of the compact space $\prod_{\lambda\in\Lambda} X_\lambda$ (Theorem 1.2.7). We show that $\mathfrak{S}$ has the finite intersection property. Let $n \in \mathbb{N}$ and let $\mathcal{S}^{\alpha_1}_{\lambda_1}, \ldots, \mathcal{S}^{\alpha_n}_{\lambda_n}$ be elements of $\mathfrak{S}$. Since Λ is a directed set, there exists $\gamma \in \Lambda$ such that $\alpha_j \leq \gamma$ for every $j \in \{1, \ldots, n\}$. Let $x_\gamma \in X_\gamma$ and for each $j \in \{1, \ldots, n\}$, let $x_{\lambda_j} = f^{\gamma}_{\lambda_j}(x_\gamma)$ and $x_{\alpha_j} = f^{\gamma}_{\alpha_j}(x_\gamma)$. Consider the point $\{z_\lambda\}_{\lambda\in\Lambda} \in \prod_{\lambda\in\Lambda} X_\lambda$ given by $z_{\lambda_j} = x_{\lambda_j}$, $z_{\alpha_j} = x_{\alpha_j}$ for every $j \in \{1, \ldots, n\}$, $z_\gamma = x_\gamma$ and $z_\lambda \in X_\lambda$ for each $\lambda \in \Lambda \setminus \{\lambda_1, \ldots, \lambda_n, \alpha_1, \ldots, \lambda_n, \gamma\}$. Since $f^{\alpha_j}_{\lambda_j}(z_{\alpha_j}) = f^{\alpha_j}_{\lambda_j} \circ f^{\gamma}_{\lambda_j}(z_\gamma) = f^{\alpha_j}_{\lambda_j} \circ f^{\gamma}_{\lambda_j}(x_\gamma) = f^{\gamma}_{\lambda_j}(x_\gamma) = x_{\lambda_j} = z_{\lambda_j}$, we have that $\{z_\lambda\}_{\lambda\in\Lambda} \in \bigcap_{j=1}^{n} \mathcal{S}^{\alpha_j}_{\lambda_j}$. Thus, $\mathfrak{S}$ has the finite intersection property. Hence, $\bigcap\{\mathcal{S}^{\alpha}_{\lambda_0} \mid \lambda_0 \text{ and } \alpha \text{ belong to } \Lambda \text{ and } \lambda_0 < \alpha\} \neq \emptyset$ [36, Theorem 1.3, p. 223]. By Lemma 1.2.14, $\mathcal{X}_\infty$ is a closed subset of $\prod_{\lambda\in\Lambda} X_\lambda$ and $\bigcap\{\mathcal{S}^{\alpha}_{\lambda_0} \mid \lambda_0 \text{ and } \alpha \text{ belong to } \Lambda \text{ and } \lambda_0 < \alpha\} = \mathcal{X}_\infty$. Therefore, $\mathcal{X}_\infty$ is a nonempty compact space. The fact that $\mathcal{X}_\infty$ is a Hausdorff space follows from [36, Theorem 1.3, p. 138] □

1.2.16 Theorem. *Let $\{X_\lambda, f^{\alpha}_{\lambda}, \Lambda\}$ be an inverse system of compacta whose inverse limit is $\mathcal{X}_\infty$. Then the bonding maps are surjective if and only if the projection maps are surjective.*

Proof. Suppose that for every λ and α in Λ, with $\lambda \leq \alpha$, f^{α}_{λ} is surjective. Let μ be an element of Λ. We show that f_μ is surjective. Let x_μ be a point of X_μ and let $\mathcal{K} = \{x_\mu\} \times \prod_{\lambda\in\Lambda\setminus\{\mu\}} X_\lambda$. Recall that, by Lemma 1.2.14, $\mathcal{X}_\infty = \bigcap\{\mathcal{S}^{\alpha}_{\lambda_0} \mid \lambda_0 \text{ and } \alpha \text{ belong to } \Lambda \text{ and } \lambda_0 < \alpha\}$. By the proof of Theorem 1.2.15, $\mathfrak{S} = \{\mathcal{S}^{\alpha}_{\lambda_0} \mid \lambda_0 \text{ and } \alpha \text{ belong to } \Lambda \text{ and } \lambda_0 < \alpha\}$ has the finite intersection property. We prove that $\mathfrak{S} \cup \{\mathcal{K}\}$ also has the finite intersection property. Let $n \in \mathbb{N}$ and let $\mathcal{S}^{\alpha_1}_{\lambda_1}, \ldots, \mathcal{S}^{\alpha_n}_{\lambda_n}$ be elements of $\mathfrak{S}$. We see that $\mathcal{K} \cap \left(\bigcap_{j=1}^{n} \mathcal{S}^{\alpha_j}_{\lambda_j}\right) \neq \emptyset$. Since Λ is a directed set, there exists $\gamma \in \Lambda$ such that $\mu \leq \gamma$ and $\alpha_j \leq \gamma$ for all

$j \in \{1, \ldots, n\}$. Since f_μ^γ is surjective, there exists x_γ in X_γ such that $f_\mu^\gamma(x_\gamma) = x_\mu$. Let $\{z_\lambda\}_{\lambda\in\Lambda}$ in $\prod_{\lambda\in\Lambda} X_\lambda$ be such that $z_\gamma = x_\gamma$, $z_\mu = x_\mu$ and for each $j \in \{1, \ldots, n\}$, $z_{\lambda_j} = f_{\lambda_j}^\gamma(z_\gamma)$ and $z_{\alpha_j} = f_{\alpha_j}^\gamma(z_\gamma)$. Clearly, $\{z_\lambda\}_{\lambda\in\Lambda}$ belongs to $\mathcal{K}$. Let $j \in \{1, \ldots, n\}$. Note that $f_{\lambda_j}^{\alpha_j}(z_{\alpha_j}) = f_{\lambda_j}^{\alpha_j} \circ f_{\alpha_j}^\gamma(z_\gamma) = f_{\lambda_j}^\gamma(z_\gamma) = z_{\lambda_j}$. Hence, $\{z_\lambda\}_{\lambda\in\Lambda}$ belongs to $\mathcal{S}_{\lambda_j}^{\alpha_j}$. Thus, $\{z_\lambda\}_{\lambda\in\Lambda}$ is in $\mathcal{K} \cap \left(\bigcap_{j=1}^n \mathcal{S}_{\lambda_j}^{\alpha_j}\right)$. Since $\prod_{\lambda\in\Lambda} X_\lambda$ is compact (Theorem 1.2.7), $\{\mathcal{K}\} \cap (\bigcap \mathfrak{S}) \neq \emptyset$. Therefore, by Lemma 1.2.14, there exists $\{w_\lambda\}_{\lambda\in\Lambda}$ in $\mathcal{X}_\infty$ such that $f_\mu(\{w_\lambda\}_{\lambda\in\Lambda}) = x_\mu$, and f_μ is surjective. The inverse implication is clear by Lemma 1.2.12. □

1.2.17 Theorem. *Let $\{X_\lambda, f_\lambda^\alpha, \Lambda\}$ be an inverse system of compacta with surjective bonding maps, whose inverse limit is $\mathcal{X}_\infty$. If $\mathcal{A}$ is a nonempty closed subset of $\mathcal{X}_\infty$, then $\{f_\lambda(\mathcal{A}), f_\lambda^\alpha|_{f_\alpha(\mathcal{A})}, \Lambda\}$ is an inverse system with surjective bonding maps and*

$$\varprojlim\{f_\lambda(\mathcal{A}), f_\lambda^\alpha|_{f_\alpha(\mathcal{A})}, \Lambda\} = \mathcal{A} = \left[\prod_{\lambda\in\Lambda} f_\lambda(\mathcal{A})\right] \cap \mathcal{X}_\infty. \quad (*)$$

Proof. Given two elements λ and α of Λ such that $\lambda \leq \alpha$, we know that $f_\lambda^\alpha \circ f_\alpha = f_\lambda$. Then $f_\lambda^\alpha \circ f_\alpha(\mathcal{A}) = f_\lambda(\mathcal{A})$. Thus, we have that $f_\lambda^\alpha|_{f_\alpha(\mathcal{A})} \colon f_\alpha(\mathcal{A}) \to f_\lambda(\mathcal{A})$ is surjective. Hence, $\{f_\lambda(\mathcal{A}), f_\lambda^\alpha|_{f_\alpha(\mathcal{A})}, \Lambda\}$ is an inverse system with surjective bonding maps.

Next, we prove the equalities are true. Let

$$\{x_\lambda\}_{\lambda\in\Lambda} \in \varprojlim\{f_\lambda(\mathcal{A}), f_\lambda^\alpha|_{f_\alpha(\mathcal{A})}, \Lambda\}.$$

Then $\{x_\lambda\}_{\lambda\in\Lambda} \in \prod_{\lambda\in\Lambda} f_\lambda(\mathcal{A}) \subset \prod_{\lambda\in\Lambda} X_\lambda$ and for each pair of elements λ_0 and α of Λ, with $\lambda_0 \leq \alpha$, $f_{\lambda_0}^\alpha|_{f_\alpha(\mathcal{A})}(x_\alpha) = x_{\lambda_0}$. Therefore, $\{x_\lambda\}_{\lambda\in\Lambda} \in \left[\prod_{\lambda\in\Lambda} f_\lambda(\mathcal{A})\right] \cap \mathcal{X}_\infty$.

Now, let $\{x_\lambda\}_{\lambda\in\Lambda} \in \left[\prod_{\lambda\in\Lambda} f_\lambda(\mathcal{A})\right] \cap \mathcal{X}_\infty$. Hence, $\{x_\lambda\}_{\lambda\in\Lambda} \in \prod_{\lambda\in\Lambda} f_\lambda(\mathcal{A})$ and if $\lambda_0 \leq \alpha$, then $f_{\lambda_0}^\alpha(x_\alpha) = x_{\lambda_0}$. This implies that $f_{\lambda_0}^\alpha|_{f_\alpha(\mathcal{A})}(x_\alpha) = x_{\lambda_0}$. Thus, $\{x_\lambda\}_{\lambda\in\Lambda} \in \varprojlim\{f_\lambda(\mathcal{A}), f_\lambda^\alpha|_{f_\alpha(\mathcal{A})}, \Lambda\}$.

Let $\{x_\lambda\}_{\lambda\in\Lambda} \in \left[\prod_{\lambda\in\Lambda} f_\lambda(\mathcal{A})\right] \cap \mathcal{X}_\infty$. We prove that $\{x_\lambda\}_{\lambda\in\Lambda} \in \mathcal{A}$. For each $\lambda_0 \in \Lambda$, let $\mathcal{K}_{\lambda_0} = \mathcal{A} \cap f_{\lambda_0}^{-1}(x_{\lambda_0})$. Note that $\mathcal{K}_{\lambda_0}$ is a nonempty compact subset of $\mathcal{A}$. We see that $\mathfrak{K} = \{\mathcal{K}_\lambda\}_{\lambda\in\Lambda}$ has the finite intersection property. Let $\mathcal{K}_{\lambda_1}, \ldots, \mathcal{K}_{\lambda_n}$ be a finite subfamily of $\mathfrak{K}$. Since Λ is a directed set, there exists $\gamma \in \Lambda$ such that $\lambda_j \leq \gamma$ for each $j \in \{1, \ldots, n\}$. Let $j \in \{1, \ldots, n\}$. We show that $\mathcal{K}_\gamma \subset \mathcal{K}_{\lambda_j}$. Let $\{y_\lambda\}_{\lambda\in\Lambda} \in \mathcal{K}_\gamma = \mathcal{A} \cap f_\gamma^{-1}(x_\gamma)$. Then $f_{\lambda_j}^\gamma \circ f_\gamma(\{y_\lambda\}_{\lambda\in\Lambda}) = f_{\lambda_j}(\{y_\lambda\}_{\lambda\in\Lambda})$. Hence, $f_{\lambda_j}^\gamma(y_\gamma) = f_{\lambda_j}^\gamma(x_\gamma) = x_{\lambda_j}$ and $\mathcal{K}_\gamma \subset \mathcal{K}_{\lambda_j}$. Thus, $\bigcap_{j=1}^n \mathcal{K}_{\lambda_j} \neq \emptyset$. Therefore, $\bigcap_{\lambda\in\Lambda} \mathcal{K}_\lambda \neq \emptyset$.

Let $\{z_\lambda\}_{\lambda\in\Lambda} \in \bigcap_{\lambda\in\Lambda} \mathcal{K}_\lambda$. Then $f_\lambda(\{z_\lambda\}_{\lambda\in\Lambda}) = x_\lambda$ for all $\lambda \in \Lambda$. Hence, $z_\lambda = x_\lambda$ for every $\lambda \in \Lambda$. Therefore, $\{x_\lambda\}_{\lambda\in\Lambda} \in \mathcal{A}$. Clearly, $\mathcal{A} \subset \left[\prod_{\lambda\in\Lambda} f_\lambda(\mathcal{A})\right] \cap \mathcal{X}_\infty$. □

1.2.18 Theorem. *Let $\{X_\lambda, f_\lambda^\alpha, \Lambda\}$ be an inverse system of compacta whose inverse limit is $\mathcal{X}_\infty$. Let Λ_0 be a cofinal subset of Λ. For each $\lambda \in \Lambda_0$, let A_λ be* *a nonempty closed subset of* X_λ *and suppose that for every* λ *and* α *in* Λ_0, *with* $\lambda \leq \alpha$, $f_\lambda^\alpha(A_\alpha) \subset A_\lambda$. *Then* $\{A_\lambda, f_\lambda^\alpha|_{A_\alpha}, \Lambda_0\}$ *is and inverse system and* $\underleftarrow{\lim}\{A_\lambda, f_\lambda^\alpha|_{A_\alpha}, \Lambda_0\} = \bigcap_{\lambda\in\Lambda_0} f_\lambda^{-1}(A_\lambda)$.

Proof. Observe that $\{A_\lambda, f_\lambda^\alpha|_{A_\alpha}, \Lambda_0\}$ is an inverse system. Let λ and α be elements of Λ_0 such that $\lambda \leq \alpha$. Since $f_\lambda^\alpha(A_\alpha) \subset A_\lambda$, we have $f_\alpha^{-1}(A_\alpha) \subset f_\lambda^{-1}(A_\lambda)$. Hence, $\{f_\lambda^{-1}(A_\lambda)\}_{\lambda\in\Lambda_0}$ is a directed set by inclusion.

Let $\{a_\lambda\}_{\lambda\in\Lambda_0} \in \underleftarrow{\lim}\{A_\lambda, f_\lambda^\alpha|_{A_\alpha}, \Lambda_0\}$ and let $\gamma \in \Lambda_0$. Since

$$f_\gamma(\{a_\lambda\}_{\lambda\in\Lambda_0}) = a_\gamma \in A_\gamma,$$

$\{a_\lambda\}_{\lambda\in\Lambda_0} \in f_\gamma^{-1}(A_\gamma)$. Thus, $\{a_\lambda\}_{\lambda\in\Lambda_0} \in \bigcap_{\lambda\in\Lambda_0} f_\lambda^{-1}(A_\lambda)$. Therefore, $\underleftarrow{\lim}\{A_\lambda, f_\lambda^\alpha|_{A_\alpha}, \Lambda_0\} \subset \bigcap_{\lambda\in\Lambda_0} f_\lambda^{-1}(A_\lambda)$.

Let $\{b_\lambda\}_{\lambda\in\Lambda_0} \in \bigcap_{\lambda\in\Lambda_0} f_\lambda^{-1}(A_\lambda)$. Then $f_\gamma(\{b_\lambda\}_{\lambda\in\Lambda_0}) = b_\gamma \in A_\gamma$, for each $\gamma \in \Lambda_0$. Hence, $\{b_\lambda\}_{\lambda\in\Lambda_0} \in \prod_{\lambda\in\Lambda_0} A_\lambda$ and given λ and α in Λ_0, $f_\lambda^\alpha(b_\alpha) = b_\lambda$. This implies that $\{b_\lambda\}_{\lambda\in\Lambda_0} \in \underleftarrow{\lim}\{A_\lambda, f_\lambda^\alpha|_{A_\alpha}, \Lambda_0\}$. Therefore, $\bigcap_{\lambda\in\Lambda_0} f_\lambda^{-1}(A_\lambda) \subset \underleftarrow{\lim}\{A_\lambda, f_\lambda^\alpha|_{A_\alpha}, \Lambda_0\}$. □

As a consequence of the Theorems 1.2.17 and 1.2.18, we obtain:

1.2.19 Corollary. *Let* $\{X_\lambda, f_\lambda^\alpha, \Lambda\}$ *be an inverse system of compacta whose inverse limit is* $\mathcal{X}_\infty$. *Let* Λ_0 *be a cofinal subset of* Λ. *If* $\mathcal{A}$ *is a nonempty closed subset* $\mathcal{X}_\infty$, *then*

$$\mathcal{A} = \bigcap_{\lambda\in\Lambda_0} f_\lambda^{-1}(f_\lambda(\mathcal{A})).$$

1.2.20 Theorem. *Let* $\{X_\lambda, f_\lambda^\alpha, \Lambda\}$ *be an inverse system of connected compacta with surjective projection maps, whose inverse limit is* $\mathcal{X}_\infty$. *Then* $\mathcal{X}_\infty$ *is a connected compactum.*

Proof. By Theorem 1.2.15, $\mathcal{X}_\infty$ is a nonempty compactum. We show that if $\mathcal{X}_\infty = \mathcal{A} \cup \mathcal{B}$, where $\mathcal{A}$ and $\mathcal{B}$ are nonempty closed subsets of $\mathcal{X}_\infty$, then $\mathcal{A} \cap \mathcal{B} \neq \emptyset$. Since $\mathcal{X}_\infty$ is compact, both $\mathcal{A}$ and $\mathcal{B}$ are compact subsets of $\mathcal{X}_\infty$. Hence, for every $\lambda \in \Lambda$, $f_\lambda(\mathcal{A})$ and $f_\lambda(\mathcal{B})$ are closed subsets of X_λ. Since f_λ is surjective, $X_\lambda = f_\lambda(\mathcal{A}) \cup f_\lambda(\mathcal{B})$. Let $\lambda \in \Lambda$ and let $Z_\lambda = f_\lambda(\mathcal{A}) \cap f_\lambda(\mathcal{B})$. Since X_λ is connected, $Z_\lambda \neq \emptyset$. If $\alpha \in \Lambda$ is such that $\lambda \leq \alpha$, then define $g_\lambda^\alpha = f_\lambda^\alpha|_{Z_\alpha}$. Since $f_\lambda^\alpha(Z_\alpha) = f_\lambda^\alpha(f_\alpha(\mathcal{A}) \cap f_\alpha(\mathcal{B})) \subset f_\lambda^\alpha \circ f_\alpha(\mathcal{A}) \cap f_\lambda^\alpha \circ f_\alpha(\mathcal{B}) = f_\lambda(\mathcal{A}) \cap f_\lambda(\mathcal{B}) = Z_\lambda$, we have that $g_\lambda^\alpha(Z_\alpha) \subset Z_\lambda$. Hence, $\{Z_\lambda, g_\lambda, \Lambda\}$ is an inverse system of compacta, by Theorem 1.2.15, $\mathcal{Z}_\infty \neq \emptyset$. Let $\{z_\lambda\}_{\lambda\in\Lambda} \in \mathcal{Z}_\infty$. Then for each $\alpha \in \Lambda$, $f_\alpha(\{z_\lambda\}_{\lambda\in\Lambda}) \in f_\alpha(\mathcal{A}) \cap f_\alpha(\mathcal{B})$. By Theorem 1.2.17, $\{z_\lambda\}_{\lambda\in\Lambda} \in \mathcal{A} \cap \mathcal{B}$. Therefore, $\mathcal{X}_\infty$ is connected. □

Now, we turn our attention to generalized inverse limits of compacta with a single bonding function.

1.2.21 Definition. Let X and Y be compacta. Then a function $f\colon X \to 2^Y$ (Definition 1.6.1) is *upper semicontinuous* if for every point x of X and each open subset U of Y such that $f(x) \subset U$, there exists an open subset V of X such that if $x' \in V$, then $f(x') \subset U$. The *image of* f, denoted $\mathrm{Im}(f)$, is the set $\{y \in Y \mid \text{there exists } x \in X \text{ such that } y \in f(x)\}$. If $\mathrm{Im}(f) = Y$, then we say that f is *surjective*.

1.2.22 Definition. Let X, Y and Z be compacta and let $f\colon X \to 2^Y$ and $g\colon Y \to 2^Z$ be upper semicontinuous functions. Then the *composition* of f and g, denoted $g \circ f$, is defined by $(g \circ f)(x) = \{z \in Z \mid \text{there exists } y \in Y \text{ such that } y \in f(x) \text{ and } z \in g(y)\}$.

1.2.23 Notation. Let X be a compactum. Then X^ω denotes the product of countably many copies of X with the product topology. The symbol Δ_X^ω denotes the diagonal of X^ω; that is, $\Delta_X^\omega = \{(x, x, x, \ldots) \mid x \in X\}$.

1.2.24 Definition. Let X be a compactum. If $f\colon X \to 2^X$ is an upper semicontinuous function, then its *generalized inverse limit* is

$$\varprojlim f = \{(x_n)_{n=1}^{\omega} \in X^\omega \mid x_n \in f(x_{n+1}) \text{ for each positive integer } n\}.$$

The function f is called the *bonding function*. For each positive integer n, $\pi_n\colon \varprojlim f \to X$ is the projection map.

1.2.25 Notation. Let X be a compactum and let $\mathcal{G}$ be an upper semicontinuous decomposition of X. Then $f_{\mathcal{G}}\colon X \to 2^X$ is defined by $f_{\mathcal{G}}(x) = G_x$, where G_x is the unique element of $\mathcal{G}$ such that $x \in G_x$.

1.2.26 Lemma. *If X is a compactum and $\mathcal{G}$ is an upper semicontinuous decomposition of X, then $f_{\mathcal{G}}$ is an upper semicontinuous function.*

Proof. Let $x_0 \in X$ and let U be an open subset of X such that $f_{\mathcal{G}}(x_0) \subset U$. Since $\mathcal{G}$ is upper semicontinuous, there exists an open subset V of X such that $f_{\mathcal{G}}(x_0) \subset V$ and if $G' \in \mathcal{G}$ and $G' \cap V \neq \emptyset$, then $G' \subset U$. Thus, if $x \in V$, then $f_{\mathcal{G}}(x) \subset U$. Therefore, $f_{\mathcal{G}}$ is upper semicontinuous. □

1.2.27 Lemma. *If X is a compactum and $\mathcal{G} = \{G_x \mid x \in X\}$ is an upper semicontinuous decomposition of X, then $\varprojlim f_{\mathcal{G}} = \bigcup\{G_x^\omega \mid x \in X\}$. Moreover, $\Delta_X^\omega \subset \varprojlim f_{\mathcal{G}}$.*

Proof. By Lemma 1.2.26, $f_{\mathcal{G}}$ is upper semicontinuous. Let $(x_n)_{n=1}^{\omega} \in \varprojlim f_{\mathcal{G}}$. Then $x_1 \in f_{\mathcal{G}}(x_2) = G_{x_2}$, $x_2 \in f_{\mathcal{G}}(x_3) = G_{x_3}$, etc. Since $\mathcal{G}$ is a decomposition, $G_{x_1} = G_{x_2} = \cdots$. Hence, $(x_n)_{n=1}^{\omega} \in G_{x_1}^\omega$. The other inclusion is clear.

If $x \in X$, then $(x, x, x, , \ldots) \in G_x^\omega$. Therefore, $\Delta_X^\omega \subset \varprojlim f_{\mathcal{G}}$. □

1.3 Uniformities

We introduce the notion of uniformity of a topological space and give some of its basic properties.

Let X be a Hausdorff space. If V and W are subsets of $X \times X$, then

$$-V = \{(x', x) \in X \times X \mid (x, x') \in V\}$$

and

$$V + W = \{(x, x'') \mid \text{there exists } x' \in X \text{ such that}$$
$$(x, x') \in V \text{ and } (x', x'') \in W\}.$$

We write $1V = V$ and for each positive integer n, $(n+1)V = nV + 1V$.

The diagonal of X is the set $\Delta_X = \{(x, x) \mid x \in X\}$. An *entourage* of Δ_X is a subset V of $X \times X$ such that $\Delta_X \subset V$ and $V = -V$. The family of entourages of the diagonal of X is denoted by $\mathfrak{D}_X$. If $V \in \mathfrak{D}_X$ and $x \in X$, then $B(x, V) = \{x' \in X \mid (x, x') \in V\}$. By [40, 8.1.3], $\mathrm{Int}(B(x, V))$ is an open neighbourhood of x. If A is a subset of X and $V \in \mathfrak{D}_X$, then $B(A, V) = \bigcup\{B(a, V) \mid a \in A\}$. If $V \in \mathfrak{D}_X$ and $(x, x') \in V$, then we write $\rho(x, x') < V$. If $(x, x') \notin V$, then we write $\rho(x, x') \geq V$. We have that if x, x' and x'' are points of X, and V and W belong to $\mathfrak{D}_X$ then the following hold [40, p. 426]:

(*i*) $\rho(x, x) < V$.
(*ii*) $\rho(x, x') < V$ if and only if $\rho(x', x) < V$.
(*iii*) If $\rho(x, x') < V$ and $\rho(x', x'') < W$, then $\rho(x, x'') < V + W$.

Let X be a nonempty set. A *uniformity* on X is a subfamily $\mathfrak{U}$ of $\mathfrak{D}_X$ such that:

(1) If $V \in \mathfrak{U}$, $W \in \mathfrak{D}_X$ and $V \subset W$, then $W \in \mathfrak{U}$.
(2) If V and W both belong to $\mathfrak{U}$, then $V \cap W \in \mathfrak{U}$.
(3) For every $V \in \mathfrak{U}$, there exists $W \in \mathfrak{U}$ such that $2W \subset V$.
(4) $\bigcap\{V \mid V \in \mathfrak{U}\} = \Delta_X$.

1.3.1 Definition. A *uniform space* is a pair $(X, \mathfrak{U})$ consisting of a nonempty set X and a uniformity on the set X.

1.3.2 Definition. For any uniformity $\mathfrak{U}$ on a set X, the family $\mathfrak{O} = \{G \subset X \mid \text{for every } x \in G, \text{ there exists } V \in \mathfrak{U} \text{ such that } B(x, V) \subset G\}$ is a topology on the set X [40, 8.1.1]. The topology $\mathfrak{O}$ is called the *topology induced by the uniformity* $\mathfrak{U}$.

1.3.3 Notation. Let X be a Tychonoff space and let $\mathfrak{U}$ be a uniformity of X that induces its topology. If $V \in \mathfrak{U}$, then we define the cover $\mathfrak{C}(V) = \{B(x, V) \mid x \in X\}$ of X.

1.3.4 Remark. Note that by [40, 8.3.13], for every compactum X, there exists a unique uniformity $\mathfrak{U}_X$ on X that induces the original topology of X.

1.3.5 Notation. Let X be a compactum and let $\mathfrak{U}_X$ be the unique uniformity of X that induces its topology. If Y is a subspace of X and $U \in \mathfrak{U}_X$, then $U_Y = U \cap (Y \times Y)$.

We need the next result [40, 8.3.G].

1.3.6 Theorem. *If X is a compactum, then for every open cover $\mathcal{U}$ of X, there exists $V \in \mathfrak{U}_X$ such that $\mathfrak{C}(V)$ refines $\mathcal{U}$.*

1.3.7 Remark. It is known that if X is a compactum and $\mathcal{H}(X)$ is the group of homeomorphisms of X, then $\mathcal{H}(X)$ is a topological group with the compact-open topology [40, p. 441].

1.3.8 Notation. Let X be a compactum. For every $U \in \mathfrak{U}_X$ (Remark 1.3.4), let $\widehat{U}$ be the entourage of the diagonal $\Delta_{\mathcal{H}(X)} \subset \mathcal{H}(X) \times \mathcal{H}(X)$ defined by

$$\widehat{U} = \{(g_1, g_2) \in \mathcal{H}(X) \times \mathcal{H}(X) \mid \rho(g_1(x), g_2(x)) < U \text{ for all } x \in X\}.$$

It is known that the family $\widehat{\mathcal{U}} = \{\widehat{U} \mid U \in \mathfrak{U}_X\}$ is a base for a uniformity on $\mathcal{H}(X)$; the uniformity generated by this family is called the *uniformity of uniform convergence* induced by $\mathfrak{U}_X$ and is denoted by $\widehat{\mathfrak{U}_X}$ [40, p. 440].

We also need a special case of [40, 8.2.7]:

1.3.9 Theorem. *Let X be a compactum. Then the topology on $\mathcal{H}(X)$ induced by the uniformity $\widehat{\mathfrak{U}_X}$ of uniform convergence coincides with the compact-open topology on $\mathcal{H}(X)$, and depends only on the topology induced on X by the uniformity $\mathfrak{U}_X$ (Remark 1.3.4).*

The following simple lemmas are useful.

1.3.10 Lemma. *Let Z be a Tychonoff space, let $\mathfrak{U}$ be a uniformity of Z that induces its topology, let $V \in \mathfrak{U}$ and let $z \in Z$. Then $B(z, nV) \subset \mathrm{Int}_Z(B(z, (n+1)V))$ for all $n \in \mathbb{N}$.*

Proof. Let $n \in \mathbb{N}$ and let $z' \in B(z, nV)$ and let $z'' \in B(z', V)$. Since $\rho(z'', z') < V$ and $\rho(z', z) < nV$, we have that $\rho(z'', z) < (n+1)V$. Thus, $B(z', V) \subset B(z, (n+1)V)$. Therefore, by [40, 8.1.2], $B(z, nV) \subset \mathrm{Int}_Z(B(z, (n+1)V))$. □

1.3.11 Lemma. *Let Z be a compactum and let $\mathcal{U}$ be an open cover of Z. If $U_0 = \bigcup\{U \times U \mid U \in \mathcal{U}\}$, then $U_0 \in \mathfrak{U}_Z$.*

Proof. Let $U_0 = \bigcup\{U \times U \mid U \in \mathcal{U}\}$. Note that $\Delta_Z \subset U_0$ and $-U_0 = U_0$. Thus, $U_0 \in \mathfrak{D}_Z$. Since Z is a compact Hausdorff space, by Theorem 1.3.6, there exists $W \in \mathfrak{U}_Z$ such that $\mathfrak{C}(W)$ refines $\mathcal{U}$. Since $\mathfrak{C}(W)$ refines $\mathcal{U}$, we have that $W \subset U_0$. This implies that $U_0 \in \mathfrak{U}_Z$. □

The following theorems are [111, Theorem 2.1 and Theroem 2.2], respectively:

1.3.12 Theorem. *Let f be a function from a uniform space $(X, \mathfrak{U})$ onto a nonempty set Y. Then $\mathfrak{U}(f) = \{V \subset Y \times Y \mid (f \times f)^{-1}(V) \in \mathfrak{U}\}$ is a uniformity for Y.*

1.3.13 Theorem. *Let f be a function from a uniform space $(X, \mathfrak{U})$ onto the uniform space $(Y, \mathfrak{U}(f))$. Then*

$$\mathfrak{U}(f) = \{(f \times f)(U) \mid U \in \mathfrak{U}\}.$$

1.3.14 Definition. Let $(Z, \mathfrak{V})$ and $(W, \mathfrak{U})$ be uniform spaces and let $g\colon Z \to W$ be a function. Then g is *uniformly continuous* if for each $U \in \mathfrak{U}$, there exists $V \in \mathfrak{V}$ such that for each point z of Z, $g(B(z, V)) \subset B(g(z), U)$.

Note the following:

1.3.15 Theorem. *Let Z and W be compacta and let $g\colon Z \to W$ be a function. Then g is continuous if and only if g is uniformly continuous*

Proof. By [129, 35.11], every uniformly continuous function is continuous. By [129, 36.20], each continuous function between compacta is uniformly continuous. □

1.4 Continua

We begin with some basic properties of compacta and then we present many properties of continua. We consider the notion of generalized continua and end the section with some results of generalized inverse limits with a single bonding map.

1.4.1 Lemma. *Let X be a compactum, and let A be a closed subset of X with a finite number of components. If $x \in \text{Int}(A)$ and C is the component of A such that $x \in C$, then $x \in \text{Int}(C)$.*

Proof. Let $C, C_1, \ldots, C_n$ be the components of A. Since $x \in \text{Int}(A)$, there exists an open subset U of X such that $x \in U \subset A$. Since A is closed in X, each of $C, C_1, \ldots, C_n$ is closed in X. Let $V = U \cap \left(X \setminus \bigcup_{j=1}^{n} C_j\right)$. Then V is an open subset of X such that $x \in V \subset A$ and $V \cap \left(\bigcup_{j=1}^{n} C_j\right) = \emptyset$. Thus, $V \subset C$. Therefore, $x \in \text{Int}(C)$. □

1.4.2 Definition. Let X be a compactum, and let A and B be two nonempty disjoint subsets of X. A subset C of X is a *separator of A and B* (or *separates X between A and B*) if $X \setminus C = U \cup V$, where U and V are separated (i.e., $\text{Cl}(U) \cap V = \emptyset$ and $U \cap \text{Cl}(V) = \emptyset$), $A \subset U$ and $B \subset V$. If $C = \emptyset$, we say that A and B are *separated in X*.

1.4.3 Lemma. *Let X be a compactum, let a be a point of X and let B be a compact subset of X such that for each b in B, a and b are separated in X. Then a and B are separated in X.*

Proof. For each b in B, there exist two open and closed subsets U_b and V_b such that $X = U_b \cup V_b$, $a \in U_b$ and $V \in V_b$. Since B is compact, there exist $b_1, \ldots, b_n$ in B

such that $B \subset \bigcup_{j=1}^{n} V_{b_j}$. Let $V = \bigcup_{j=1}^{n} V_{b_j}$. Then V is an open and closed subset of X, $B \subset V$ and $a \in X \setminus V$. Therefore, a and B are separated. □

1.4.4 Lemma. *Let X be a compactum and let A and B be compact subsets of X. If each point of A and each point of B are separated in X, then A and B are separated.*

Proof. By Lemma 1.4.3, B and each point of A are separated. Repeating the argument of Lemma 1.4.3, we obtain that A and B are separated. □

1.4.5 Lemma. *Let X be a compactum, let a and b be two distinct points of X and let $\{H_\lambda\}_{\lambda \in \Lambda}$ be a set theoretic chain of closed subsets of X ordered by inclusion. If each H_λ contains both a and b but is not the union of two separated sets, one containing a and the other containing b, then the intersection $\bigcap_{\lambda \in \Lambda} H_\lambda$ also has this property.*

Proof. Let $H = \bigcap_{\lambda \in \Lambda} H_\lambda$ and suppose that H is the union of two separated sets A and B such that $a \in A$ and $b \in B$. Since H is a closed subset of X, we have that A and B are closed subsets of X. Since X is normal, there exist two disjoint open subsets U and V of X such that $A \subset U$ and $B \subset V$. For each $\lambda \in \Lambda$, $H_\lambda \cap U \neq \emptyset$ and $H_\lambda \cap V \neq \emptyset$. Let $\lambda \in \Lambda$. If $K_\lambda = H_\lambda \cap [X \setminus (U \cup V)]$ were empty, then $H_\lambda = (H_\lambda \cap U) \cup (H_\lambda \cap V)$ would be a separation of H_λ with $a \in H_\lambda \cap U$ and $b \in H_\lambda \cap V$, a contradiction. Hence, $K_\lambda \neq \emptyset$. Also, the family $\{K_\lambda\}_{\lambda \in \Lambda}$ is a set theoretic chain ordered by inclusion; for, given any subset R of X, if H_λ is contained in $H_{\lambda'}$, then $H_\lambda \cap R$ lies in $H_{\lambda'} \cap R$. Thus, $\{K_\lambda\}_{\lambda \in \Lambda}$ has the finite intersection property. Since X is compact, $\bigcap_{\lambda \in \Lambda} K_\lambda \neq \emptyset$. But $\bigcap_{\lambda \in \Lambda} K_\lambda \subset \bigcap_{\lambda \in \Lambda} H_\lambda$, which implies that $H \cap [X \setminus (U \cup V)] \neq \emptyset$, a contradiction. □

1.4.6 Definition. A *continuum* is a connected compactum. A *subcontinuum* is a continuum contained in a space.

1.4.7 Theorem. *Let X be compactum and let a and b be two distinct points of X. If X is not the union of two disjoint open sets, one containing a and the other containing b, then X contains a subcontinuum containing both a and b.*

Proof. Let $\mathcal{H} = \{H_\lambda\}_{\lambda \in \Lambda}$ be the collection of all closed subsets of X, each of which contains $\{a, b\}$ but in none of which a and b are separated. Since $X \in \mathcal{H}$, $\mathcal{H} \neq \emptyset$. Partially order $\mathcal{H}$ by inclusion. Let $\mathcal{L} = \{L_\gamma\}_{\gamma \in \Gamma}$ be a set theoretic chain of elements of $\mathcal{H}$, we show that $\mathcal{L}$ has a lower bound. Let $L = \bigcap_{\gamma \in \Gamma} L_\gamma$. Then L is a nonempty closed subset of X and, by Lemma 1.4.5, L is not the union of two separated sets. Hence, $L \in \mathcal{H}$. By the Kuratowsky-Zorn Lemma, $\mathcal{H}$ has a minimal element, say H. We show that H is connected. Suppose H is not connected. Then there exist two disjoint closed subsets A and B of X such that $H = A \cup B$. Since $H \in \mathcal{H}$, H is not the union of two separated sets. Thus, either $\{a, b\} \subset A$ or $\{a, b\} \subset B$. Either case contradicts the minimality of H. Therefore, H is connected. □

The next result is known as *The Cut Wire Theorem*; it is very useful in continuum theory.

1.4.8 Theorem. *Let X be a compactum and let A and B be nonempty closed subsets of X. If no connected subset of X intersects both A and B, then there exist two disjoint closed subsets X_1 and X_2 of X such that $A \subset X_1$, $B \subset X_2$ and $X = X_1 \cup X_2$.*

Proof. By hypothesis, no component, K, of X intersects both A and B. Note that the components of X are subcontinua of X. Hence, by Theorem 1.4.7, each point of A and each point of B are separated in X. Hence, by Lemma 1.4.4, A and B are separated in X. Therefore, there exist two disjoint closed subsets X_1 and X_2 of X such that $A \subset X_1$, $B \subset X_2$ and $X = X_1 \cup X_2$. □

1.4.9 Example. Let

$$X = \{0\} \times [-1, 1] \cup \left\{ \left(x, \sin\left(\frac{1}{x}\right) \right) \in \mathbb{R}^2 \,\middle|\, x \in \left(0, \frac{2}{\pi}\right] \right\}.$$

Then X is called the *topologist sine curve* (Figure 1.7).

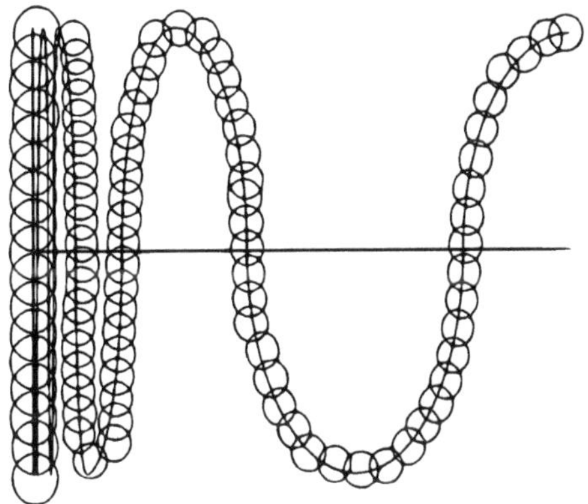

Fig. 1.7 Topologist sine curve

The following definition is due to Davis and Doyle and they used it in their study of invertible continua [31]:

1.4.10 Definition. Let X be a continuum, and let $x \in X$. We say that X *is almost connected im kleinen at* x provided that for each open subset U of X containing x, there exists a subcontinuum W of X such that $\text{Int}(W) \neq \emptyset$ and $W \subset U$. We say that X is *almost connected im kleinen* if it is almost connected im kleinen at each of its points.

1.4.11 Example. Let X be the cone over the closure of the harmonic sequence $\{0\} \cup \left\{\frac{1}{n}\right\}_{n=1}^{\infty}$. Then X is called the *harmonic fan*. Note that X is almost connected im kleinen at each of its points (Figure 1.8).

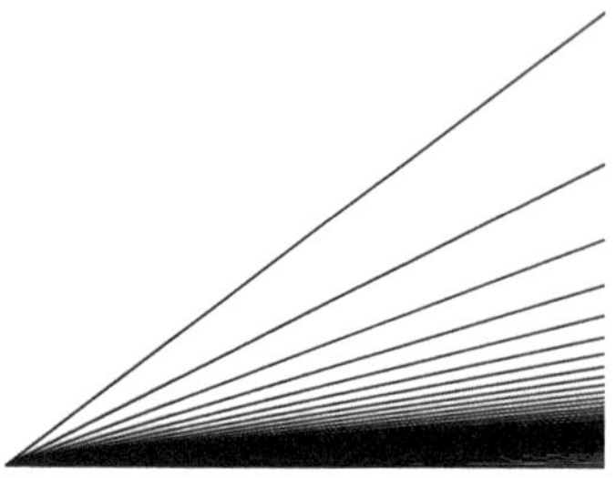

Fig. 1.8 Almost connected im kleinen

1.4.12 Definition. Let X be a continuum, and let $x \in X$. We say that X *is connected im kleinen at* x if for each closed subset F of X such that $F \subset X \setminus \{x\}$, there exists a subcontinuum W of X such that $x \in \text{Int}(W) \subset W \subset X \setminus F$.

1.4.13 Example. Let X be a sequence of harmonic fans converging to a point p; see Figure 1.9. Then X is connected im kleinen at p but X is not locally connected at that point (Definition 1.4.16).

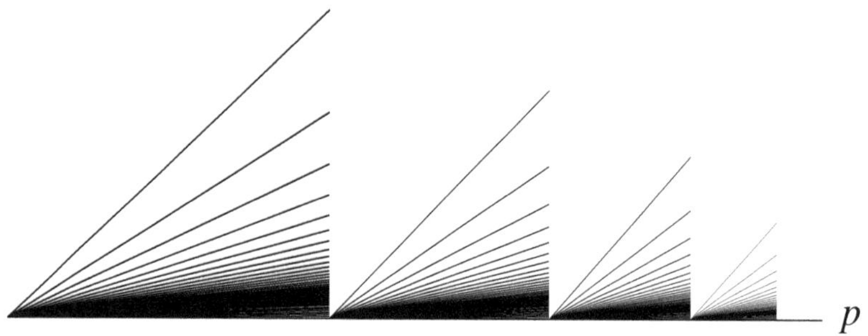

Fig. 1.9 Connected im kleinen at p

1.4.14 Remark. Note that we do not define connectedness im kleinen globally because, by Theorem 1.4.18, a continuum is connected im kleinen globally if and only if it is locally connected (Definition 1.4.16).

The way we define connectedness im kleinen in Definition 1.4.12 is not the usual one. This definition provides the author the answer to the fact that aposyndesis (Definition 1.4.21) is a generalization of connectedness im kleinen. The next theorem presents the usual definition of this concept:

1.4.15 Theorem. *If X is a continuum and $x \in X$, then the following are equivalent:*

(a) X is connected im kleinen at x.

(*b*) *For each open subset U of X such that $x \in U$, there exists an open subset V of X such that $x \in V \subset U$ and with the property that for each $y \in V$, there exists a connected subset C_y of X such that $\{x, y\} \subset C_y \subset U$.*
(*c*) *For each open subset U of X such that $x \in U$, there exists a subcontinuum W of X such that $x \in \operatorname{Int}(W) \subset W \subset U$.*

Proof. Suppose X is connected im kleinen at x. We show (b). Let U be an open subset of X such that $x \in U$. Then $X \setminus U$ is a closed subset of X not containing x. By hypothesis, there exists a subcontinuum W of X such that $x \in \operatorname{Int}(W) \subset W \subset X \setminus (X \setminus U) = U$. Thus, $V = \operatorname{Int}(W)$ satisfies the required properties.

Next, suppose (b). We prove (c). Let U be an open subset of X such that $x \in U$. Let U' be an open subset of X such that $x \in U' \subset \operatorname{Cl}(U') \subset U$. By hypothesis, there exists an open subset V of X such that $x \in V \subset U'$ with the property that for each $y \in V$, there exists a connected subset C_y of X such that $\{x, y\} \subset C_y \subset U'$. Let $W = \operatorname{Cl}\left(\bigcup_{y \in V} C_y\right)$. Then W is a subcontinuum of X such that $x \in V \subset W \subset \operatorname{Cl}(U') \subset U$.

Finally, suppose (c). We show X is connected im kleinen at x. Let F be a closed subset of X such that $F \subset X \setminus \{x\}$. Then $X \setminus F$ is an open subset of X containing x. By hypothesis, there exists a subcontinuum W of X such that $x \in \operatorname{Int}(W) \subset W \subset X \setminus F$. Therefore, X is connected im kleinen at x. □

1.4.16 Definition. Let X be a continuum, and let $x \in X$. We say that X *is locally connected at x* provided that for each open subset U of X such that $x \in U$, there exists a connected open subset V of X such that $x \in V \subset U$. We say X *is locally connected* if it is locally connected at each of its points.

1.4.17 Lemma. *A continuum X is locally connected if and only if the components of the open subsets of X are open.*

Proof. Suppose X is locally connected. Let U be an open subset of X, and let C be a component of U. For each point $x \in C$, there exists an open connected set V_x such that $x \in V_x \subset U$. Then $C \cup V_x$ is a connected subset of U. Hence, $V_x \subset C$. Therefore, each point of C is an interior point of C. Thus, C is open.

Next, suppose the components of open sets are open. Let $x \in X$, and let U be an open subset of X such that $x \in U$. By hypothesis, the component, C, of U containing x is open. Then C is an open connected set such that $x \in C \subset U$. Therefore, X is locally connected. □

1.4.18 Theorem. *A continuum X is connected im kleinen at each of its points if and only if X is locally connected.*

Proof. Clearly, if X is locally connected, then X is connected im kleinen at each of its points.

Suppose X is connected im kleinen at each of its points. Let U be an open subset of X, and let C be a component of U. If $x \in C$, then there exists a subcontinuum W of X such that $x \in \operatorname{Int}(W) \subset W \subset U$, Theorem 1.4.15. Since W is a connected subset of U and $W \cap C \neq \emptyset$, $W \subset C$. Then x is an interior point of C. Thus, C is open. Therefore, by Lemma 1.4.17, X is locally connected. □

1.4.19 Definition. Let X be a continuum, and let p and q be points of X. We say that X *is semi-aposyndetic at p and q* provided that there exists a subcontinuum W of X such that $\{p, q\} \cap \operatorname{Int}(W) \neq \emptyset$ and $\{p, q\} \setminus W \neq \emptyset$. X is *semi-aposyndetic* if it is semi-aposyndetic at each pair of its points.

1.4.20 Example. Let X be the harmonic fan (Example 1.4.11). Then X is semi-aposyndetic at each pair of its points (Figure 1.10).

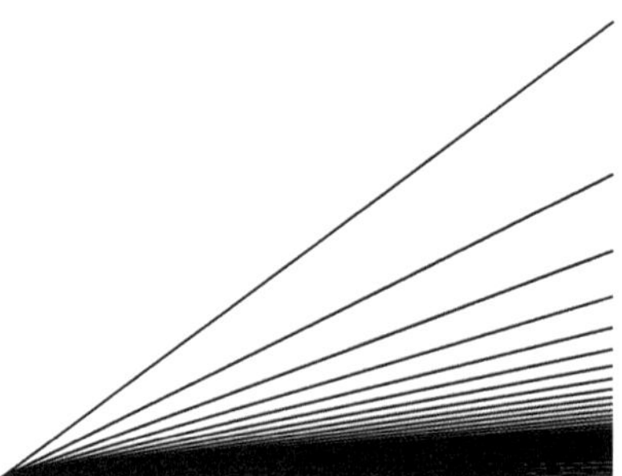

Fig. 1.10 Semi-aposyndetic continuum

1.4.21 Definition. Let X be a continuum, and let p and q be points of X. We say that X *is aposyndetic at p with respect to q* provided that there exists a subcontinuum W of X such that $p \in \operatorname{Int}(W) \subset W \subset X \setminus \{q\}$. Now, X *is aposyndetic at p* if X is aposyndetic at p with respect to each point of $X \setminus \{p\}$. We say that X *is aposyndetic* provided that X is aposyndetic at each of its points (Figure 1.11).

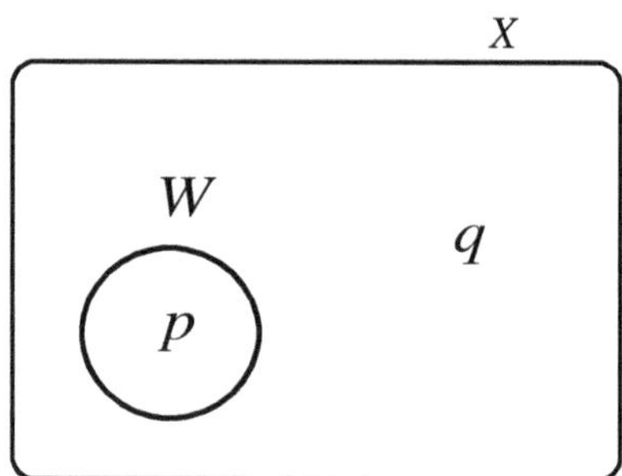

Fig. 1.11 Aposyndetic at p with respect to q

1.4.22 Definition. Let X be a continuum and let p and q be points of X. We say that X is *mutually aposyndetic at p and q* if there exist two disjoint subcontinua K and L of X such that $p \in \operatorname{Int}(K)$ and $q \in \operatorname{Int}(L)$. X is *mutually aposyndetic* if it is mutually aposyndetic at each pair of its points.

1.4.23 Example. Let X be the suspension over the closure of the harmonic sequence $\{0\} \cup \left\{\frac{1}{n}\right\}_{n=1}^{\infty}$, with vertexes a and b (Figure 1.12). Then X is mutually aposyndetic.

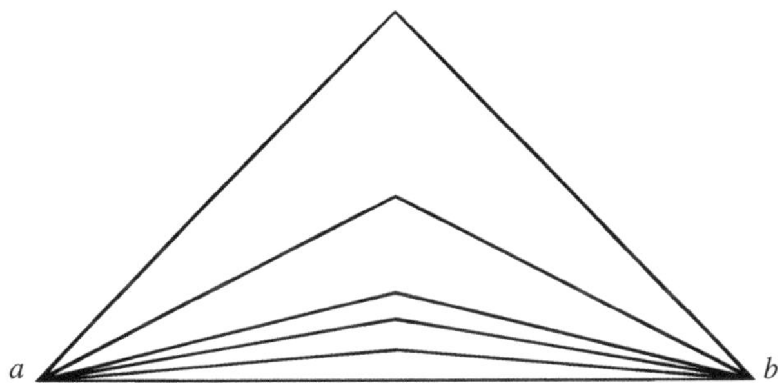

Fig. 1.12 Mutually aposyndetic

1.4.24 Remark. The notions of semi-aposyndesis, aposyndesis and mutual aposyndesis resemble those of T_0, T_1 and T_2 topological spaces, respectively.

1.4.25 Definition. Let X be a continuum, and let $p \in X$. We say that X *is semi-locally connected at* p provided that for each open subset U of X such that $p \in U$, there exists an open subset V of X such that $p \in V \subset U$ and $X \setminus V$ only has a finite number of components. We say X *is semi-locally connected* if X is semi-locally connected at each of its points (Figure 1.13).

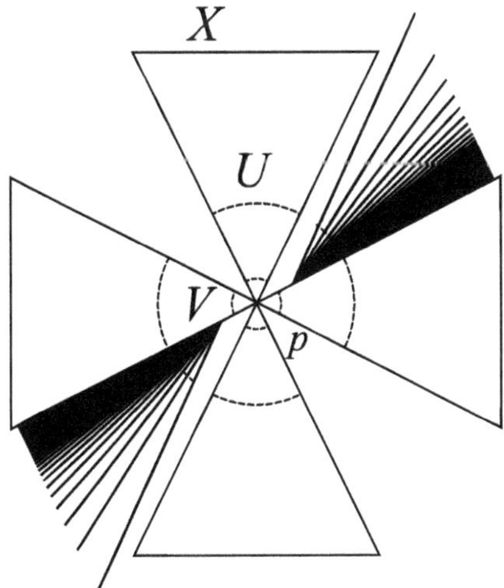

Fig. 1.13 Semi-locally connected continuum

Even though aposyndesis and semi-local connectedness seem to be different concepts, it turns out that globally they are equivalent:

1.4.26 Theorem. *A continuum X is aposyndetic if and only if it is semi-locally connected.*

Proof. Suppose X is aposyndetic. Let $x \in X$. We show that X is semi-locally connected at x. Let U be an open subset of X such that $x \in U$. Since X is aposyndetic, for each $y \in X \setminus U$, there exists a subcontinuum W_y of X such that $y \in \text{Int}(W_y) \subset W_y \subset X \setminus \{x\}$. Since $X \setminus U$ is compact, there exist $y_1, \ldots, y_m$ in $X \setminus U$ such that $X \setminus U \subset \bigcup_{j=1}^{m} \text{Int}(W_{y_j})$. Let $V = X \setminus \left(\bigcup_{j=1}^{m} W_{y_j}\right)$. Then V is an open subset of X such that $x \in V \subset U$ and $X \setminus V$ only has a finite number of components. Therefore, X is semi-locally connected, since x is an arbitrary point of X.

Now, suppose X is semi-locally connected. Let x and y be points of X. We show that X is aposyndetic at x with respect to y. Let U be an open subset of X such that $y \in U \subset X \setminus \{x\}$. Let U' be an open subset of X such that $y \in U' \subset \text{Cl}(U') \subset U$. Since X is semi-locally connected, there exists an open subset V of X such that $y \in V \subset U'$ and $X \setminus V$ only has a finite number of components. By Lemma 1.4.1, x is contained in the interior of the component of $X \setminus V$ containing x. Hence, X is aposyndetic at x with respect to y. Therefore, X is aposyndetic. □

Another concept related to aposyndesis is free decomposability.

1.4.27 Definition. A continuum X is *freely decomposable* if for each pair of distinct points p and q of X, there exist two subcontinua P and Q of X such that $X = P \cup Q$, $p \in P \setminus Q$ and $q \in Q \setminus P$.

The next theorem shows that the concepts of free decomposability and aposyndesis coincide.

1.4.28 Theorem. *A continuum X is aposyndetic if and only if X is freely decomposable.*

Proof. Suppose X is freely decomposable. Let p and q be two distinct points of X. Since X is freely decomposable, there exist two subcontinua P and Q of X such that $X = P \cup Q$, $p \in P \setminus Q$ and $q \in Q \setminus P$. Hence, X is an aposyndetic continuum.

Now, assume X is an aposyndetic continuum. Let p and q be two distinct points of X. Since X is aposyndetic, there exist two subcontinua A and B such that $p \in \text{Int}(A) \subset A \subset X \setminus \{q\}$ and $q \in \text{Int}(B) \subset B \subset X \setminus \{p\}$. Let U and V be open subsets of X such that $U \subset A$, $V \subset B$, $p \in U \subset X \setminus B$ and $q \in V \subset X \setminus A$. Let A_p be the component of $X \setminus V$ that contains p and let B_q be the component of $X \setminus U$ that contains q. Note that A_p and B_q are subcontinua of X, $p \in A_p \setminus B_q$ and $q \in B_q \setminus A_p$. We show that $X = A_p \cup B_q$. Suppose this is not true and let $L = X \setminus (A_p \cup B_q)$. Since $p \in U \subset A \subset X \setminus V$ and $q \in V \subset B \subset X \setminus U$, we have that $A \subset A_p$ and $B \subset B_q$. Hence, $U \cup V \subset A_p \cup B_q$. Thus, $L \subset X \setminus (U \cup V) = (X \setminus U) \cap (X \setminus V)$. This implies that $\text{Cl}(L) \subset X \setminus V$. Note that $A_p \cup \text{Cl}(L)$ is a closed subset of $X \setminus V$ that contains A_p properly. Since A_p is a component of $X \setminus V$, $A_p \cup \text{Cl}(L)$ is not connected. Hence, there exist two disjoint closed subsets R_1 and S_1 of X such that $A_p \cup \mathrm{Cl}(L) = R_1 \cup S_1$ (Theorem 1.4.8). Without loss of generality, we assume that $A_p \subset R_1$. Thus, $S_1 \subset \mathrm{Cl}(L)$. Since $\mathrm{Cl}(L) \subset X \setminus U$, we have that $B_q \cup S_1 \subset X \setminus U$. Observe that $B_q \cup S_1$ is a closed subset of $X \setminus U$ containing B_q properly. Since B_q is a component of $X \setminus U$, $B_q \cup S_1$ is not connected. Thus, there exist two disjoint closed subsets R_2 and S_2 of X such that $B_q \cup S_1 = R_2 \cup S_2$ (Theorem 1.4.8). Without loss of generality, we assume that $B_q \subset R_2$. Hence, $S_2 \subset S_1$. Since $R_1 \cap S_1 = \emptyset$ and $R_1 \cap S_2 = \emptyset$, we obtain that $(R_1 \cup R_2) \cap S_2 = \emptyset$. Since $X = A_p \cup B_q \cup L$, we have that $X = A_p \cup B_q \cup \mathrm{Cl}(L)$. Also, since $A_p \cup \mathrm{Cl}(L) = R_1 \cup S_1$, $X = R_1 \cup S_1 \cup B_q$. Moreover, since $S_1 \cup B_q = R_2 \cup S_2$, we obtain that $X = (R_1 \cup R_2) \cup S_2$. A contradiction to the connectedness of X. Therefore, $X = A_p \cup B_q$ and X is freely decomposable. □

The notion of aposyndesis may be extended as follows:

1.4.29 Definition. A continuum X is *aposyndetic at p with respect to the subset K of X* if there exists a subcontinuum W of X such that $p \in \mathrm{Int}(W) \subset W \subset X \setminus K$.

1.4.30 Definition. A continuum X is *continuum aposyndetic at p* if X is aposyndetic at p with respect to each subcontinuum of X not containing p. X is *continuum aposyndetic* provided that X is continuum aposyndetic at each of its points.

1.4.31 Definition. A continuum X is *freely decomposable with respect to points and continua* if for each subcontinuum C of X and each point $a \in X \setminus C$, there exist two subcontinua A and B of X such that $X = A \cup B$, $a \in A \setminus B$ and $C \subset B \setminus A$.

1.4.32 Lemma. *Let X be a continuum, and let A be a subcontinuum of X such that $X \setminus A$ is not connected. If U and V are nonempty disjoint open subsets of X such that $X \setminus A = U \cup V$, then $A \cup U$ and $A \cup V$ are subcontinua of X.*

Proof. Since $X \setminus (A \cup U) = V$, $A \cup U$ is closed, hence, compact. Similarly $A \cup V$ is compact.

We show $A \cup U$ is connected. To see this, suppose $A \cup U$ is not connected. Then there exist two nonempty disjoint closed subsets K and L of X such that $A \cup U = K \cup L$. Since A is connected, without loss of generality, we assume that $A \subset K$. Note that, in this case, $L \subset U$. Hence, $L \cap \mathrm{Cl}(V) = \emptyset$. Thus, $X = L \cup (K \cup \mathrm{Cl}(V))$, a contradiction, since L and $K \cup \mathrm{Cl}(V)$ are disjoint closed subsets of X. Therefore, $A \cup U$ is connected. Similarly, $A \cup V$ is connected. □

1.4.33 Definition. A continuum X is *decomposable* provided that it can be written as the union of two of its proper subcontinua. We say X is *indecomposable* if it is not decomposable. We say X is *hereditarily decomposable* (*indecomposable*) if each nondegenerate subcontinuum of X is decomposable (indecomposable, respectively).

1.4.34 Lemma. *A continuum X is decomposable if and only if X contains a proper subcontinuum with nonempty interior.*

Proof. Suppose X is a decomposable continuum. Then there exist two proper subcontinua, A and B, of X such that $X = A \cup B$. Note that $X \setminus B$ is an open set contained in A. Therefore, $\text{Int}(A) \neq \emptyset$.

Now, assume A is a proper subcontinuum of X with nonempty interior. If $X \setminus A$ is connected, then $\text{Cl}(X \setminus A)$ is a proper subcontinuum of X and $X = A \cup \text{Cl}(X \setminus A)$. Hence, X is decomposable.

Suppose that $X \setminus A$ is not connected. Then there exist two nonempty disjoint open subsets U and V of X such that $X \setminus A = U \cup V$. By Lemma 1.4.32, $A \cup U$ and $A \cup V$ are subcontinua of X, and $X = (A \cup U) \cup (A \cup V)$. Therefore, X is decomposable. □

1.4.35 Corollary. *A continuum X is indecomposable if and only if each proper subcontinuum of X has empty interior.*

The next result is known as the *Boundary Bumping Theorem.*

1.4.36 Theorem. *Let X be a continuum and let U be a nonempty, proper, not dense open subset of X. If K is a component of* $\text{Cl}(U)$, *then* $K \cap \text{Bd}(U) \neq \emptyset$.

Proof. Let U be a not dense open subset of X and let K be a component of $\text{Cl}(U)$. Assume that $K \cap \text{Bd}(U) = \emptyset$. Hence, $K \subset U$. Since X is normal, there exists an open subset V of X such that $K \subset V \subset \text{Cl}(V) \subset U$. Clearly, $\text{Cl}(V)$ is a compactum. Since no connected subset of $\text{Cl}(V)$ intersects both K and $\text{Bd}(V)$ (K is a component of U), by Theorem 1.4.8, there exists two disjoint closed subsets L and M of X such that $\text{Cl}(V) = L \cup M$, $K \subset L$ and $\text{Bd}(V) \subset M$. Then $X = L \cup [M \cup (X \setminus V)]$, where L and $[M \cup (X \setminus V)]$ are disjoint closed subset of X, a contradiction to the connectedness of X. Therefore, $K \cap \text{Bd}(U) \neq \emptyset$. □

The following corollary is very useful:

1.4.37 Corollary. *Let X be a nondegenerate continuum. If A is a proper subcontinuum of X and U is a proper open subset of X such that $A \subset U$, then there exists a subcontinuum B of X such that $A \subsetneq B \subset U$.*

Proof. Let A be a proper subcontinuum of X and let U be a proper open subset of X. Since X is normal, there exists an open subset V of X such that $A \subset V \subset \text{Cl}(V) \subset U$. Let B be the component of $\text{Cl}(V)$ that contains A. By Theorem 1.4.36, $B \cap \text{Bd}(V) \neq \emptyset$. Therefore, B is a subcontinuum of X and $A \subsetneq B \subset U$. □

1.4.38 Lemma. *If X is a continuum such that each of its proper subcontinua is indecomposable, then X is indecomposable. Hence, X is hereditarily indecomposable.*

Proof. Suppose X is decomposable. Then there exist two proper subcontinua A and B of X such that $X = A \cup B$. Note that $X \setminus B$ is an open subset of X contained in A. On the other hand, there exists a proper subcontinuum, H, of X containing A (Corollary 1.4.37). Hence, H is an indecomposable continuum containing a proper subcontinuum with nonempty interior, which is impossible (Corollary 1.4.35). Therefore, X is indecomposable. □

1.4.39 Definition. A continuum X *is irreducible between two of its points* if no proper subcontinuum of X contains both points. A continuum is *irreducible* if it is irreducible between two of its points.

The following results present some of the properties of irreducible continua.

1.4.40 Theorem. *Let X be an irreducible continuum between a and b. If C is a subcontinuum of X such that $X \setminus C$ is not connected, then $X \setminus C$ is the union of two open and connected sets, one containing a and the other containing b. Moreover, if $a \in C$, then $X \setminus C$ is connected.*

Proof. Suppose $X \setminus C$ is not connected. Then there exist two nonempty disjoint open subsets U and V of X such that $X \setminus C = U \cup V$. By Lemma 1.4.32, $A = C \cup U$ and $B = C \cup V$ are subcontinua of X such that $X = A \cup B$, $A \cap B = C$, $A \neq X$ and $B \neq X$.

Since X is irreducible between a and b, $\{a, b\} \cap C = \emptyset$. If $\{a, b\} \cap C \neq \emptyset$, then either A or B is a proper subcontinuum of X containing $\{a, b\}$, a contradiction. Therefore, $\{a, b\} \cap C = \emptyset$. We assume that $a \in U$ and $b \in V$.

Since A and B are proper subcontinua of X, neither A nor B may contains $\{a, b\}$.

Since A is a proper subcontinuum of X and $a \in A$, we assert that $V = X \setminus A$ is connected. To see this, suppose $X \setminus A$ is not connected. Then there exist two nonempty disjoint open subsets K and L of X such that $X \setminus A = K \cup L$. Since $b \in X \setminus A$, we may assume that $b \in K$. Then, by Lemma 1.4.32, $A \cup K$ is subcontinuum of X which is proper and satisfies that $\{a, b\} \subset A \cup K$, a contradiction. Therefore, $V = X \setminus A$ is connected. Similarly, $U = X \setminus B$ is connected.

A similar argument shows that if $a \in C$, then $X \setminus C$ is connected. □

1.4.41 Lemma. *Let X be an irreducible decomposable continuum. If Y and Z are two disjoint subcontinua of X, then $X \setminus (Y \cup Z)$ has at most three components.*

Proof. Since X is irreducible, by Theorem 1.4.40, $X \setminus Y$ has at most two components, U and V, such that they are connected open subsets of X, $Z \subset U$ and V may be empty.

Similarly, we assume that $X \setminus Z = H \cup K$, where H and K are connected open subsets of X such that $Y \subset H$ and K may be empty. Let $R = (U \setminus Z) \cap (H \setminus Y)$. Note that R is an open subset of X, $\mathrm{Cl}(R) \cap Y \neq \emptyset$ and $\mathrm{Cl}(R) \cap Z \neq \emptyset$. We assert that R is connected. To show this, assume R is not connected. Let C be a component of R. By Theorem 1.4.36, $\mathrm{Cl}(C) \cap (Y \cup Z) \neq \emptyset$. If $\mathrm{Cl}(C) \cap Y \neq \emptyset$ and $\mathrm{Cl}(C) \cap Z \neq \emptyset$, then $V \cup Y \cup \mathrm{Cl}(C) \cup Z \cup K$ is a subcontinuum of X containing the points of irreducibility of X. Thus, $X = V \cup Y \cup \mathrm{Cl}(C) \cup Z \cup K$. Since $R \cap (V \cup Y \cup Z \cup K) = \emptyset$, it follows that $R \subset \mathrm{Cl}(C)$. Hence, R is connected ($C \subset R \subset \mathrm{Cl}(C)$), a contradiction. Therefore, either $\mathrm{Cl}(C) \cap Y = \emptyset$ or $\mathrm{Cl}(C) \cap Z = \emptyset$. Let

$$\mathcal{A} = \{\mathrm{Cl}(C) \mid C \text{ is a component of } R \text{ and } \mathrm{Cl}(C) \cap Y \neq \emptyset\}$$

and

$$\mathcal{B} = \{\mathrm{Cl}(C) \mid C \text{ is a component of } R \text{ and } \mathrm{Cl}(C) \cap Z \neq \emptyset\}.$$

We claim that $\mathcal{A}$ and $\mathcal{B}$ are both nonempty. Suppose, to the contrary, that $\mathcal{B}$ is empty. Note that $V \cup Y \cup (\bigcup \mathcal{A})$ is a connected set and that $\bigcup \mathcal{A}$ is not connected. Then there exist two separated subsets J and L of $V \cup Y \cup (\bigcup \mathcal{A})$ such that $\bigcup \mathcal{A} = J \cup L$. Note that $V \cup Y \cup J$ and $V \cup Y \cup L$ are connected sets (Lemma 1.4.32). Hence, either $(V \cup Y \cup \mathrm{Cl}(J)) \cap Z \neq \emptyset$ or $(V \cup Y \cup \mathrm{Cl}(L)) \cap Z \neq \emptyset$. Suppose $(V \cup Y \cup \mathrm{Cl}(J)) \cap Z \neq \emptyset$. Then $(V \cup Y \cup \mathrm{Cl}(J)) \cup Z \cup K$ is a proper subcontinuum of X containing the points of irreducibility of X, a contradiction. Therefore, $\mathcal{B} \neq \emptyset$. Similarly, $\mathcal{A} \neq \emptyset$.

Since X is a continuum, $\mathrm{Cl}\,(\bigcup \mathcal{A}) \cap \mathrm{Cl}\,(\bigcup \mathcal{B}) \neq \emptyset$. Let $x \in \mathrm{Cl}\,(\bigcup \mathcal{A}) \cap \mathrm{Cl}\,(\bigcup \mathcal{B})$. Since $x \in \mathrm{Cl}\,(\bigcup \mathcal{A})$, there exists a net $\{a_\lambda\}_{\lambda \in \Lambda}$ of elements of $\bigcup \mathcal{A}$ converging to x [40, 1.6.3]. For each $\lambda \in \Lambda$, let $\mathrm{Cl}(C_\lambda) \in \mathcal{A}$ be such that $a_\lambda \in \mathrm{Cl}(C_\lambda)$. By Theorem 1.6.7, we assume that the net $\{\mathrm{Cl}(C_\lambda)\}_{\lambda \in \Lambda}$ of subcontinua of X converges (in the Vietoris topology) to a subcontinuum T of X [104, Theorem 4]. Note that $T \cap Y \neq \emptyset$ and $x \in T$. Similarly, there exists a subcontinuum T' of X such that $T' \cap Z \neq \emptyset$ and $x \in T'$. Consequently, $V \cup Y \cup T \cup T' \cup Z \cup K$ is a proper subcontinuum of X containing its points of irreducibility, a contradiction. Therefore, R is connected.

Now, observe that

$$\begin{aligned} X \setminus [(V \cup Y) \cup (Z \cup K)] &= (X \setminus V) \cap (X \setminus Y) \cap (X \setminus Z) \cap (X \setminus K) \\ &= (U \cap Y) \cap (X \setminus Y) \cap (X \setminus Z) \cap (H \cap Z) \\ &= U \cap H = U \cap (X \setminus Y) \cap (X \setminus Z) \cap H \\ &= U \cap (X \setminus Z) \cap (X \setminus Y) \cap H \\ &= (U \setminus Z) \cap (H \setminus Y) = R. \end{aligned}$$

Since $V \cap (Y \cup Z \cup K) = \emptyset$ and $K \cap (Z \cup Y \cup V) = \emptyset$, $X \setminus (Y \cup Z) = V \cup R \cup K$. □

1.4.42 Definition. A continuum X *is weakly irreducible* provided that the complement of each finite union of subcontinua of X only has a finite number of components.

1.4.43 Theorem. *If X is an irreducible continuum, then X is weakly irreducible.*

Proof. Let $Z_1, \ldots, Z_n$ be a finite family of subcontinua of X such that $Z_j \cap Z_k = \emptyset$ if $j \neq k$. For each $j \in \{1, \ldots, n\}$, by Theorem 1.4.40, we assume that $X \setminus Z_j = U_j \cup V_j$, where U_j and V_j are open connected subsets of X. Without loss of generality, we suppose that $\bigcup_{j=2}^{n} Z_j \subset V_1$ and $\bigcup_{j=1}^{n-1} Z_j \subset U_n$. Hence, U_1 and V_n may be empty. We assume also that, for each $j \in \{2, \ldots, n-1\}$, $\bigcup_{k=1}^{j-1} Z_k \subset U_j$ and $\bigcup_{k=j+1}^{n} Z_k \subset V_j$. For every $j \in \{1, \ldots, n-1\}$, let $R_j = (V_j \setminus Z_{j+1}) \cap (U_{j+1} \setminus Z_j)$.

By the proof of Lemma 1.4.41, R_j is a connected open subset of X. Note that $X \setminus \left(\bigcup_{j=1}^n Z_j\right) = U_1 \cup \left(\bigcup_{j=1}^{n-1} R_j\right) \cup V_n$. Thus, $X \setminus \left(\bigcup_{j=1}^n Z_j\right)$ has, at most, $n+1$ components. Therefore, X is weakly irreducible. □

1.4.44 Definition. A continuum X *is unicoherent* provided that for every pair, A and B, of subcontinua of X such that $X = A \cup B$, $A \cap B$ is connected. We say that X *is hereditarily unicoherent* if each subcontinuum of X is unicoherent (Figure 1.14).

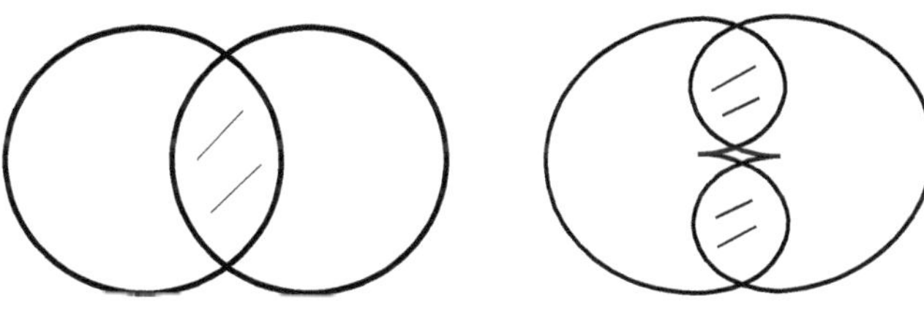

Fig. 1.14 *Unicoherent* *Not Unicoherent*

1.4.45 Definition. Let X and Y be continua. A surjective map $f: X \twoheadrightarrow Y$ is said to be *monotone* if $f^{-1}(y)$ is connected for each $y \in Y$.

The next lemma gives a very useful characterization of monotone maps.

1.4.46 Lemma. *Let X and Y be continua, and let $f: X \twoheadrightarrow Y$ be a surjective map. Then f is monotone if and only if $f^{-1}(C)$ is connected for each connected subset C of Y.*

Proof. Suppose f is monotone and let C be a subset of Y such that $f^{-1}(C)$ is not connected. Then there exist two nonempty subsets A and B of X such that $f^{-1}(C) = A \cup B$, $\mathrm{Cl}_X(A) \cap B = \emptyset$ and $A \cap \mathrm{Cl}_X(B) = \emptyset$. Observe that if $y \in C$ and $f^{-1}(y) \cap A \neq \emptyset$, then $f^{-1}(y) \subset A$ ($f^{-1}(y)$ is connected). Let $M = \{y \in C \mid f^{-1}(y) \subset A\}$. Hence, $A = f^{-1}(M)$. Similarly, if $N = \{y \in C \mid f^{-1}(y) \subset B\}$, then $B = f^{-1}(N)$. Note that $C = M \cup N$. Suppose there exists $y \in \mathrm{Cl}_Y(M) \cap N$. Since $y \in N$, $f^{-1}(y) \subset B$. Since $y \in \mathrm{Cl}_Y(M)$, there exists a net $\{y_\lambda\}_{\lambda \in \Lambda}$ of points of M converging to y [40, 1.6.3]. This implies that $f^{-1}(y_\lambda) \subset A$ for every $\lambda \in \Lambda$. Let $x_\lambda \in f^{-1}(y_\lambda)$. Since X is a compactum, without loss of generality, we assume that $\{x_\lambda\}_{\lambda \in \Lambda}$ converges to a point x [40, 3.1.23 and 1.6.1]. Note that $x \in \mathrm{Cl}_X(A)$ and, by the continuity of f, $f(x) = y$ [97, Proposition 3.38]. Thus, $x \in \mathrm{Cl}_X(A) \cap f^{-1}(y) \subset \mathrm{Cl}_X(A) \cap B$, a contradiction. Hence, $\mathrm{Cl}_Y(M) \cap N = \emptyset$. Similarly, $M \cap \mathrm{Cl}_Y(N) = \emptyset$. Therefore, C is not connected. The other implication is clear. □

1.4.47 Definition. An *arc* is a continuum with the property that in contains exactly two points whose complement is connected. A *metric arc* is a metric continuum homeomorphic to $[0, 1]$.

Now, we present the definition of an arc-smooth continuum. This notion is used to study continua which are strictly point $\mathcal{T}$-asymmetric (Definition 7.3.1).

1.4.48 Definition. An arcwise connected continuum X is *arc-smooth at* $p \in X$ if there exists a map $\alpha\colon X \to \mathcal{C}_1(X)$ (Definition 1.6.1) such that $\alpha(p) = \{p\}$, $\alpha(x)$ is an arc in X joining p and x and if $y \in \alpha(x)$, then $\alpha(y) \subset \alpha(x)$. The continuum X is *arc-smooth* if there exists a point at which it is arc-smooth.

1.4.49 Theorem. *If X is an arc-smooth continuum, then for each closed subset H of X, the set $\bigcup_{x\in H}\alpha(x)$ is a subcontinuum of X.*

Proof. Suppose X is arc-smooth at p and let $M = \bigcup_{x\in H}\alpha(x)$. Note that for each $x \in H$, $p \in \alpha(x)$. Hence, M is a connected subset of X. Let $z \in \mathrm{Cl}(M)$. Then there exists a net $\{z_\lambda\}_{\lambda\in\Lambda}$ of points of M converging to z [40, 1.6.3]. Thus, for each $\lambda \in \Lambda$, there exists $x_\lambda \in H$ such that $z_\lambda \in \alpha(x_\lambda)$. Since H is closed in X, without loss of generality, we assume that the net $\{x_\lambda\}_{\lambda\in\Lambda}$ converges to a point $x \in H$ [40, 3.1.23 and 1.6.1]. Since X is arc-smooth, α is continuous. Hence, the net of arcs $\{\alpha(x_\lambda)\}_{\lambda\in\Lambda}$ converges to the arc $\alpha(x)$ [97, Proposition 3.38]. Note that, by the properties of α, the net of arcs $\{\alpha(z_\lambda)\}_{\lambda\in\Lambda}$ converges to the arc $\alpha(z)$ and $z \in \alpha(x)$. Thus, $z \in M$ and M is closed. Therefore, M is a subcontinuum of X. □

1.4.50 Definition. Let X be a continuum and let K be a subcontinuum of X. Then K is a *terminal subcontinuum* of X if for each subcontinuum L of X such that $L \cap K \neq \emptyset$, then either $L \subset K$ or $K \subset L$.

1.4.51 Lemma. *Let X be a continuum. If W is a proper terminal subcontinuum of X, then* $\mathrm{Int}(W) = \emptyset$.

Proof. Suppose W is a proper terminal subcontinuum of X and $\mathrm{Int}(W) \neq \emptyset$. Note that $\mathrm{Bd}(\mathrm{Int}(W)) \subset W$. Let $x \in X \setminus W$ and let C be the component of $X \setminus \mathrm{Int}(W)$ containing x. By Theorem 1.4.36, $C \cap \mathrm{Bd}(\mathrm{Int}(W)) \neq \emptyset$. Hence, $C \cap W \neq \emptyset$, and $C \setminus W \neq \emptyset$. Since W is a terminal subcontinuum of X, $W \subset C$, a contradiction to the fact that $C \cap \mathrm{Int}(W) = \emptyset$. Therefore, $\mathrm{Int}(W) = \emptyset$. □

1.4.52 Lemma. *Let X be a continuum. If X is aposyndetic, then X does not contain nondegenerate proper terminal subcontinua.*

Proof. Suppose Y is a nondegenerate proper terminal subcontinuum of X. Let $y \in Y$ and let $y' \in Y \setminus \{y\}$. Then since X is aposyndetic, there exists a subcontinuum W of X such that $y \in \mathrm{Int}(W) \subset W \subset X \setminus \{y'\}$. Since Y is terminal and $Y \setminus W \neq \emptyset$, $W \subset Y$. Thus, y is an interior point of Y. Since y is an arbitrary point of Y, all the points of Y are interior points. Hence, Y is a nonempty open and closed proper subset of X. This contradicts the fact that X is connected. Therefore, X does not contain nondegenerate proper terminal subcontinua. □

1.4.53 Lemma. *Let X and Y be continua. If $f\colon X \twoheadrightarrow Y$ is a monotone map and W is a terminal subcontinuum of X, then $f(W)$ is a terminal subcontinuum of Y.*

Proof. Let W be a terminal subcontinuum of X. Let K be a subcontinuum of Y such that $K \cap f(W) \neq \emptyset$. Since f is a monotone map, $f^{-1}(K)$ is a subcontinuum of X (Lemma 1.4.46) such that $f^{-1}(K) \cap W \neq \emptyset$. Hence, since W is a terminal subcontinuum of X, either $W \subset f^{-1}(K)$ or $f^{-1}(K) \subset W$. This implies that either $f(W) \subset K$ or $K \subset f(W)$ (f is surjective). Therefore, $f(W)$ is a terminal subcontinuum of Y. □

1.4.54 Corollary. *Let X and Y be continua. If Z is a terminal subcontinuum of X and $h\colon X \twoheadrightarrow Y$ is a homeomorphism, then $h(Z)$ is a terminal subcontinuum of Y.*

1.4.55 Definition. A continuum X is *homogeneous* provided that for every pair of points x_1 and x_2 of X, there exists a homeomorphism $h\colon X \twoheadrightarrow X$ such that $h(x_1) = x_2$.

1.4.56 Notation. Let X be a homogeneous continuum, let $\mathcal{H}(X)$ be its group of homeomorphisms. If $x \in X$, then define $\gamma_x\colon \mathcal{H}(X) \to X$ by $\gamma_x(h) = h(x)$.

1.4.57 Definition. A homogeneous continuum X is said to be an *Effros continuum*, if γ_x is an open map (Notation 1.4.56) for all $x \in X$.

1.4.58 Definition. Let Z be a Tychonoff space and let $\mathfrak{U}$ be a uniformity that induces the topology of Z. Then Z has the *uniform property of Effros with respect to* $\mathfrak{U}$, provided that for each $U \in \mathfrak{U}$, there exists $V \in \mathfrak{U}$ such that if z_1 and z_2 are two points of Z with $\rho(z_1, z_2) < V$, there exists a homeomorphism $h\colon Z \twoheadrightarrow Z$ such that $h(z_1) = z_2$ and $\rho(z, h(z)) < U$ for all $z \in Z$. The entourage V is called an *Effros entourage for* U. A homeomorphism $h\colon Z \twoheadrightarrow Z$ satisfying $\rho(z, h(z)) < U$, for all $z \in Z$, is called a *U-homeomorphism*.

1.4.59 Theorem. *Let Z be a connected Tychonoff space and let $\mathfrak{U}$ be a uniformity that induces the topology of Z. If Z has the uniform property of Effros with respect to $\mathfrak{U}$, then Z is a homogeneous space.*

Proof. Let z_1 and z_2 be two points of Z and let $U \in \mathfrak{U}$ be such that $\rho(z_1, z_2) \geq U$. Since Z has the uniform property of Effros with respect to $\mathfrak{U}$, there exists an Effros entourage V for U. Let $V' \in \mathfrak{U}$ be such that $2V' \subset V$. Note that $\{\mathrm{Int}(B(z, V')) \mid z \in Z\}$ is an open cover of Z. Since Z is connected, by [24, (2F2)], there exist $w_1, \ldots, w_n$ in Z such that $z_1 \in \mathrm{Int}(B(w_1, V'))$, $z_2 \in \mathrm{Int}(B(w_n, V'))$ and $\mathrm{Int}(B(w_j, V')) \cap \mathrm{Int}(B(w_k, V')) \neq \emptyset$ if and only if $|j - k| \leq 1$. Let $x_0 = z_1$ and let $x_{n+1} = z_2$. For each $j \in \{1, \ldots, n-1\}$, let $x_j \in \mathrm{Int}(B(w_j, V')) \cap \mathrm{Int}(B(w_{j+1}, V'))$. Let $j \in \{0, \ldots, n\}$. Since $\rho(x_j, x_{j+1}) < V$, there exists a U-homeomorphism $h_j\colon Z \twoheadrightarrow Z$ such that $h_j(x_j) = x_{j+1}$. Let $h = h_n \circ \cdots \circ h_0$. Then $h\colon Z \twoheadrightarrow Z$ is a homeomorphism such that $h(z_1) = z_2$. Therefore, Z is a homogeneous space. □

1.4.60 Theorem. *If X is an Effros continuum, then X has the uniform property of Effros.*

Proof. By Theorem 1.3.9, the topology induced by $\widehat{\mathfrak{U}_X}$ on $\mathcal{H}(X)$ coincides with the compact-open topology. Also, $\widehat{\mathcal{U}}$ is a base for the uniformity $\widehat{\mathfrak{U}_X}$. Let x_0 be a point of X and let $U \in \mathfrak{U}_X$ (Remark 1.3.4). Let $U' \in \mathfrak{U}_X$ be such that $2U' \subset U$. Since X is an Effros continuum, the map γ_{x_0} is an open map. Then $\gamma_{x_0}(\mathrm{Int}(B(1_X, \widehat{U'})))$

is an open neighbourhood of x_0 in X. Then there exists $V_{x_0} \in \mathfrak{U}_X$ such that $B(x_0, V_{x_0}) \subset \gamma_{x_0}(\text{Int}(B(1_X, \widehat{U'})))$. Thus, if $x \in B(x_0, V_{x_0})$, then there exists $h \in \text{Int}(B(1_X, \widehat{U'}))$ such that $h(x_0) = x$ and $\rho(z, h(z)) < U'$ for every $z \in X$. Observe that $\{\text{Int}(B(x, V_x)) \mid x \in X\}$ is an open cover of X. Since X is compact, by Theorem 1.3.6, there exists $V \in \mathfrak{U}_X$ such that $\mathfrak{C}(V)$ refines $\{\text{Int}(B(x, V_x)) \mid x \in X\}$. Let x_1 and x_2 be points of X such that $\rho(x_1, x_2) < V$. Then there exists x_3 in X such that x_1 and x_2 both belong to $\text{Int}(B(x_3, V_{x_3}))$. Let h_1 and h_2 be elements of $\text{Int}(B(1_X, \widehat{U'}))$ such that $h_1(x_3) = x_1$ and $h_2(x_3) = x_2$. Let $h = h_2 \circ h_1^{-1}$. Then h is a homeomorphism of X and $h(x_1) = x_2$. Let z be a point of X. Since $\rho(z, h(z)) = \rho(z, h_2 \circ h_1^{-1}(z))$, $\rho(z, h_1^{-1}(z)) = \rho(h_1 \circ h_1^{-1}(z), h_1^{-1}(z)) < U'$, $\rho(h_1^{-1}(z), h_2 \circ h_1^{-1}(z)) < U'$ and $2U' \subset U$, we obtain that $\rho(z, h(z)) < U$.

Therefore, X has the uniform property of Effros. □

As a first consequence of the uniform property of Effros, we have:

1.4.61 Theorem. *Let X be a continuum with the uniform property of Effros. Then the following are equivalent:*

(1) *X is locally connected;*
(2) *X is locally connected at some point;*
(3) *X is connected im kleinen at some point;*
(4) *X is almost connected im kleinen at every point of X;*
(5) *X is almost connected im kleinen at some point of X.*

Proof. By Theorem 1.4.59, X is homogeneous. We only prove that (5) implies (1), the other implications are clear.

Suppose X is almost connected im kleinen at x_0. Let A be an open subset of X such that $x_0 \in A$. Let $U \in \mathfrak{U}_X$ be such that $B(x_0, U) \subset A$. Let $U' \in \mathfrak{U}_X$ be such that $2U' \subset U$. Since X has the uniform property of Effros, there exists an Effros entourage V for U'. Without loss of generality, we assume that $V \subset U'$. Since X is almost connected im kleinen at x_0, there exists a subcontinuum W of X such that $W \subset \text{Int}(B(x_0, V)$ and $\text{Int}(W) \neq \emptyset$. Let $w_0 \in \text{Int}(W)$. Then $\rho(x_0, w_0) < V$. Hence, there exists a U'-homeomorphism $h\colon X \twoheadrightarrow X$ such that $h(w_0) = x_0$. Thus, $x_0 \in \text{Int}(h(W))$. Since for each $w \in W$, $\rho(x_0, w) < V$, $\rho(w, h(w)) < U'$ and $V \subset U'$, we obtain that for every $w \in W$, $\rho(x_0, h(w)) < 2U'$. Hence, since $2U' \subset U$, we have that $h(W) \subset B(x_0, U) \subset A$. Thus, X is connected im kleinen at x_0. Since X is homogeneous, X is connected im kleinen at each of its points. Therefore, X is locally connected, Theorem 1.4.18. □

1.4.62 Definition. A metric continuum X is a *graph* provided that it can be written as the union of finitely many metric arcs any two of which are either disjoint or intersect only in one or both of their end points.

1.4.63 Definition. A metric continuum X is of *type* λ provided that X is irreducible and each indecomposable subcontinuum of X has empty interior.

1.4.64 Remark. By [119, Theorem 10, p. 15], a metric continuum X is of type λ if and only if it admits a monotone upper semicontinuous decomposition $\mathcal{G}$ such that each element of $\mathcal{G}$ is nowhere dense and $X/\mathcal{G}$ is a metric arc. Each element of $\mathcal{G}$ is called a *layer* of X.

1.4.65 Definition. A metric continuum X of type λ for which $\mathcal{G}$ (Remark 1.4.64) is continuous is a *continuously irreducible continuum.*

1.4.66 Definition. A continuum X is a θ*-continuum* (θ_n*-continuum* for some $n \in \mathbb{N}$) provided that for each subcontinuum K of X, $X \setminus K$ only has finitely many components ($X \setminus K$ has at most n components).

1.4.67 Definition. Let X be a metric θ-continuum (metric θ_n-continuum). We say that X is of *type A* provided that it admits a monotone upper semicontinuous decomposition $\mathcal{D}$ whose quotient space is a graph. X is of *type A′* if, in addition, the elements of the decomposition have empty interior.

1.4.68 Definition. A metric θ-continuum (metric θ_n-continuum) of type A' for which the decomposition $\mathcal{D}$ is continuous is a *continuously type A′ θ-continuum* (θ_n*-continuum*).

1.4.69 Lemma. *Let X be a continuously type A' θ-continuum and let $q\colon X \twoheadrightarrow D$ be the quotient map, where D is a graph. If K is a proper subcontinuum of X, then $q(K) \neq D$.*

Proof. Let K be a proper subcontinuum of X. If $\mathrm{Int}_X(K) = \emptyset$, then, since q is an open map and D is a finite graph, $q(K)$ is a degenerate subcontinuum of D. Thus, $q(K) \neq D$.

Assume that $\mathrm{Int}_X(K) \neq \emptyset$. Since X is θ-continuum, $X \setminus K$ only has a finite number of components, say $C_1, \ldots, C_\ell$. Note that each C_j is an open subset of X, Lemma 1.4.1. Suppose $q(K) = D$. Since C_1 is open, $q(C_1)$ is an open connected subset of D. Let $\chi_0 \in \mathrm{Int}_D(q(C_1))$ and let $\{\chi_j\}_{j=1}^{\infty}$ be a sequence of distinct elements of $\mathrm{Int}_D(q(C_1))$ converging to χ_0. Since D is a graph, $D \setminus \{\chi_j\}_{j=0}^{\infty}$ has infinitely many components. Let $L = K \cup \left(\bigcup_{j=0}^{\infty} q^{-1}(\chi_k)\right)$. Then, since $q(K) = D$, L is a subcontinuum of X. Also, since $K \cap C_1 = \emptyset$ and, for each $j \in \mathbb{N} \cup \{0\}$, $C_1 \cap q^{-1}(\chi_j) \neq \emptyset$, we obtain that $X \setminus L$ has infinitely many components, a contradiction to the fact that X is a θ-continuum. Therefore, $q(K) \neq D$. □

A proof of the next result may be found in [45, Theorem 3.4].

1.4.70 Theorem. *A continuum X is weakly irreducible if and only if X is a θ-continuum.*

Next, we consider the notion of *generalized metric continuum.*

1.4.71 Definition. A *generalized metric continuum* is a locally compact, connected, metric space

1.4.72 Definition. A generalized metric continuum Z is an *exhausted metric σ-continuum* provided that there exists a sequence $\{K_n\}_{n=1}^{\infty}$ of subcontinua of Z such that $K_n \subset \mathrm{Int}(K_{n+1})$, for all $n \in \mathbb{N}$, and $Z = \bigcup_{n=1}^{\infty} K_n$. The sequence $\{K_n\}_{n=1}^{\infty}$ is an *exhaustive sequence* of subcontinua of Z.

1.4.73 Example. Let X be the topologist sine curve, Example 1.4.9, and let $Z = X \setminus \{(0,0)\}$. Then Z is a generalized metric continuum that is not an exhaustive σ-continuum.

1.4.74 Definition. A Hausdorff space Z is *continuumwise connected* provided that for each pair of points z_1 and z_2 of Z, there exists a subcontinuum W of Z such that $\{z_1, z_2\} \subset W$.

1.4.75 Theorem. *Let Z be a generalized metric continuum. Then the following are equivalent:*

(1) *Z is an exhausted metric σ-continuum;*
(2) *For every pair of points p and q of Z, there exists a subcontinuum K of Z such that $\{p, q\} \subset \mathrm{Int}(K)$;*
(3) *Z is continuumwise connected and for each point p in Z, there exists a subcontinuum K of Z such that $p \in \mathrm{Int}(K)$;*
(4) *For every point p of Z, there exists a subcontinuum K of Z such that $p \in \mathrm{Int}(K)$;*
(5) *For each compact subset L of Z, there exists a subcontinuum K of Z such that $L \subset \mathrm{Int}(K)$;*
(6) *For every compact subset L of Z, there exists a subcontinuum K of Z such that $L \subset K$;*
(7) *The hyperspace of all compact subsets of Z is arcwise connected;*
(8) *The hyperspace of all compact subsets of Z is continuumwise connected.*

Proof. Note that if Z is an exhausted metric σ-continuum, then, by definition, for every pair of points p and q of Z, there exists a subcontinuum K of Z such that $\{p, q\} \subset \mathrm{Int}(K)$. Now, suppose (2) holds. Then Z is continuumwise connected and for each point p of Z, there exists a subcontinuum K of Z such that $p \in \mathrm{Int}(K)$. Assume (3) is true. Then for every point p of Z, there exists a subcontinuum K of Z such that $p \in \mathrm{Int}(K)$.

Now suppose (4) holds, we show (3) is true. Let p be a point of Z and let $\kappa(p)$ be the union of all subcontinua of Z containing p. Note that (4) implies that $\kappa(p)$ is an open subset of Z. Since the family $\{\kappa(z) \mid z \in Z\}$ forms a decomposition of Z, $\kappa(p)$ is a closed subset of Z. Since Z is connected, $\kappa(p) = Z$. Thus, Z is continuumwise connected.

Next, we assume (4) and prove (5) holds. Let L be a compact subset of Z. For each point $l \in L$, let K_l be a subcontinuum of Z such that $l \in \mathrm{Int}(K_l)$. Since L is compact, there exist $l_1, \ldots, l_n$ in L such that $L \subset \bigcup_{j=1}^{n} \mathrm{Int}(K_{l_j})$. Let z be a point of Z. Since (3) is true, Z is continuumwise connected. Hence, for each $j \in \{1, \ldots, n\}$, there exists a subcontinuum M_j such that $\{z, l_j\} \subset M_j$. Let $K = \bigcup_{j=1}^{n}(K_{l_j} \cup M_j)$. Then K is a subcontinuum of Z and $L \subset \mathrm{Int}(K)$. Clearly, if (5) is true, then (6) holds.

Assume (6), we show (1). We use mathematical induction. Let $\{L_n\}_{n=1}^{\infty}$ be a sequence of compact subsets of Z such that $L_n \subset \mathrm{Int}(L_{n+1})$ for all $n \in \mathbb{N}$ and $Z = \bigcup_{n=1}^{\infty} L_n$. By (6), there exists a subcontinuum K_1 of Z such that $L_1 \subset \mathrm{Int}(K_1)$. Suppose $K_1, \dots, K_n$ are subcontinua of Z such that $K_j \subset \mathrm{Int}(K_{j+1})$ for each $j \in \{1, \dots, n-1\}$. By (6), there exists a subcontinuum K_{n+1} of Z such that $L_{n+1} \cup K_n \subset \mathrm{Int}(K_{n+1})$. Hence, $\{K_n\}_{n=1}^{\infty}$ is an exhaustive sequence of subcontinua of Z. Therefore, Z is an exhausted metric σ-continuum.

Next, suppose (6) and we prove (7). Let L_1 and L_2 be compact subsets of Z. By (6), there exists a subcontinuum K of Z such that $L_1 \cup L_2 \subset K$. By [105, (1.8)], there exist two arcs $\alpha, \beta \colon [0,1] \to \mathcal{C}_1(K)$ such that $\alpha(0) = L_1$, $\alpha(1) = K$, $\beta(0) = L_2$ and $\beta(1) = K$. Thus, $\alpha([0,1]) \cup \beta([0,1])$ contains an arc joining L_1 and L_2. Clearly, (7) implies (8).

To finish, we show that if (8) is true, then (6) holds. Let L be a compact subset of Z and let z be a point of Z. Since the hyperspace of compact subsets of Z is continuumwise connected, there exists a subcontinuum $\mathcal{K}$ of this space containing L and $\{z\}$. Let $K = \bigcup \mathcal{K}$. By [105, (1.49)], K is a subcontinuum of Z containing L. □

We end this section with some results of generalized inverse limits of metric continua with a single bonding function.

1.4.76 Theorem. *If X is a metric continuum and $\mathcal{G}$ is a monotone upper semicontinuous decomposition of X, then $\varprojlim f_{\mathcal{G}}$ is an infinite-dimensional decomposable metric continuum.*

Proof. By [65, Theorem 3.2] and Lemma 1.2.27, $\varprojlim f_{\mathcal{G}}$ is compact and connected. By [92, Lemma 1.1.7 and 1.1.8], $\varprojlim f_{\mathcal{G}}$ is a metric space. Since, by Lemma 1.2.27, if $G \in \mathcal{G}$, then $G^{\omega} \subset \varprojlim f_{\mathcal{G}}$, we have that $\varprojlim f_{\mathcal{G}}$ is infinite-dimensional.

To see that $\varprojlim f_{\mathcal{G}}$ is decomposable let $(x_n)_{n=1}^{\omega}$ and $(y_n)_{n=1}^{\omega}$ be two points of $\varprojlim f_{\mathcal{G}}$. Then there exist two points z_1 and z_2 of X such that $(x_n)_{n=1}^{\omega} \in G_{z_1}^{\omega}$ and $(y_n)_{n=1}^{\omega} \in G_{z_2}^{\omega}$. Then $G_{z_1}^{\omega} \cup \Delta_X^{\omega} \cup G_{z_2}^{\omega}$ is a proper subcontinuum of $\varprojlim f_{\mathcal{G}}$ containing $(x_n)_{n=1}^{\omega}$ and $(y_n)_{n=1}^{\omega}$. Therefore, $\varprojlim f_{\mathcal{G}}$ is decomposable. □

1.4.77 Remark. Let X be a metric continuum and let $\mathcal{G}$ be an upper semicontinuous decomposition of X. If $\mathcal{G}^{\star} = \{K \mid K \text{ is a component of } G, \text{ for some } G \in \mathcal{G}\}$, then $\mathcal{G}^{\star}$ is an upper semicontinuous decomposition of X (Theorem 1.1.35) and $\varprojlim f_{\mathcal{G}^{\star}}$ is a metric subcontinuum of $\varprojlim f_{\mathcal{G}}$.

1.4.78 Theorem. *If X is a metric continuum and $\mathcal{G} = \{G_x \mid x \in X\}$ is a continuous decomposition of X, then $\mathcal{G}^{*} = \{G_x^{\omega} \mid x \in X\}$ is a continuous decomposition of $\varprojlim f_{\mathcal{G}}$. Moreover, $\varprojlim f_{\mathcal{G}}/\mathcal{G}^{*}$ is homeomorphic to $X/\mathcal{G}$.*

Proof. Let $q\colon X \twoheadrightarrow X/\mathcal{G}$ be the quotient map, and let $\pi_1\colon \varprojlim f_{\mathcal{G}} \to X$ be the projection map to the first factor space. Note that $\mathcal{G}^* = \{(q\circ\pi_1)^{-1}(\chi) \mid \chi \in X/\mathcal{G}\}$. Hence, $\mathcal{G}^*$ is an upper semicontinuous decomposition of $\varprojlim f_{\mathcal{G}}$ (Theorem 1.1.18).

Let $(x_n)_{n=1}^{\omega}$ and $(y_n)_{n=1}^{\omega}$ be two points of G_x^{ω}, and let $\mathcal{U}$ be an open subset of $\varprojlim f_{\mathcal{G}}$ such that $(x_n)_{n=1}^{\omega} \in \mathcal{U}$. Let $\bigcap_{j=1}^{m} \pi_j^{-1}(U_j)$ be a basic open set such that $(x_n)_{n=1}^{\omega} \in \bigcap_{j=1}^{m} \pi_j^{-1}(U_j) \subset \mathcal{U}$, where $\pi_j\colon \varprojlim f_{\mathcal{G}} \to X$ is the projection map to the jth factor space. Note that for each $j \in \{1,\ldots,m\}$, $x_j \in G_x \cap U_j$. Since $\mathcal{G}$ is a continuous decomposition, for each $j \in \{1,\ldots,m\}$, there exists an open subset V_j such that $y_j \in V_j$ and if $G \in \mathcal{G}$ is such that $G \cap V_j \neq \emptyset$, we have that $G \cap U_j \neq \emptyset$. Let $\mathcal{V} = \bigcap_{j=1}^{m} \pi_j^{-1}(V_j)$. Let $G^{\omega} \in \mathcal{G}^*$ be such that $G^{\omega} \cap \mathcal{V} \neq \emptyset$. Then $G \cap V_j \neq \emptyset$ for every $j \in \{1,\ldots,m\}$. By the construction of the V_j's, we have that $G \cap U_j \neq \emptyset$ for all $j \in \{1,\ldots,m\}$. For each $j \in \{1,\ldots,m\}$, let $z_j \in G \cap U_j$, and for each $j \geq m+1$, let $z_j \in G$. Then $(z_n)_{n=1}^{\omega} \in G^{\omega} \cap \left(\bigcap_{j=1}^{m} \pi_j^{-1}(U_j)\right)$. Hence $G^{\omega} \cap \mathcal{U} \neq \emptyset$. Therefore, $\mathcal{G}^*$ is continuous.

Let $q^*\colon \varprojlim f_{\mathcal{G}} \to \varprojlim f_{\mathcal{G}}/\mathcal{G}^*$ be the quotient map. Note that $q\circ\pi_1$ is constant on each of the fibres of q^* and q^* is constant on each of the fibres of $q\circ\pi_1$. Then, by [36, 3.2, p. 123], there exist two maps $h\colon \varprojlim f_{\mathcal{G}}/\mathcal{G}^* \to X/\mathcal{G}$ and $h'\colon X/\mathcal{G} \to \varprojlim f_{\mathcal{G}}/\mathcal{G}^*$ such that $h\circ q^* = q\circ\pi_1$ and $h'\circ q\circ\pi_1 = q^*$. Note that h and h' are homeomorphisms. Therefore, $\varprojlim f_{\mathcal{G}}/\mathcal{G}^*$ is homeomorphic to $X/\mathcal{G}$. □

1.5 Uniformly Completely Regular Maps

We extend the notion of completely regular maps [37] to uniform spaces and present the properties needed to prove the F. B. Jones' Aposyndetic Decomposition Theorem (Theorem 3.3.8).

1.5.1 Definition. Let X and Y be Tychonoff spaces, let $\mathfrak{U}$ and $\mathfrak{U}'$ be uniformities inducing the topologies of X and Y, respectively. Let $g\colon X \twoheadrightarrow Y$ be a uniformly continuous and surjective function. Then g is *uniformly completely regular with respect to* $\mathfrak{U}$ *and* $\mathfrak{U}'$ if for each $U \in \mathfrak{U}$, there exists $V \in \mathfrak{U}'$ such that if y and y' are two points of Y such that $\rho_Y(y,y') < V$, there exists a homeomorphism $h\colon g^{-1}(y) \twoheadrightarrow g^{-1}(y')$ such that $\rho_X(z,h(z)) < U$ for all $z \in g^{-1}(y)$.

1.5.2 Theorem. *Let X and Y be compacta. If $g\colon X \twoheadrightarrow Y$ is a uniformly completely regular map, then g is open.*

Proof. Let W be an open subset of X and let $y \in g(W)$. Then there exists $x \in W$ such that $g(x) = y$. Also, there exists $U \in \mathfrak{U}_X$ such that $B_X(x,U) \subset W$. Since g is uniformly completely regular, there exists $V \in \mathfrak{U}_Y$ such that if y' is a point of Y and $\rho_Y(y,y') < V$, there exists a homeomorphism $h\colon g^{-1}(y) \twoheadrightarrow g^{-1}(y')$ such that $\rho_X(z,h(z)) < U$ for every $z \in g^{-1}(y)$. In particular, $\rho_X(x,h(x)) < U$. Hence,

$h(x) \in W$ and $g(h(x)) = y'$. Thus, $B_Y(y, V) \subset g(W)$. Since y is an arbitrary point of $g(W)$, $g(W)$ is an open subset of Y. Therefore, g is an open map. □

1.5.3 Definition. Let $\mathfrak{D}$ be the class of metric spaces or the class of normal spaces. By an *absolute neighbourhood retract for the class* $\mathfrak{D}$ we mean a space Y in $\mathfrak{D}$ such that for each $Z \in \mathfrak{D}$ for which there exists an embedding $i\colon Y \to Z$ such that $i(Y)$ is a closed subset of Z, there exist an open subset U of Z containing $i(Y)$ and a map $r\colon U \twoheadrightarrow i(Y)$ such that $r(r(u)) = r(u)$ for all $u \in U$.

1.5.4 Remark. Note that an absolute neighbourhood retract Y for the class of metric spaces is an absolute retract for the class of normal spaces if and only if Y is separable and topologically complete [61, Theorem 4.1, p. 86].

1.5.5 Theorem. *Let X be a continuum. If A is a terminal subcontinuum of X, if B is a nonempty subset of X disjoint from A, and if $f\colon A \to Y$ is a map from A into a compact metric absolute neighbourhood retract Y, then there exists a map $F\colon X \to Y$ such that $F|_A = f$ and $F|_B$ is homotopic to a constant map.*

Proof. Since Y is a metric absolute neighbourhood retract, there exist an open subset U of X containing A and a map $g\colon U \to Y$ such that $g|_A = f$ [99, 1.5.2]. Without loss of generality, we assume that $U \cap B = \emptyset$. By [61, Theorem 7.1, p. 96], Y is locally contractible. Hence, there exists an open neighbourhood V of a point a of A such that $V \subset U$ and $g|_V$ is homotopic to a constant map.

Note that $A \setminus V$ and $X \setminus U$ are two closed subsets of $X \setminus V$ such that no connected subset of $X \setminus V$ intersects both $A \setminus V$ and $X \setminus U$. (If K is a connected subset of $X \setminus V$ such that $K \cap (A \setminus V) \neq \emptyset$ and $K \cap (X \setminus U) \neq \emptyset$, then $\mathrm{Cl}(K)$ is a continuum in $X \setminus V$ intersecting A and $X \setminus A$. Since A is terminal, $A \subset \mathrm{Cl}(K)$. This implies that $V \cap \mathrm{Cl}(K) \neq \emptyset$, a contradiction.) By Theorem 1.4.8, there exist two disjoint closed subsets X_1 and X_2 of X such that $X \setminus V = X_1 \cup X_2$, $A \setminus V \subset X_1$ and $X \setminus U \subset X_2$.

Note that $X_1 \cup A$ and X_2 are two disjoint closed subsets of X. Then, by Urysohn's Lemma, there exists a map $h\colon X \twoheadrightarrow [0,1]$ such that $h(X_1 \cup A) = \{0\}$ and $h(X_2) = \{1\}$. Let $M = h^{-1}\left(\left[0, \frac{1}{2}\right]\right)$ and let $N = h^{-1}\left(\left[\frac{1}{2}, 1\right]\right)$. Then $X = M \cup N$, $A \subset M$, $X \setminus U \subset N$ and $M \cap N \subset V \setminus A$. Since $g|_{M \cap N}$ is homotopic to a constant map (because $M \cap N \subset V$ and $g|_V$ is homotopic to a constant map), by [24, (15.A.1)], there exists a map $k\colon N \to Y$ such that k is homotopic to a constant map and $k|_{M \cap N} = g|_{M \cap N}$.

Let $F\colon X \to Y$ be given by

$$F(x) = \begin{cases} g(x), & \text{if } x \in M, \\ k(x), & \text{if } x \in N. \end{cases}$$

Then F is well defined and continuous. Note that $F|_A = g|_A = f$ and $F|_B = k|_B$. Since $k|_B$ is homotopic to a constant map (because $B \subset X \setminus U \subset N$ and k is homotopic to a constant map), F is the desired extension of f. □

1.5.6 Definition. A continuum X is *cell-like* if each map of X into a compact metric absolute neighbourhood retract is homotopic to a constant map.

1.5.7 Definition. Let X and Y be continua and let $f\colon X \twoheadrightarrow Y$ be a surjective map. Then f a *cell-like map* if for each $y \in Y$, $f^{-1}(y)$ is a cell-like subcontinuum of X.

1.5.8 Theorem. *Let X and Z be nondegenerate continua. Suppose $g\colon X \twoheadrightarrow Z$ is a monotone uniform completely regular map. If z_1 is a point of Z such that $g^{-1}(z_1)$ is a terminal subcontinuum of X, then g is a cell-like map.*

Proof. Let Y be a compact metric absolute neighbourhood retract and let $f\colon g^{-1}(z_1) \to Y$ be a map. Let $z_2 \in Z \setminus \{z_1\}$. By Theorem 1.5.5, there exists an extension $F\colon X \to Y$ of f such that $F|_{g^{-1}(z_2)}$ is homotopic to a constant map.

Since Y is a compact metric absolute neighbourhood retract, by [61, Theorem 1.1, p. 111], there exists $\varepsilon > 0$ such that for any space A and any two maps $k, k'\colon A \to Y$ such that $d(k(a), k'(a)) < \varepsilon$, for all $a \in A$, where d is a metric compatible with Y, k and k' are homotopic.

Since F is uniformly continuous (Theorem 1.3.15), there exists $V \in \mathfrak{U}_X$ such that if $\rho_X(x, x') < V$, then $d(F(x), F(x')) < \varepsilon$.

Let $Z' = \{z \in Z \mid F|_{g^{-1}(z)}$ is homotopic to a constant map$\}$. Note that, since $z_2 \in Z'$, we have that $Z' \neq \emptyset$. We show that Z' is closed in Z. Let $z \in \mathrm{Cl}(Z')$. Since g is uniformly completely regular, there exists $W \in \mathfrak{U}_Z$ such that if z' is a point of Z such that $\rho_Z(z, z') < W$, then there exists a homeomorphism $h\colon g^{-1}(z) \to g^{-1}(z')$ such that $\rho_X(x, h(x)) < V$ for each $x \in g^{-1}(z)$. Observe that there exists a point $z' \in Z'$ such that $\rho_Z(z, z') < W$. Thus, there exists a homeomorphism h as described above. Let $x \in g^{-1}(z)$. Then $\rho_X(x, h(x)) < V$. Hence, $d(F(x), F(h(x))) = d(F|_{g^{-1}(z)}(x), (F|_{g^{-1}(z')}) \circ h(x)) < \varepsilon$. Thus, $F|_{g^{-1}(z)}$ and $(F|_{g^{-1}(z')}) \circ h$ are homotopic. Since $z' \in Z'$, $(F|_{g^{-1}(z')}) \circ h$ is homotopic to a constant map. As a consequence of this, $F|_{g^{-1}(z)}$ is homotopic to a constant map, and $z \in Z'$. Therefore, Z' is closed in Z. A similar argument proves that Z' is open in Z. Since Z' is a nonempty, closed and open subset of Z and Z is connected, $Z' = Z$. This implies that $F|_{g^{-1}(z_1)} = f$ is homotopic to a constant map. Since Y and f are arbitrary, $g^{-1}(z_1)$ is a cell-like continuum. Since the fibres of uniformly completely regular maps defined on continua are homeomorphic, g is a cell-like map. □

1.6 Hyperspaces

We give the definition of the main hyperspaces associated with a continuum. We present some of their elementary properties. We introduce the property of Kelley for continua and we relate it to the uniform property of Effros.

1.6.1 Definition. Given a compactum X, we define its *hyperspaces* as the following sets:

$$2^X = \{A \subset X \mid A \text{ is closed and nonempty}\},$$

and for each $n \in \mathbb{N}$

$$\mathcal{C}_n(X) = \{A \in 2^X \mid A \text{ has at most } n \text{ components}\},$$

$$\mathcal{F}_n(X) = \{A \in 2^X \mid A \text{ has at most } n \text{ points}\}.$$

$\mathcal{F}_n(X)$ is called *n-fold symmetric product of* X and $\mathcal{C}_n(X)$ is called *n-fold hyperspace of* X (Figure 1.15).

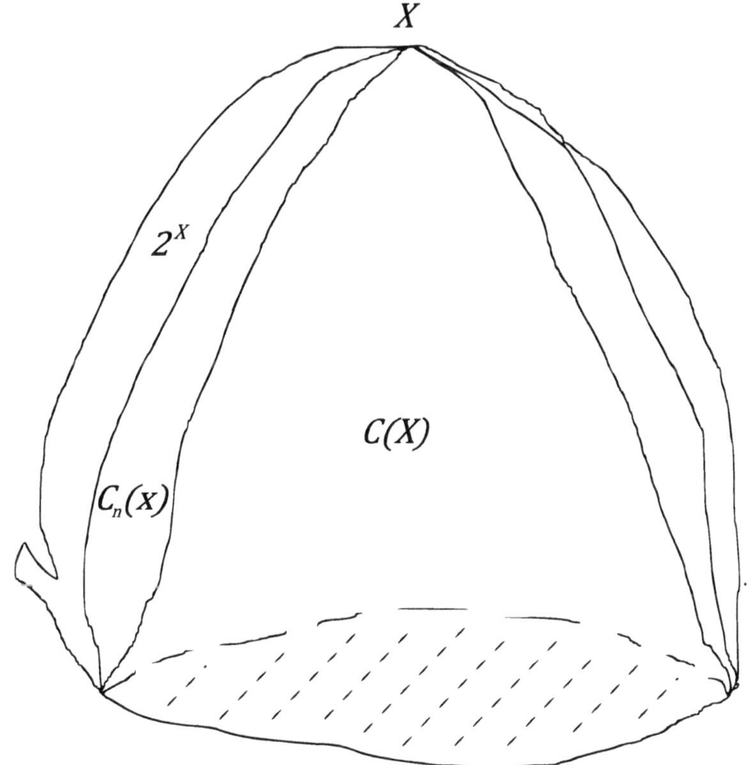

Fig. 1.15 Hyperspaces

1.6.2 Remark. Given a continuum X, let us observe that for each $n \in \mathbb{N}$,

$$\mathcal{F}_n(X) \subset \mathcal{C}_n(X),$$

$$\mathcal{C}_n(X) \subset \mathcal{C}_{n+1}(X),$$

and that

$$\mathcal{F}_n(X) \subset \mathcal{F}_{n+1}(X).$$

The n-fold symmetric products were defined by Borsuk and Ulam [15]. These hyperspaces have been studied by many people. Some recent results and references about n-fold symmetric products may be found in [79, 80] and [48]. It is not clear who defined or where the n-fold hyperspaces were defined. A study of the n-fold hyperspaces for metric continua is presented in [92, Chapter 6], see also [81].

1.6.3 Notation. Let X be a compactum. Given a finite collection of nonempty subsets of X, $U_1, \dots, U_m$, we define
$\langle U_1, \dots, U_m \rangle =$

$$\left\{ A \in 2^X \;\middle|\; A \subset \bigcup_{k=1}^{m} U_k \text{ and } A \cap U_k \neq \emptyset \text{ for each } k \in \{1, \dots, m\} \right\}$$

(Figure 1.16).

Fig. 1.16 The set $\langle U_1, \dots, U_m \rangle$

1.6.4 Notation. Let X be a compactum and let $n \in \mathbb{N}$. If $U_1, \dots, U_m$ are nonempty subsets of X, then

$$\langle U_1, \dots, U_m \rangle_n = \langle U_1, \dots, U_m \rangle \cap \mathcal{C}_n(X).$$

A proof of the next theorem may be found in [105, (0.11)]:

1.6.5 Theorem. *Let X be a compactum. If*

$$\mathcal{B} = \{\langle U_1, \dots, U_n \rangle \mid U_1, \dots, U_n \textit{ are open subsets of } X \textit{ and } n \in \mathbb{N}\},$$

then $\mathcal{B}$ is a basis for a topology of 2^X.

1.6.6 Definition. The topology for 2^X given by Theorem 1.6.5 is called the *Vietoris topology*.

By [98, 4.9.6 and 4.13.5] we have:

1.6.7 Theorem. *If X is a compactum, then 2^X and $\mathcal{C}_1(X)$ are compacta.*

1.6.8 Lemma. *Let X be a continuum. If $\mathcal{K}$ is a subcontinuum of 2^X and $K \in \mathcal{K}$, then each component of $\bigcup \mathcal{K}$ intersects K.*

Proof. First, note that $\bigcup \mathcal{K} \in 2^X$ [98, 2.5.2]. Suppose that there exists a component C of $\bigcup \mathcal{K}$ such that $K \cap C = \emptyset$. Hence, by Theorem 1.4.8, there exist two disjoint closed subsets K_1 and K_2 of $\bigcup \mathcal{K}$ such that $\bigcup \mathcal{K} = K_1 \cup K_2$, $K \subset K_1$ and $C \subset K_2$. Let $\mathcal{K}_1 = \{L \in \mathcal{K} \mid L \subset K_1\}$ and $\mathcal{K}_2 = \{L \in \mathcal{K} \mid L \cap K_2 \neq \emptyset\}$. Then $\mathcal{K}_1$ and $\mathcal{K}_2$ are disjoint closed subsets of $\mathcal{K}$. Also, $\mathcal{K} = \mathcal{K}_1 \cup \mathcal{K}_2$, a contradiction to the fact that $\mathcal{K}$ is connected. Therefore, each component of $\bigcup \mathcal{K}$ intersects K. □

1.6.9 Lemma. *If X is a compactum, then $\mathcal{F}(X) = \bigcup_{n=1}^{\infty} \mathcal{F}_n(X)$ is dense in 2^X.*

Proof. Let $\mathcal{U}$ be an open subset of 2^X and let $A \in \mathcal{U}$. By Theorem 1.6.5, there exist open subsets $U_1, \ldots, U_m$ of X such that $A \in \langle U_1, \ldots, U_m \rangle \subset \mathcal{U}$. For each $j \in \{1, \ldots, m\}$, let $x_j \in U_j$. Then $\{x_1, \ldots, x_m\} \in \langle U_1, \ldots, U_m \rangle \cap \mathcal{F}(X)$. Hence, $\{x_1, \ldots, x_m\} \in \mathcal{U} \cap \mathcal{F}(X)$. Therefore, $\mathcal{F}(X)$ is dense in 2^X. □

1.6.10 Lemma. *Let X be a continuum and let $n \in \mathbb{N}$. Then X is locally connected if and only if $\mathcal{F}_n(X)$ is locally connected.*

Proof. Suppose X is locally connected. Then X^n is locally connected. By [98, Proposition 2.4.3], the function $f_n \colon X^n \twoheadrightarrow \mathcal{F}_n(X)$ is continuous and surjective. Hence, $\mathcal{F}_n(X)$ is locally connected.

Suppose $\mathcal{F}_n(X)$ is locally connected. Since $\mathcal{F}_1(X)$ is homeomorphic to X, we may assume that $n \geq 2$. Let x_0 be a point of X and let U be an open subset of X such that $x_0 \in U$. Since $\mathcal{F}_n(X)$ is locally connected, there exists a connected open subset $\mathcal{W}$ of $\mathcal{F}_n(X)$ such that $\{x_0\} \in \mathcal{W} \subset \mathrm{Cl}(\mathcal{W}) \subset \langle U \rangle \cap \mathcal{F}_n(X)$. By Theorem 1.6.5, there exist open subsets $V_1, \ldots, V_m$ of X such that $\{x_0\} \in \langle V_1, \ldots, V_m \rangle \cap \mathcal{F}_n(X) \subset \mathcal{W}$. Since $\mathrm{Cl}(\mathcal{W})$ is a subcontinuum of 2^X and $\mathrm{Cl}(\mathcal{W}) \cap \mathcal{C}_1(X)$, by Lemma 1.6.8, $\bigcup \mathrm{Cl}(\mathcal{W})$ is a subcontinuum of X. Since $\mathrm{Cl}(\mathcal{W}) \subset \langle U \rangle$, we have that $\bigcup \mathrm{Cl}(\mathcal{W}) \subset U$ and $x_0 \in \bigcup \mathrm{Cl}(\mathcal{W})$. Let $V = \bigcap_{j=1}^{m} V_j$, and let $x \in V$. Then $\{x\} \in \langle V_1, \ldots, V_m \rangle \cap \mathcal{F}_n(X) \subset \mathcal{W}$. Hence, $x \in \bigcup \mathrm{Cl}(\mathcal{W})$, and $V \subset \bigcup \mathrm{Cl}(\mathcal{W})$. Thus, $\bigcup \mathrm{Cl}(\mathcal{W})$ is a connected neighbourhood of x_0 contained in U. Thus, X is connected im kleinen at each of its points. Therefore, X is locally connected (Theorem 1.4.18). □

1.6.11 Definition. Let X be a continuum, we define a uniformity on 2^X as follows: If $U \in \mathfrak{U}_X$ (Remark 1.3.4), then let $2^U = \{(A, A') \in 2^X \times 2^X \mid A \subset B(A', U)$ and $A' \subset B(A, U)\}$. Let $\mathfrak{B}_X = \{2^U \mid U \in \mathfrak{U}_X\}$. Then $\mathfrak{B}_X$ is a base for a uniformity, denoted by $2^{\mathfrak{U}_X}$ [40, 8.5.16]. Observe that the topology generated by $2^{\mathfrak{U}_X}$ coincides with the Vietoris topology [98, 3.3]. Note that 2^X is compact and Hausdorff, Theorem 1.6.7. Thus, $2^{\mathfrak{U}_X}$ is unique (Remark 1.3.4), and

$$2^{\mathfrak{U}_X} = \{\mathcal{W} \subset 2^X \times 2^X \mid \text{there exists } U \in \mathfrak{U}_X$$
$$\text{such that } 2^U \subset \mathcal{W}\}.$$

Given $n \in \mathbb{N}$, we consider the restriction of $2^{\mathfrak{U}_X}$ to $\mathcal{C}_n(X)$ and $\mathcal{F}_n(X)$ and denoted by $\mathcal{C}_n(\mathfrak{U}_X)$ and $\mathcal{F}_n(\mathfrak{U}_X)$, respectively.

1.6.12 Remark. Let X be a continuum. We need to consider the hyperspace 2^{2^X} with the corresponding uniformity given by Definition 1.6.11, $2^{2^{\mathfrak{U}_X}}$.

1.6.13 Lemma. *Let X be a continuum and let $\mathcal{A}$ and $\mathcal{B}$ be elements of 2^{2^X}. If $U \in \mathfrak{U}_X$ and $\rho_{2^{2^X}}(\mathcal{A}, \mathcal{B}) < 2^{2^U}$, then $\rho_{2^X}(\bigcup \mathcal{A}, \bigcup \mathcal{B}) < 2^U$.*

Proof. Let $U \in \mathfrak{U}_X$ and let $\mathcal{A}$ and $\mathcal{B}$ be elements of 2^{2^X} such that $\rho_{2^{2^X}}(\mathcal{A}, \mathcal{B}) < 2^{2^U}$. Observe that, by [98, 2.5.2], $\bigcup \mathcal{A}$ and $\bigcup \mathcal{B}$ both belong to 2^X. Let $a \in \bigcup \mathcal{A}$. Then there exists $A \in \mathcal{A}$ such that $a \in A$. Since $\rho_{2^{2^X}}(\mathcal{A}, \mathcal{B}) < 2^{2^U}$, $\mathcal{A} \subset B(\mathcal{B}, 2^{2^U})$. Hence, there exists $B \in \mathcal{B}$ such that $\rho_{2^X}(A, B) < 2^U$. This implies that $A \subset B(B, U)$. Thus, there exists $b \in B$ such that $\rho_X(a, b) < U$. Hence, $\bigcup \mathcal{A} \subset B(\bigcup \mathcal{B}, U)$. Similarly, $\bigcup \mathcal{B} \subset B(\bigcup \mathcal{A}, U)$. Therefore, $\rho_{2^X}(\bigcup \mathcal{A}, \bigcup \mathcal{B}) < 2^U$. □

1.6.14 Definition. Let $f : X \to Y$ be a map between compacta. Then $2^f : 2^X \to 2^Y$ given by $2^f(A) = f(A)$ is called the *induced map between the hyperspaces of closed subsets of X and Y*. For each $n \in \mathbb{N}$, the functions $\mathcal{C}_n(f) : \mathcal{C}_n(X) \to \mathcal{C}_n(Y)$ and $\mathcal{F}_n(f) : \mathcal{F}_n(X) \to \mathcal{F}_n(Y)$ given by $\mathcal{C}_n(f) = 2^f|_{\mathcal{C}_n(X)}$ and $\mathcal{F}_n(f) = 2^f|_{\mathcal{F}_n(X)}$ are called the *induced map between the n-fold hyperspaces of X and Y* and the *induced map between the n-fold symmetric products of X and Y*, respectively.

A proof of the following two theorems may be found in [98, 5.10].

1.6.15 Theorem. *Let $f : X \to Y$ be a map between compacta. Then the functions 2^f, $\mathcal{C}_n(f)$ and $\mathcal{F}_n(f)$ ($n \in \mathbb{N}$) are continuous.*

1.6.16 Theorem. *Let $f : X \to Y$ be a map between compacta. If f is open, then the function $\Im(f) : 2^Y \to 2^X$ given by $\Im(f)(B) = f^{-1}(B)$ is continuous.*

1.6.17 Definition. A continuum X has the *property of Kelley* provided that for each $U \in \mathfrak{U}_X$, there exists $V \in \mathfrak{U}_X$ such that for any two points p and q of X with $\rho_X(p, q) < V$, and each $P \in \mathcal{C}_1(X)$ with $p \in P$, there exists $Q \in \mathcal{C}_1(X)$ such that $q \in Q$ and $\rho_{\mathcal{C}_1(X)}(P, Q) < \mathcal{C}_1(U)$. The entourage V is called a *Kelley entourage* for $\mathcal{C}_1(U)$.

W. J. Charatonik [23] and W. Makuchowski [95] defined the pointwise version of the property of Kelley as follows:

1.6.18 Definition. A continuum X has the *property of Kelley at a point x_0* if for every subcontinuum L of X containing x_0 and each open subset $\mathcal{A}$ of $\mathcal{C}_1(X)$ containing L, there exists an open subset $K(x_0, L, \mathcal{A})$ of X such that $x_0 \in K(x_0, L, \mathcal{A})$ and if $x \in K(x_0, L, \mathcal{A})$, then there exists a subcontinuum M of X

such that $x \in M$ and $M \in \mathcal{A}$. The open set $K(x_0, L, \mathcal{A})$ is called a *Kelley set* for $\mathcal{A}$, L and x_0.

1.6.19 Theorem. *A continuum X has the property of Kelley if and only if X has the property of Kelley at each of its points.*

Proof. Suppose X has the property of Kelley. Let p be an element of X, let $P \in \mathcal{C}_1(X)$ be such that $p \in P$ and let $\mathcal{A}$ be an open subset of $\mathcal{C}_1(X)$ such that $P \in \mathcal{A}$. Note that there exists $U \in \mathfrak{U}_X$ such that $B(P, \mathcal{C}_1(U)) \subset \mathcal{A}$. Let V be a Kelley entourage for $\mathcal{C}_1(U)$. Hence, $\mathrm{Int}_X(B(p, V))$ is an open subset of X. Let $q \in \mathrm{Int}_X(B(p, V))$. Thus, $\rho_X(p, q) < V$. Then there exists $Q \in \mathcal{C}_1(X)$ such that $q \in Q$ and $\rho_{\mathcal{C}_1(X)}(P, Q) < \mathcal{C}_1(U)$. This implies that $Q \in \mathcal{A}$. Since p is an arbitrary point of X, X has the property of Kelley at each of its points.

Now, assume that X has the property of Kelley at each of its points and let $U \in \mathfrak{U}_X$. Since X has the property of Kelley at each of its points, for every point z of X and each $Z \in \mathcal{C}_1(X)$, there exists a Kelley set $K(z, Z, \mathrm{Int}_{\mathcal{C}_1(X)}(B(Z, \mathcal{C}_1(U))))$ for $\mathrm{Int}_{\mathcal{C}_1(X)}(B(Z, \mathcal{C}_1(U)))$, Z and z. Note that

$$\{K(z, Z, \mathrm{Int}_{\mathcal{C}_1(X)}(B(Z, \mathcal{C}_1(U)))) \mid z \in X \text{ and } Z \in \mathcal{C}_1(X)\}$$

is an open cover of X. By Lemma 1.3.11,
$V =$

$$\bigcup\{K(z, Z, \mathrm{Int}_{\mathcal{C}_1(X)}(B(Z, \mathcal{C}_1(U)))) \times K(z, Z, \mathrm{Int}_{\mathcal{C}_1(X)}(B(Z, \mathcal{C}_1(U)))) \mid$$

$$z \in X \text{ and } Z \in \mathcal{C}_1(X)\}$$

belongs to $\mathfrak{U}_X$. Let p and q be points of X such that $\rho_X(p, q) < V$ and let $P \in \mathcal{C}_1(X)$ be such that $p \in P$. Then, by the definition of V, we have that $q \in K(p, P, \mathrm{Int}_{\mathcal{C}_1(X)}(B(P, \mathcal{C}_1(U))))$. Thus, there exists $Q \in \mathcal{C}_1(X)$ such that $Q \in \mathrm{Int}_{\mathcal{C}_1(X)}(B(P, \mathcal{C}_1(U)))$ and $\rho_{\mathcal{C}_1(X)}(P, Q) < \mathcal{C}_1(U)$. Therefore, X has the property of Kelley. □

For the following results we use the pointwise version of the property of Kelley.

1.6.20 Theorem. *Let X be a continuum with the property of Kelley. Let A be a nonempty closed subset of X and let W be a subcontinuum of X such that $\mathrm{Int}(W) \neq \emptyset$ and $W \cap A = \emptyset$. If $w \in W$, then there exists a subcontinuum K of X such that $w \in \mathrm{Int}(K)$ and $K \cap A = \emptyset$.*

Proof. Let $w \in W$. Since X is normal, there exists an open subset U of X such that $W \subset U$ and $\mathrm{Cl}(U) \cap A = \emptyset$. Let $\mathcal{U} = \langle U, \mathrm{Int}(W)\rangle_1$. Note that $W \in \mathcal{U}$. Also observe that if $L \in \mathcal{U}$, then $L \subset U$ and $L \cap W \neq \emptyset$. Since X has the property of Kelley at w, there exists a Kelley set V for $\mathcal{U}$ and w. Let $v \in V \setminus \{w\}$. Then there exists a subcontinuum L_v of X such that $v \in L_v$ and $L_v \in \mathcal{U}$. Let $L_w = W$, and let $K = \mathrm{Cl}\left(\bigcup\{L_v \mid v \in V\}\right)$. Then K is a subcontinuum of X, $w \in V \subset K$ and $K \subset \mathrm{Cl}(U) \subset X \setminus A$. □

1.6.21 Corollary. *Let X be a continuum with the property of Kelley, let W be a subcontinuum of X, with nonempty interior, and let A be a nonempty closed subset of X such that $W \cap A = \emptyset$. Then there exists a subcontinuum M of X such that $W \subset \mathrm{Int}(M)$ and $M \cap A = \emptyset$.*

Proof. Let $w \in W$. By Theorem 1.6.20, there exists a subcontinuum K_w such that $w \in \mathrm{Int}(K)$ and $K \cap A = \emptyset$. Note that $\{\mathrm{Int}(K_w) \mid w \in W\}$ forms an open cover of W, by compactness, there exist $w_1, \ldots, w_n$ in W such that $W \subset \bigcup_{j=1}^{n} \mathrm{Int}(K_{w_j})$. Let $M = \bigcup_{j=1}^{n} K_{w_j}$. Then M is a subcontinuum of X, $W \subset \mathrm{Int}(M)$ and $M \cap A = \emptyset$. □

Next, we show that continua with the uniform property of Effros have the property of Kelley.

1.6.22 Theorem. *If X is a continuum with the uniform property of Effros, then X has the property of Kelley.*

Proof. By Theorem 1.6.19, it is enough to show that X has the property of Kelley at each of its points. Let x_0 be a point of X, let K be a subcontinuum of X with $x_0 \in K$ and let $\mathcal{U}$ be an open subset of $\mathcal{C}_1(X)$ containing K. Then there exists $U \in \mathfrak{U}_X$ (Remark 1.3.4) such that $B(K, \mathcal{C}_1(U)) \subset \mathcal{U}$. Since X has the uniform property of Effros, there exists an Effros entourage $V \in \mathfrak{U}_X$ for U. Let $x \in \mathrm{Int}(B(x_0, V))$. Then there exists a U-homeomorphism $h\colon X \twoheadrightarrow X$ such that $h(x_0) = x$. Then $h(K)$ is a subcontinuum of X, $x \in h(K)$ and $h(K) \in B(K, \mathcal{C}_1(U)) \subset \mathcal{U}$. Therefore, X has the property of Kelley. □

As a consequence of Theorems 1.4.60 and 1.6.22, we obtain.

1.6.23 Corollary. *If X is an Effros continuum, then X has the property of Kelley.*

References for Chapter 1

Section 1.1: [24, 36, 60, 92].
Section 1.2: [24, 36, 48, 65].
Section 1.3: [40, 48, 49, 111, 129]
Section 1.4: [1, 31, 36, 40, 45, 52, 55, 65–67, 75, 90–92, 97, 104, 105, 119, 127].
Section 1.5: [24, 37, 38, 61, 92, 99].
Section 1.6: [15, 23, 24, 40, 48, 79–81, 90–92, 95, 98, 105].

Chapter 2
The Set Function $\mathcal{T}$

We prove basic results about the set function $\mathcal{T}$ defined by F. Burton Jones [68] to study the properties of metric continua. We define this function on compacta, and then we concentrate on continua. In particular, we present some of the well known properties (such as connectedness im kleinen, local connectedness, semi-local connectedness, etc.) using the set function $\mathcal{T}$. The notion of aposyndesis was the main motivation of Jones to define this function. We present some properties of a continuum when it is $\mathcal{T}$-symmetric and $\mathcal{T}$-additive. We give properties of continuum on which $\mathcal{T}$ is idempotent, idempotent on closed sets and idempotent on contina. We also present results about the set functions $\mathcal{T}^n$, when $n \in \mathbb{N}$.

2.1 Main Properties of $\mathcal{T}$

2.1.1 Definition. Given a topological space X, the *power set of* X, denoted by $\mathcal{P}(X)$, is

$$\mathcal{P}(X) = \{A \mid A \subset X\}.$$

2.1.2 Remark. Let X be a topological space. Let us note that if A is a subset of X, then its closure satisfies

$$\begin{aligned}\mathrm{Cl}(A) = \{x \in X \mid &\text{ for each open subset } U \text{ of } X \text{ such that} \\ &x \in U, \text{ we have that } U \cap A \neq \emptyset\}.\end{aligned}$$

As we see below (Definition 2.1.3), this property is similar to the definition of the function $\mathcal{T}$.

2.1.3 Definition. Let X be a compactum. Define

$$\mathcal{T}\colon \mathcal{P}(X) \to \mathcal{P}(X)$$

S. Macías, *Set Function* $\mathcal{T}$, Developments in Mathematics 67,
https://doi.org/10.1007/978-3-030-65081-0_2

by

$$\mathcal{T}(A) = \{x \in X \mid \text{for each subcontinuum } W \text{ of } X \text{ such that } x \in \operatorname{Int}(W), \text{ we have that } W \cap A \neq \emptyset\},$$

for every subset A of X. The function $\mathcal{T}$ is called *Jones' set function* $\mathcal{T}$.

2.1.4 Remark. In general, when working with the set function $\mathcal{T}$, we usually work with complements. Hence, for any compactum X and any subset A of X, we have that

$$\mathcal{T}(A) = X \setminus \{x \in X \mid \text{there exists a subcontinuum } W \text{ of } X \text{ such that } x \in \operatorname{Int}(W) \subset W \subset X \setminus A\}.$$

2.1.5 Remark. Let X be a compactum. If A is a subset of X, then $A \subset \mathcal{T}(A)$, and $\mathcal{T}(A)$ is closed in X. Hence, the range of $\mathcal{T}$ is $2^X \cup \{\emptyset\}$ (Definition 1.6.1), i.e.,

$$\mathcal{T}\colon \mathcal{P}(X) \to 2^X \cup \{\emptyset\}.$$

The next proposition gives a relation between aposyndesis and the set function $\mathcal{T}$.

2.1.6 Proposition. *Let X be a compactum, and let A be a subset of X. If $x \in X \setminus \mathcal{T}(A)$, then X is aposyndetic at x with respect to each point of A. Also, X is aposyndetic at x with respect to A.*

Proof. Let $x \in X \setminus \mathcal{T}(A)$, and let $a \in A$. Since $x \in X \setminus \mathcal{T}(A)$, by Definition 2.1.3, there exists a subcontinuum W of X such that $x \in \operatorname{Int}(W) \subset W \subset X \setminus A$. In particular, $W \subset X \setminus \{a\}$. Therefore, X is aposyndetic at x with respect to a and with respect to A. □

2.1.7 Proposition. *Let X be a compactum. If A and B are subsets of X and $A \subset B$, then $\mathcal{T}(A) \subset \mathcal{T}(B)$.*

Proof. Let $x \in X \setminus \mathcal{T}(B)$. Then there exists a subcontinuum W of X such that $x \in \operatorname{Int}(W) \subset W \subset X \setminus B$. Since $X \setminus B \subset X \setminus A$, $x \in \operatorname{Int}(W) \subset W \subset X \setminus A$. Therefore, $x \in X \setminus \mathcal{T}(A)$. □

2.1.8 Corollary. *Let X be a compactum. If A and B are subsets of X, then $\mathcal{T}(A) \cup \mathcal{T}(B) \subset \mathcal{T}(A \cup B)$.*

2.1.9 Remark. Let us note that the reverse inclusion of Corollary 2.1.8 is, in general, not true (Example 2.1.18). It is an open question to characterize continua X such that $\mathcal{T}(A \cup B) = \mathcal{T}(A) \cup \mathcal{T}(B)$ for each pair of closed subsets A and B of X.

Let us see a couple of examples.

2.1.10 Example. Let X be the Cantor set. Then $\mathcal{T}(\emptyset) = X$. To see this, suppose there exists a point $x \in X \setminus \mathcal{T}(\emptyset)$. Then there exists a subcontinuum W of X such that $x \in \text{Int}(W) \subset W \subset X \setminus \emptyset = X$. Since X is the Cantor set, X is totally disconnected and perfect. Hence, no subcontinuum of X has interior. Therefore, W cannot exist. Hence, $\mathcal{T}(\emptyset) = X$. Note that, by Proposition 2.1.7, $\mathcal{T}(A) = X$ for each subset A of X.

2.1.11 Example. Let $X = \{0\} \cup \left\{\frac{1}{n}\right\}_{n=1}^{\infty}$. Then $\mathcal{T}(\emptyset) = \{0\}$. This follows from the fact that the only subcontinuum of X with empty interior is $\{0\}$. Hence, by Remark 2.1.5 and Proposition 2.1.7, $\mathcal{T}(A) = \{0\} \cup A$ for each subset A of X.

2.1.12 Remark. Note that if X is as in Example 2.1.10 or 2.1.11, then $\mathcal{T}|_{2^X} : 2^X \to 2^X$ is continuous, where 2^X has the topology given by the Hausdorff metric [92, Theorem 1.8.3] or the Vietoris topology.

The following theorem gives us a characterization of compacta X for which $\mathcal{T}(\emptyset) = \emptyset$.

2.1.13 Theorem. *Let X be a compactum. Then $\mathcal{T}(\emptyset) = \emptyset$ if and only if X only has finitely many components.*

Proof. Suppose X only has finitely many components. Let $x \in X$, and let C be the component of X such that $x \in C$. Hence, C is a subcontinuum of X and, by Lemma 1.4.1, $x \in \text{Int}(C)$. Thus, each point of X is contained in the interior of a proper subcontinuum of X. Therefore, $\mathcal{T}(\emptyset) = \emptyset$.

Now, suppose $\mathcal{T}(\emptyset) = \emptyset$. Then for each point $x \in X$, there exists a subcontinuum W_x of X such that $x \in \text{Int}(W_x) \subset W_x \subset X \setminus \emptyset$. Hence, $\{\text{Int}(W_x) \mid x \in X\}$ is an open cover of X. Since X is compact, there exist $x_1, \ldots, x_n$ in X such that $X = \bigcup_{j=1}^{n} \text{Int}(W_{x_j}) \subset \bigcup_{j=1}^{n} W_{x_j} \subset X$. Thus, X is the union of finitely many continua. Therefore, X only has finitely many components. □

2.1.14 Corollary. *If X is a continuum, then $\mathcal{T}(\emptyset) = \emptyset$.*

Let us see more examples.

2.1.15 Example. Let X be the cone over the Cantor set, with vertex ν_X and base B (Figure 2.1). Since the Cantor set is totally disconnected and perfect, it follows

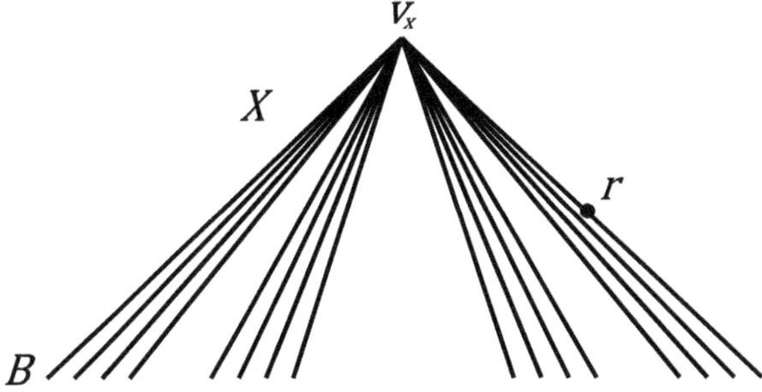

Fig. 2.1 Cantor fan

that the subcontinua of X having nonempty interior must contain ν_X. Hence, $\mathcal{T}(\{\nu_X\}) = X$. If $r \in X \setminus \{\nu_X\}$, then $\mathcal{T}(\{r\})$ is the line segment from r to the base B. Also, $\mathcal{T}(B) = B$.

2.1.16 Example. Let $Y = X \cup X'$, where X is the cone over the Cantor set, with vertex ν_X and base B, and X' is another copy of the cone over the Cantor set, with vertex $\nu_{X'} \in B$ (Figure 2.2). In this case, $\mathcal{T}(\{\nu_X\}) = X$, $\mathcal{T}(\{\nu_{X'}\}) = X'$ and $\mathcal{T}\mathcal{T}(\{\nu_X\}) = \mathcal{T}^2(\{\nu_X\}) = Y$. Note that this implies, in general, that the function $\mathcal{T}$ is not idempotent (Definition 2.3.1).

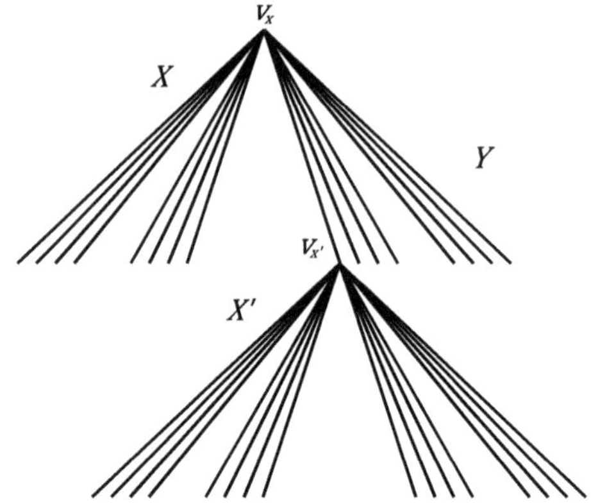

Fig. 2.2 Double Cantor fan

2.1.17 Example. Let X be the topologist sine curve, Example 1.4.9. Let $J = X \setminus \left\{\left(x, \sin\left(\frac{1}{x}\right)\right) \,\middle|\, 0 < x \leq \frac{2}{\pi}\right\}$. Since X is not locally connected at any point of J, if $p \in J$, then $\mathcal{T}(\{p\}) = J$. Also, $\mathcal{T}(\{q\}) = \{q\}$ for each $q \in X \setminus J$.

The following example shows that, in general, $\mathcal{T}(A \cup B) = \mathcal{T}(A) \cup \mathcal{T}(B)$ does not hold.

2.1.18 Example. Let X be the suspension over the closure of the harmonic sequence (Example 1.4.23). Note that for each $x \in X$, $\mathcal{T}(\{x\}) = \{x\}$. In particular,

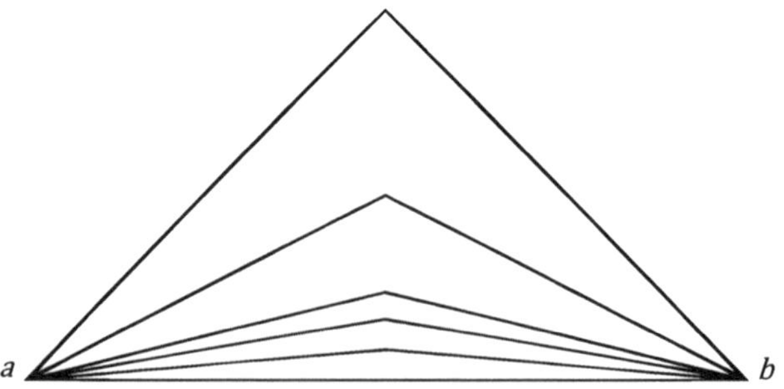

Fig. 2.3 Suspension over the closure of the harmonic sequence

$\mathcal{T}(\{a\}) = \{a\}$ and $\mathcal{T}(\{b\}) = \{b\}$; however, $\mathcal{T}(\{a, b\})$ consists of the limit segment from a to b (Figure 2.3).

2.1.19 Theorem. *Let X be a continuum. If A is a subset of X, then $\mathcal{T}(A)$ intersects each closed separator of A and some point of $\mathcal{T}(A)$.*

Proof. Suppose the theorem is not true. Then there exist $p \in \mathcal{T}(A)$ and a closed subset B of X such that $X \setminus B = H \cup K$, $A \subset H$, $p \in K$, where H and K are disjoint open subsets of X and $B \cap \mathcal{T}(A) = \emptyset$. Hence, for each $b \in B$, there exists a subcontinuum W_b of X such that $b \in \mathrm{Int}(W_b) \subset W_b \subset X \setminus A$. Since B is compact, there exist $b_1, \ldots, b_n$ in B such that $B \subset \bigcup_{j=1}^{n} \mathrm{Int}(W_{b_j})$.

Since $\mathrm{Cl}(K) \setminus K \subset B$, by Theorem 1.4.36, $Y = K \cup \bigcup_{j=1}^{n} W_{b_j}$ only has a finite number of components. Note that Y is closed in X, since it is the union of a finite number of continua and $\mathrm{Cl}(K) \setminus K \subset Y$. Since $p \in K$ and $K \subset Y$, $p \in Y$. Let C be the component of Y such that $p \in C$. By Lemma 1.4.1, $p \in \mathrm{Int}(C)$. This implies that $p \in X \setminus \mathcal{T}(A)$, a contradiction. Therefore, $B \cap \mathcal{T}(A) \neq \emptyset$. □

Now, we present several consequences of Theorem 2.1.19.

2.1.20 Corollary. *Let X be a continuum. If A is a subset of X, then each component of $\mathcal{T}(A)$ intersects* $\mathrm{Cl}(A)$.

Proof. Suppose the corollary is not true. Then there exists a component C of $\mathcal{T}(A)$ such that $C \cap \mathrm{Cl}(A) = \emptyset$. Hence, by Theorem 1.4.8, there exist two disjoint closed subsets H and K of $\mathcal{T}(A)$ such that $\mathcal{T}(A) = H \cup K$, $\mathrm{Cl}(A) \subset H$ and $C \subset K$. Since X is a normal space, there exist two disjoint open subsets U and V of X such that $H \subset U$ and $K \subset V$. Let $B = X \setminus (U \cup V)$. Then B is a closed separator of A and C and $B \cap \mathcal{T}(A) = \emptyset$, a contradiction to Theorem 2.1.19. □

2.1.21 Corollary. *Let X be a continuum. If A is a subset of X, then the cardinality of the components of $\mathcal{T}(A)$ is at most equal to the cardinality of the components of* $\mathrm{Cl}(A)$.

2.1.22 Corollary. *Let X be a continuum, and let A and B be two closed subsets of X. If K is a component of $\mathcal{T}(A \cup B)$ such that $K \setminus (\mathcal{T}(A) \cup \mathcal{T}(B)) \neq \emptyset$, then $K \cap A \neq \emptyset$ and $K \cap B \neq \emptyset$.*

Proof. Since X is a continuum, by Corollary 2.1.14, $\mathcal{T}(\emptyset) = \emptyset$. By Corollary 2.1.20, $K \cap (A \cup B) \neq \emptyset$. Now, by Theorem 2.1.47, $\mathcal{T}((A \cup B) \cap K) = K$. Since $K \setminus (\mathcal{T}(A) \cup \mathcal{T}(B)) \neq \emptyset$, $(A \cup B) \cap K$ meets both A and B. Thus, $K \cap A \neq \emptyset$ and $K \cap B \neq \emptyset$. □

2.1.23 Corollary. *Let X be a continuum, and let A and B be closed subsets of X. If $\mathcal{T}(A \cup B) \neq \mathcal{T}(A) \cup \mathcal{T}(B)$, then there exists a subcontinuum K of X such that $K \subset \mathcal{T}(A \cup B)$, $K \cap A \neq \emptyset$ and $K \cap B \neq \emptyset$.*

Proof. Let K be a component of $\mathcal{T}(A \cup B)$ such that $K \setminus (\mathcal{T}(A) \cup \mathcal{T}(B)) \neq \emptyset$ and apply Corollary 2.1.22. □

The next result is due to H. S. Davis [120, p. 18].

2.1.24 Theorem. *Let X be a compactum, and let A and B be nonempty closed subsets of X. Then the following are equivalent:*

(1) $\mathcal{T}(A) \cap B = \emptyset$.
(2) *There exist two closed subsets M and N of X such that $A \subset \mathrm{Int}(M)$, $B \subset \mathrm{Int}(N)$ and $\mathcal{T}(M) \cap N = \emptyset$.*

Proof. Assume (1), we show (2). For each $b \in B$, there exists a subcontinuum W_b of X such that $b \in \mathrm{Int}(W_b) \subset W_b \subset X \setminus A$. Since B is compact, there exist $b_1, \ldots, b_n$ in B such that $B \subset \bigcup_{j=1}^n \mathrm{Int}(W_{b_j}) \subset \bigcup_{j=1}^n W_{b_j} \subset X \setminus A$. Since X is a normal space, there exist two open sets U and V of X such that $A \subset U \subset \mathrm{Cl}(U) \subset \left(X \setminus \bigcup_{j=1}^n W_{b_j}\right)$ and $B \subset V \subset \mathrm{Cl}(V) \subset \bigcup_{j=1}^n \mathrm{Int}(W_{b_j})$.

Let $M = \mathrm{Cl}(U)$ and let $N = \mathrm{Cl}(V)$. Then M and N are closed subsets of X such that $A \subset \mathrm{Int}(M)$, $B \subset \mathrm{Int}(N)$ and $\mathcal{T}(M) \cap N = \emptyset$. The other implication follows easily from Proposition 2.1.7. □

The next corollary is a consequence of proof of Theorem 2.1.24.

2.1.25 Corollary. *Let X be a continuum, and let A be a closed subset of X. If $x \in X \setminus \mathcal{T}(A)$, then there exists an open subset U of X such that $A \subset U$ and $x \in X \setminus \mathcal{T}(\mathrm{Cl}(U))$.*

2.1.26 Theorem. *Let X be a continuum, and let A be a nonempty subset of X. Then $\mathcal{T}(A)$ is disconnected if and only if there exist a nonempty proper subset B of A and a closed subset W of X such that $W \cap A = \emptyset$, W only has finitely many components and* $\mathrm{Int}(W)$ *separates B from $A \setminus B$ in X.*

Proof. Suppose $\mathcal{T}(A)$ is disconnected. Then $\mathcal{T}(A) = K_1 \cup K_2$, where K_1 and K_2 are nonempty disjoint closed subsets of X. Let $B = A \cap K_1$. Then $A \setminus B = A \cap K_2$. By Theorem 2.1.19, B is a nonempty proper subset of A. Let U be an open subset of X such that $K_1 \subset U$ and $\mathrm{Cl}(U) \cap K_2 = \emptyset$. For each $z \in \mathrm{Bd}(U)$, there exists a subcontinuum W_z of X such that $z \in \mathrm{Int}(W_z) \subset W_z \subset X \setminus A$. Since $\mathrm{Bd}(U)$ is compact, there exist $z_1, \ldots, z_m$ in $\mathrm{Bd}(U)$ such that $\mathrm{Bd}(U) \subset \bigcup_{j=1}^m \mathrm{Int}(W_{z_j})$. Let $W = \bigcup_{j=1}^m W_{z_j}$. Then W only has finitely many components and $\mathrm{Bd}(U) \subset \mathrm{Int}(W)$. Hence, $X \setminus \mathrm{Int}(W) = [U \cap (X \setminus \mathrm{Int}(W))] \cup [(X \setminus \mathrm{Cl}(U)) \cap (X \setminus \mathrm{Int}(W))]$. Note that $B \subset U \cap (X \setminus \mathrm{Int}(W))$ and $A \setminus B \subset (X \setminus \mathrm{Cl}(U)) \cap (X \setminus \mathrm{Int}(W))$.

Now, assume that there exist a nonempty proper subset B of A and a closed subset W of X such that $W \cap A = \emptyset$, W only has finitely many components and $\mathrm{Int}(W)$ separates B from $A \setminus B$ in X. Let $x \in \mathrm{Int}(W)$, and let W_1 be the component of W containing x. Then $x \in \mathrm{Int}(W_1) \subset W_1 \subset X \setminus A$ (Lemma 1.4.1). Thus, $x \in X \setminus \mathcal{T}(A)$. Hence, $\mathcal{T}(A) \subset X \setminus \mathrm{Int}(W) = L_1 \cup L_2$, where L_1 and L_2 are nonempty disjoint closed subsets of X, with $B \subset L_1 \cap \mathcal{T}(A)$ and $A \setminus B \subset L_2 \cap \mathcal{T}(A)$. Therefore, $\mathcal{T}(A)$ is not connected. □

2.1.27 Theorem. *Let X be a continuum. If W is a subcontinuum of X, then $\mathcal{T}(W)$ is also a subcontinuum of X.*

Proof. By Remark 2.1.5, $\mathcal{T}(W)$ is a closed subset of X.

Suppose $\mathcal{T}(W)$ is not connected. Then there exist two disjoint closed subsets A and B of X such that $\mathcal{T}(W) = A \cup B$. Since W is connected, without loss of generality, we assume that $W \subset A$. Since X is a normal space, there exists an open subset U of X such that $A \subset U$ and $\text{Cl}(U) \cap B = \emptyset$.

Note that $\text{Bd}(U) \cap \mathcal{T}(W) = \emptyset$. Hence, for each $z \in \text{Bd}(U)$, there exists a subcontinuum K_z of X such that $z \in \text{Int}(K_z) \subset K_z \subset X \setminus W$. Since $\text{Bd}(U)$ is compact, there exist $z_1, \ldots, z_n$ in $\text{Bd}(U)$ such that $\text{Bd}(U) \subset \bigcup_{j=1}^{n} \text{Int}(K_{z_j}) \subset \bigcup_{j=1}^{n} K_{z_j}$ (Figure 2.4). Let $V = U \setminus \left(\bigcup_{j=1}^{n} K_{z_j}\right)$. Let $Y = X \setminus V = (X \setminus U) \bigcup \left(\bigcup_{j=1}^{n} K_{z_j}\right)$. By Theorem 1.4.36, Y only has a finite number of components. Observe that $B \subset X \setminus \text{Cl}(U) \subset X \setminus U \subset Y$. Hence, $B \subset \text{Int}(Y)$. Let $b \in B$, and let C be the component of Y such that $b \in C$. By Lemma 1.4.1, $b \in \text{Int}(C)$. By construction, $C \cap W = \emptyset$. Thus, $b \in X \setminus \mathcal{T}(W)$, a contradiction. Therefore, $\mathcal{T}(W)$ is connected. □

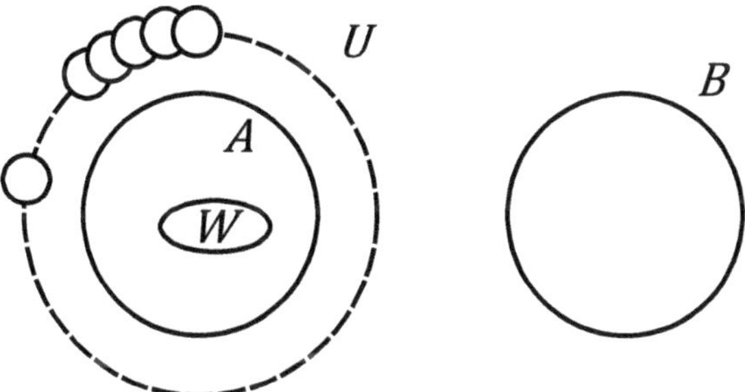

Fig. 2.4 Covering the boundary

The set function $\mathcal{T}$ may be used to define some of the properties of continua like connectedness im kleinen or local connectedness. In the following theorems we show the equivalence between both types of definitions.

The next theorem gives us a local definition of almost connectedness im kleinen using $\mathcal{T}$.

2.1.28 Theorem. *Let X be a continuum. If $p \in X$, then X is almost connected im kleinen at p if and only if for each subset A of X such that $p \in \text{Int}(\mathcal{T}(A))$, $p \in \text{Cl}(A)$.*

Proof. Suppose X is almost connected im kleinen at p. Let A be a subset of X such that $p \in \text{Int}(\mathcal{T}(A))$. Let $\mathcal{N} = \{U_\lambda\}_{\lambda \in \Lambda}$ be an open local base about p. Without loss of generality, we assume that Λ is a directed set. Then there exists a cofinal subset Λ' of Λ such that for every $\lambda \in \Lambda'$, $p \in U_\lambda \subset \text{Cl}(U_\lambda) \subset \mathcal{T}(A)$ and $\bigcap_{\lambda \in \Lambda'} U_\lambda = \{p\}$. Since X is almost connected im kleinen at p, for each $\lambda \in \Lambda'$, there exists a subcontinuum W_λ of X such that $\text{Int}(W_\lambda) \neq \emptyset$ and $W_\lambda \subset U_\lambda \subset \text{Cl}(U_\lambda) \subset \mathcal{T}(A)$. Hence, $W_\lambda \cap A \neq \emptyset$ for each $\lambda \in \Lambda'$. Let $x_\lambda \in W_\lambda \cap A$, for every $\lambda \in \Lambda'$. Note that, by construction, the net $\{x_\lambda\}_{\lambda\in\Lambda'}$ converges to p and $\{x_\lambda\}_{\lambda\in\Lambda'} \subset A$. Therefore, $p \in \text{Cl}(A)$.

Now suppose that $p \in X$ satisfies that for each subset A of X such that $p \in \text{Int}(\mathcal{T}(A))$, $p \in \text{Cl}(A)$. Let U be an open subset of X such that $p \in U$. Let V be an open subset of X such that $p \in V \subset \text{Cl}(V) \subset U$. If some component of $\text{Cl}(V)$ has nonempty interior, then X is almost connected im kleinen at p. Suppose, then, that all the components of $\text{Cl}(V)$ have empty interior. Let $A = \text{Bd}(V)$. Then A is a closed subset of X and $p \in X \setminus A$. We show that $V \subset \mathcal{T}(A)$. To this end, suppose there exists $x \in V \setminus \mathcal{T}(A)$. Thus, there exists a subcontinuum W of X such that $x \in \text{Int}(W) \subset W \subset X \setminus A$. Since all the components of $\text{Cl}(V)$ have empty interior, and W is a subcontinuum with nonempty interior, we have that $W \cap (X \setminus V) \neq \emptyset$ and, of course, $W \cap V \neq \emptyset$. Since W is connected, $W \cap \text{Bd}(V) = W \cap A \neq \emptyset$, a contradiction. Hence, $V \subset \mathcal{T}(A)$. Since $p \in V \subset \mathcal{T}(A)$, by hypothesis, $p \in \text{Cl}(A) = A$, a contradiction. Therefore, $\text{Cl}(V)$ has a component with nonempty interior, and X is almost connected im kleinen at p. □

The next theorem gives us a global definition of almost connectedness im kleinen using $\mathcal{T}$.

2.1.29 Theorem. *A continuum X is almost connected im kleinen if and only if for each closed subset, F, of X,* $\text{Int}(F) = \text{Int}(\mathcal{T}(F))$.

Proof. Suppose X is almost connected im kleinen. Let F be a closed subset of X. Since $F \subset \mathcal{T}(F)$ (Remark 2.1.5), $\text{Int}(F) \subset \text{Int}(\mathcal{T}(F))$. Let $x \in \text{Int}(\mathcal{T}(F))$. Then, by Theorem 2.1.28, $x \in \text{Cl}(F) = F$. Hence, $\text{Int}(\mathcal{T}(F)) \subset F$. Therefore, $\text{Int}(\mathcal{T}(F)) \subset \text{Int}(F)$ and $\text{Int}(\mathcal{T}(F)) = \text{Int}(F)$.

Now, suppose that $\text{Int}(F) = \text{Int}(\mathcal{T}(F))$ for each closed subset F of X. Let $x \in X$, and let U be an open subset of X such that $x \in U$. Since $\text{Int}(X \setminus U) \cap U = \emptyset$, by hypothesis, $\text{Int}(\mathcal{T}(X \setminus U)) \cap U = \emptyset$. Hence, there exists $y \in U$ such that $y \in X \setminus \mathcal{T}(X \setminus U)$. Thus, there exists a subcontinuum K of X such that $y \in \text{Int}(K) \subset K \subset X \setminus (X \setminus U) = U$. Therefore, X is almost connected im kleinen at x. □

2.1.30 Theorem. *Let X be continuum and let p be a point of X. Then X is connected im kleinen at p if and only if for each subset A of X such that $p \in \mathcal{T}(A)$,* $p \in \text{Cl}(A)$.

Proof. Let $p \in X$. Assume X is connected im kleinen at p. Let A be a subset of X and suppose $p \in X \setminus \text{Cl}(A)$. Hence, $\text{Cl}(A)$ is a closed subset of $X \setminus \{p\}$. Since X is connected im kleinen at p, there exists a subcontinuum W of X such that $p \in \text{Int}(W) \subset W \subset X \setminus A$. Therefore, $p \in X \setminus \mathcal{T}(A)$.

Now, suppose that $p \in X$ satisfies that for each subset A of X such that $p \in \mathcal{T}(A)$, $p \in \text{Cl}(A)$. Let D be a closed subset of X such that $D \subset X \setminus \{p\}$. Since $p \in X \setminus D$, by hypothesis, $p \in X \setminus \mathcal{T}(D)$. Then there exists a subcontinuum K of X such that $p \in \text{Int}(K) \subset K \subset X \setminus D$. Therefore, X is connected im kleinen at p. □

2.1.31 Corollary. *Let X be continuum and let p be a point of X. Then X is connected im kleinen at p if and only if for each closed subset A of X such that $p \in \mathcal{T}(A)$, $p \in A$.*

The following theorem follows from Definition 1.4.19.

2.1.32 Theorem. *Let X be a continuum, and let p and q be points of X. Then X is semi-aposyndetic at p and q if and only if either $p \in X \setminus \mathcal{T}(\{q\})$ or $q \in X \setminus \mathcal{T}(\{p\})$.*

The next theorem follows from the definition of aposyndesis.

2.1.33 Theorem. *Let X be a continuum, and let p and q be points of X. Then X is aposyndetic at p with respect to q if and only if $p \in X \setminus \mathcal{T}(\{q\})$.*

The following theorem is a global version of Theorem 2.1.33.

2.1.34 Theorem. *A continuum X is aposyndetic if and only if $\mathcal{T}(\{p\}) = \{p\}$, for each $p \in X$.*

Proof. Suppose X is aposyndetic. Let $p \in X$. Then for each $q \in X \setminus \{p\}$, X is aposyndetic at q with respect to p. Hence, by Remark 2.1.5 and Theorem 2.1.33, $q \in X \setminus \mathcal{T}(\{p\})$. Therefore, $\mathcal{T}(\{p\}) = \{p\}$.

Now, suppose $\mathcal{T}(\{x\}) = \{x\}$ for each $x \in X$. Let p and q be points of X such that $p \neq q$. Since $\mathcal{T}(\{q\}) = \{q\}$, $p \in X \setminus \mathcal{T}(\{q\})$. Thus, by Theorem 2.1.33, X is aposyndetic at p with respect to q. Since p and q are arbitrary points of X, X is aposyndetic. □

2.1.35 Theorem. *Let X be a continuum, and let p be a point of X. Then X is semi-locally connected at p if and only if $\mathcal{T}(\{p\}) = \{p\}$.*

Proof. Suppose X is semi-locally connected at p. By Remark 2.1.5, $\{p\} \subset \mathcal{T}(\{p\})$. Let $q \in X \setminus \{p\}$, and let U be an open subset of X such that $p \in U$ and $q \in X \setminus \mathrm{Cl}(U)$. Since X is semi-locally connected at p, there exists an open subset V of X such that $p \in V \subset U$ and $X \setminus V$ only has finitely many components. By Lemma 1.4.1, q belongs to the interior of the component of $X \setminus V$ containing q. Hence, $q \in X \setminus \mathcal{T}(\{p\})$. Therefore, $\mathcal{T}(\{p\}) = \{p\}$.

Now, suppose $\mathcal{T}(\{p\}) = \{p\}$. Let U be an open subset of X such that $p \in U$. Since $\mathcal{T}(\{p\}) = \{p\}$, for each $q \in X \setminus U$, there exists a subcontinuum W_q of X such that $q \in \mathrm{Int}(W_q) \subset W_q \subset X \setminus \{p\}$. Note that $\{\mathrm{Int}(W_q) \mid q \in X \setminus U\}$ is an open cover of $X \setminus U$. Since this set is compact, there exist $q_1, \ldots, q_n$ in $X \setminus U$ such that $X \setminus U \subset \bigcup_{j=1}^{n} \mathrm{Int}(W_{q_j})$. Let $V = X \setminus \left(\bigcup_{j=1}^{n} W_{q_j}\right)$. Then V is an open subset of X such that $p \in V \subset U$ and $X \setminus V$ only has finitely many components. Therefore, X is semi-locally connected at p. □

As a consequence of Theorems 2.1.34 and 2.1.35, we have the following.

2.1.36 Corollary. *A continuum X is aposyndetic if and only if it is semi-locally connected.*

The next theorem gives us a characterization of local connectedness in terms of the set function $\mathcal{T}$.

2.1.37 Theorem. *A continuum X is locally connected if and only if for each closed subset A of X, $\mathcal{T}(A) = A$.*

Proof. Suppose X is locally connected. Let A be a closed subset of X. Then $X \setminus A$ is an open subset of X. Let $p \in X \setminus A$. Then there exists an open subset U of X such that $p \in U \subset \mathrm{Cl}(U) \subset X \setminus A$. Since X is locally connected, there exists an open connected subset V of X such that $p \in V \subset U$. Hence, $\mathrm{Cl}(V)$ is a subcontinuum of X such that $p \in V \subset \mathrm{Cl}(V) \subset \mathrm{Cl}(U) \subset X \setminus A$. Consequently, $p \in X \setminus \mathcal{T}(A)$. Thus, $\mathcal{T}(A) \subset A$. By Remark 2.1.5, $A \subset \mathcal{T}(A)$. Therefore, $\mathcal{T}(A) = A$.

Now, suppose $\mathcal{T}(A) = A$ for each closed subset A of X. Let $p \in X$, and let U be an open subset of X such that $p \in U$. Then $X \setminus U$ is a closed subset of X such that $p \notin X \setminus U$. Since $X \setminus U = \mathcal{T}(X \setminus U)$, there exists a subcontinuum W of X such that $p \in \mathrm{Int}(W) \subset W \subset X \setminus (X \setminus U) = U$. Thus, by Theorem 1.4.15, X is connected im kleinen at p. Since p is an arbitrary point of X, by Theorem 1.4.18, X is locally connected. □

Theorem 2.1.37 may be strengthened as follows.

2.1.38 Theorem. *Let X be a continuum. Then X is locally connected if and only if $\mathcal{T}(Y) = Y$ for each subcontinuum Y of X.*

Proof. If X is locally connected, by Theorem 2.1.37, $\mathcal{T}(Y) = Y$ for each subcontinuum Y of X.

Now suppose that for every subcontinuum Y of X, $\mathcal{T}(Y) = Y$. We show that X is locally connected. To this end, by Lemma 1.4.17, it is enough to see the components of open subsets of X are open. Let V be an open subset of X. Let $p \in V$. Since $\mathcal{T}(\{p\}) = \{p\}$ and $X \setminus V$ is compact, there exist finitely many subcontinua contained in $X \setminus \{p\}$ whose interiors cover $X \setminus V$. Let $\mathcal{W} = \{W_1, \ldots, W_m\}$ be such a collection of smallest possible cardinality. Let $U = X \setminus \left(\bigcup_{j=1}^m W_j\right)$, and let P be the component of $\mathrm{Cl}(U)$ such that $p \in P$. We show that $p \in \mathrm{Int}(P) \subset P \subset V$. By Theorem 1.4.36, each component of $\mathrm{Cl}(U)$ intersects $\bigcup_{j=1}^m W_j$. By the minimality of m, we assert that no component of $\mathrm{Cl}(U)$, except P, can intersect more than one of the W_j's. To see this, let P' be a component of $\mathrm{Cl}(U)$, different from P, and suppose that P' intersects both W_j and W_k, where j and k are in $\{1, \ldots, m\}$ and $j \neq k$. Note that, in this case, $P' \cup W_j \cup W_j$ is a continuum. Let $\mathcal{W}' = \{W_1, \ldots, W_m, P' \cup W_j \cup W_k\} \setminus \{W_j, W_k\}$. Then $\mathcal{W}'$ has $m - 1$ elements and $X \setminus V \subset \left(\bigcup_{\substack{l \neq j \\ l \neq k}} W_l\right) \cup (P' \cup W_j \cup W_k)$, a contradiction to the minimality of m.

For every $j \in \{1, \ldots, m\}$, let A_j be the union of all the components of $\mathrm{Cl}(U)$, which intersect W_j. Then $A_j \cap A_k \subset P$, for each j and k in $\{1, \ldots, m\}$ such that $j \neq k$.

Now, since $p \notin \mathcal{T}(W_j) = W_j$, for each $j \in \{1, \ldots, m\}$, there exists a subcontinuum K_j of X such that $p \in \mathrm{Int}(K_j) \subset K_j \subset X \setminus W_j$. Then $K_j \cap (A_j \setminus P) = \emptyset$. To see this, suppose there exists $x \in K_j \cap (A_j \setminus P)$. Let L be the component of $K_j \cap \mathrm{Cl}(U)$ such that $x \in L$. By Theorem 1.4.36, L intersects the boundary of $K_j \cap \mathrm{Cl}(U)$ in K_j. This boundary is contained in $\bigcup_{\ell=1}^m W_\ell$. (If $y \in \mathrm{Bd}_{K_j}(K_j \cap \mathrm{Cl}(U))$, then $y \in \mathrm{Cl}(U)$. If $y \in U$, then

$y \in K_j \cap U \subset K_j \cap \mathrm{Cl}(U)$ and $(K_j \cap U) \cap (K_j \setminus (K_j \cap \mathrm{Cl}(U))) = \emptyset$, a contradiction. Therefore, $y \in \mathrm{Cl}(U) \setminus U \subset \bigcup_{\ell=1}^{m} W_\ell$.) Hence, there exists $k \in \{1, \ldots, m\} \setminus \{j\}$ such that $L \cap W_k \neq \emptyset$. But $L \subset M$, for some component M of $\mathrm{Cl}(U)$, $M \neq P$ and $M \subset A_j$, so that $M \cap W_k = \emptyset$, a contradiction.

Thus, $\left(\bigcap_{j=1}^{m} K_j\right) \cap \left(\bigcup_{j=1}^{m} (A_j \setminus P)\right) = \emptyset$, and it follows that $\bigcap_{j=1}^{m} K_j \subset P$. Since $p \in \mathrm{Int}\left(\bigcap_{j=1}^{m} K_j\right)$, $p \in \mathrm{Int}(P)$. Therefore, P is a connected neighborhood of p in X contained in V. Since p is an arbitrary point of V, every component of V is open. □

Our next three results are the consequence of Theorem 2.1.38.

2.1.39 Theorem. *Let X be an aposyndetic continuum. If for each subcontinuum W of X, $X \setminus W$ only has finitely many components, then X is locally connected.*

Proof. To show that X is locally connected, by Theorem 2.1.38, it suffices to prove that $\mathcal{T}(W) = W$ for every subcontinuum W of X. Let W be a subcontinuum of X, we know that $W \subset \mathcal{T}(W)$ (Remark 2.1.5). Let $x \in X \setminus W$. Since X is aposyndetic, for each $w \in W$, there exists a subcontinuum K_w of X such that $w \in \mathrm{Int}(K_w) \subset K_w \subset X \setminus \{x\}$. Since W is compact, there exist $w_1, \ldots, w_n$ in W such that $W \subset \bigcup_{j=1}^{n} \mathrm{Int}(K_{w_j}) \subset \bigcup_{j=1}^{n} K_{w_j} \subset X \setminus \{x\}$. Let $K = \bigcup_{j=1}^{n} K_{w_j}$. Then K is a subcontinuum of X such that $W \subset \mathrm{Int}(K) \subset K \subset X \setminus \{x\}$. By hypothesis, $X \setminus K$ only has finitely many components. Let C be the component of $X \setminus K$ containing x. Then, since $X \setminus K$ is open and only has finitely many components, C is open in X. Thus, $\mathrm{Cl}(C)$ is a subcontinuum of X such that $x \in \mathrm{Int}\,(\mathrm{Cl}(C)) \subset X \setminus \mathrm{Int}(K) \subset X \setminus W$. Hence, $x \in X \setminus \mathcal{T}(W)$. Therefore, $\mathcal{T}(W) = W$, and X is locally connected. □

2.1.40 Corollary. *Let X be an aposyndetic metric continuum such that $\dim(\mathcal{C}_1(X)) < \infty$. Then X is locally connected.*

Proof. Since $\dim(\mathcal{C}_1(X)) < \infty$, it follows that there exists $n \in \mathbb{N}$ such that $\mathcal{C}_1(X)$ does not contain an n-cell. Hence, using a similar proof to the one given for [92, Theorem 6.1.11], we have that X does not contain n-ods. Note that this implies that the complement of each subcontinuum of X has at most $n-1$ components. The corollary now follows from Theorem 2.1.39. □

2.1.41 Theorem. *Let X be an aposyndetic continuum. If $\mathcal{T}(W) = W$ for each subcontinuum W of X with nonempty interior, then X is locally connected.*

Proof. Let A be a subcontinuum of X, and let $x \in X \setminus A$. Since X is aposyndetic, for each $a \in A$, there exists a subcontinuum W_a of X such that $a \in \mathrm{Int}(W_a) \subset W_a \subset X \setminus \{x\}$. Since A is compact, there exist $a_1, \ldots, a_n$ in A such that $A \subset \bigcup_{j=1}^{n} \mathrm{Int}(W_{a_j})$. Let $W = \bigcup_{j=1}^{n} W_{a_j}$. Note that $A \subset W$, $\mathrm{Int}(W) \neq \emptyset$ and $x \in X \setminus W$. Hence, by hypothesis, $\mathcal{T}(W) = W$. Since $A \subset W$, by Proposition 2.1.7, $\mathcal{T}(A) \subset \mathcal{T}(W) = W$. Thus, $x \in X \setminus \mathcal{T}(A)$. Therefore, $\mathcal{T}(A) = A$. Now the theorem follows from Theorem 2.1.38. □

2.1.42 Theorem. *A continuum X is continuum aposyndetic if and only if X is locally connected.*

Proof. Suppose X is continuum aposyndetic. It follows from the definition that if W is a subcontinuum of X, then $\mathcal{T}(W) = W$. Hence, by Theorem 2.1.38, X is locally connected. The reverse implication is clear. □

As a consequence of Theorem 2.1.42, we obtain the following.

2.1.43 Theorem. *A continuum X is freely decomposable with respect to points and continua if and only if X is locally connected.*

Proof. Suppose X is freely decomposable with respect to points and continua. Then X is continuum aposyndetic. Thus, by Theorem 2.1.42, X is locally connected.

Now, assume X is locally connected. Let C be a subcontinuum of X, and let $a \in X \setminus C$. Consider the quotient space X/C. Then X/C is a locally connected continuum. Since locally connected continua are aposyndetic, by Theorem 1.4.28, X/C is freely decomposable. Let $q\colon X \twoheadrightarrow X/C$ be the quotient map, and let χ be the point of X/C corresponding to $q(C)$. Note that q is a monotone map. Since X/C is freely decomposable, there exist two subcontinua Γ_1 and Γ_2 of X/C such that $X/C = \Gamma_1 \cup \Gamma_2$, $q(a) \in \Gamma_1 \setminus \Gamma_2$ and $\chi \in \Gamma_2 \setminus \Gamma_1$. Hence, $X = q^{-1}(\Gamma_1) \cup q^{-1}(\Gamma_2)$, $a \in q^{-1}(\Gamma_1) \setminus q^{-1}(\Gamma_2)$ and $C = q^{-1}(\chi) \subset q^{-1}(\Gamma_2) \setminus q^{-1}(\Gamma_1)$. Therefore, X is freely decomposable with respect to points and continua. □

2.1.44 Theorem. *Let X be a continuum. Then X is indecomposable if and only if $\mathcal{T}(\{p\}) = X$ for each $p \in X$. Hence, $\mathcal{T}(A) = X$ for each nonempty closed subset A of X.*

Proof. Suppose X is indecomposable. By Corollary 1.4.35, each proper subcontinuum of X has empty interior. Therefore, $\mathcal{T}(\{p\}) = X$ for each $p \in X$.

Now, suppose X is decomposable. Then there exist two proper subcontinua A and B of X such that $X = A \cup B$. Let $a \in A \setminus B$, and let $b \in B \setminus A$. Then $b \in \text{Int}(B) \subset B \subset X \setminus \{a\}$. Hence, $b \in X \setminus \mathcal{T}(\{a\})$. Therefore, $\mathcal{T}(\{a\}) \neq X$. □

2.1.45 Lemma. *Let X be a compactum, and let A be a nonempty closed subset of X. Then $p \in X \setminus \mathcal{T}(A)$ if and only if there exist a subcontinuum W and an open subset Q of X such that $p \in \text{Int}(W) \cap Q$, $\text{Bd}(Q) \cap \mathcal{T}(A) = \emptyset$ and $W \cap A \cap Q = \emptyset$.*

Proof. Let $p \in X \setminus \mathcal{T}(A)$. Then there exists a subcontinuum W of X such that $p \in \text{Int}(W) \subset W \subset X \setminus A$. Since X is a normal space, there exists an open subset Q of X such that $p \in Q \subset \text{Cl}(Q) \subset \text{Int}(W)$. Hence, $\text{Bd}(Q) \cap \mathcal{T}(A) = \emptyset$ and $W \cap A \cap Q = \emptyset$.

Now, suppose there exist a continuum W and an open subset Q of X such that $p \in \text{Int}(W) \cap Q$, $\text{Bd}(Q) \cap \mathcal{T}(A) = \emptyset$ and $W \cap A \cap Q = \emptyset$. Since $\text{Bd}(Q)$ is a compact set and $\text{Bd}(Q) \cap \mathcal{T}(A) = \emptyset$, there exist a finite collection, $W_1, \ldots, W_n$, of subcontinua of X such that $\text{Bd}(Q) \subset \bigcup_{j=1}^{n} \text{Int}(W_j) \subset \bigcup_{j=1}^{m} W_j \subset X \setminus A$. If $W \subset Q$, then, clearly, $p \in X \setminus \mathcal{T}(A)$.

Assume $W \setminus Q \neq \emptyset$. By Theorem 1.4.36, the closure of each component of $W \cap Q$ must intersect at least one of the W_j's, since $\text{Bd}(Q) \subset \bigcup_{j=1}^{m} W_j$. Hence, $H = (W \cap Q) \cup \left(\bigcup_{j=1}^{m} W_j\right)$ only has a finite number of components. Let K be the component of H such that $p \in K$. Since $p \in \text{Int}(W) \cap Q$, by Lemma 1.4.1, $p \in \text{Int}(K)$ and, of course, $K \cap A \subset H \cap A = \emptyset$. Thus, $p \in X \setminus \mathcal{T}(A)$. □

2.1.46 Lemma. *Let X be a compactum, and let A be a subset of X. If $\mathcal{T}(A) = M \cup N$, where M and N are disjoint closed sets, then $\mathcal{T}(A \cap M) = M \cup \mathcal{T}(\emptyset)$.*

Proof. Suppose $p \in \mathcal{T}(A \cap M) \setminus (M \cup \mathcal{T}(\emptyset))$. Since $p \in X \setminus \mathcal{T}(\emptyset)$, there exists a subcontinuum W of X such that $p \in \text{Int}(W)$. Since X is a normal space, there exists an open subset Q of X such that $N \subset Q$ and $\text{Cl}(Q) \cap M = \emptyset$. Note that $p \in \text{Int}(W) \cap Q$ and $\text{Bd}(Q) \cap \mathcal{T}(A \cap M) \subset \text{Bd}(Q) \cap \mathcal{T}(A) = \emptyset$, and $W \cap (A \cap M) \cap Q \subset Q \cap M = \emptyset$. Then, by Lemma 2.1.45, $p \in X \setminus \mathcal{T}(A \cap M)$, thus contradicting the assumption.

Now, suppose $p \in (M \cup \mathcal{T}(\emptyset)) \setminus \mathcal{T}(A \cap M)$. Since $p \in X \setminus \mathcal{T}(A \cap M)$ and $\emptyset \subset A \cap M$, $p \in X \setminus \mathcal{T}(\emptyset)$. Hence, $p \in M$. Since X is a normal space, there exists an open subset Q of X such that $M \subset Q$ and $\text{Cl}(Q) \cap N = \emptyset$. Since $p \in X \setminus \mathcal{T}(A \cap M)$, there exists a subcontinuum W of X such that $p \in \text{Int}(W) \subset W \subset X \setminus (A \cap M)$. Observe that $p \in \text{Int}(W) \cap Q$ and $\text{Bd}(Q) \cap \mathcal{T}(A) = \emptyset$. Since $Q \cap N = \emptyset$, $W \cap A \cap Q = W \cap (A \cap M) = \emptyset$. Hence, by Lemma 2.1.45, $p \in X \setminus \mathcal{T}(A \cap M) \subset X \setminus M$, thus contradicting our assumption. □

2.1.47 Theorem. *Let X be a compactum, and let A be a closed subset of X. If K is a component of $\mathcal{T}(A)$, then $\mathcal{T}(A \cap K) = K \cup \mathcal{T}(\emptyset)$.*

Proof. Let K be a component of $\mathcal{T}(A)$. Let $\Gamma(K) = \{L \subset \mathcal{T}(A) \mid K \subset L$ and L is open and closed in $\mathcal{T}(A)\}$. Note that the collection of $\{A \cap L \mid L \in \Gamma(K)\}$ fails to be a filterbase if for some $L \in \Gamma(K)$, $A \cap L = \emptyset$. In this case, the conclusion of Lemma 2.2.9 holds (an argument similar to the one given in the proof of Theorem 2.2.10 shows this). Note that Lemma 2.1.46 remains true even if $A \cap M = \emptyset$. Hence, by Lemma 2.1.46, for each $L \in \Gamma(K)$, $\mathcal{T}(A \cap L) = L \cup \mathcal{T}(\emptyset)$. Therefore, by Lemma 2.2.9, the following sequence of equalities establishes the theorem:

$\mathcal{T}(A \cap K) = \mathcal{T}(\bigcap\{A \cap L \mid L \in \Gamma(K)\}) = \bigcap\{\mathcal{T}(A \cap L) \mid L \in \Gamma(K)\} = \bigcap\{L \cup \mathcal{T}(\emptyset) \mid L \in \Gamma(K)\} = \bigcap\{L \mid L \in \Gamma(K)\} \cup \mathcal{T}(\emptyset) = K \cup \mathcal{T}(\emptyset)$. (The fact that $\bigcap\{L \mid L \subset \Gamma(K)\} = K$ follows from Theorem 1.4.7.) □

The following corollary says that for each nonempty closed subset A of a continuum X, the components of $\mathcal{T}(A)$ are also in the image of $\mathcal{T}$.

2.1.48 Corollary. *Let X be a continuum, and let A be a closed subset of X. If K is a component of $\mathcal{T}(A)$, then $K = \mathcal{T}(A \cap K)$.*

Proof. Let A be a closed subset of X, and let K be a component of $\mathcal{T}(A)$. By Theorem 2.1.47, $\mathcal{T}(A \cap K) = K \cup \mathcal{T}(\emptyset)$. Since $\mathcal{T}(\emptyset) = \emptyset$, Corollary 2.1.14, $K = \mathcal{T}(A \cap K)$. □

2.1.49 Corollary. *Let X be a continuum, and let W_1 and W_2 be subcontinua of X. If $\mathcal{T}(W_1 \cup W_2) \neq \mathcal{T}(W_1) \cup \mathcal{T}(W_2)$, then $\mathcal{T}(W_1 \cup W_2)$ is a continuum.*

Proof. Suppose $\mathcal{T}(W_1 \cup W_2)$ is not connected. Then there exist two disjoint closed subsets A and B of X such that $\mathcal{T}(W_1 \cup W_2) = A \cup B$. By Lemma 2.1.46 and Corollary 2.1.14, $\mathcal{T}((W_1 \cup W_2) \cap A) = A$ and $\mathcal{T}((W_1 \cup W_2) \cap B) = B$. Suppose $W_1 \subset A$. If $W_2 \subset A$, then $A = \mathcal{T}((W_1 \cup W_2) \cap A) = \mathcal{T}(W_1 \cup W_2)$, a contradiction. Thus, $W_2 \subset B$. Hence, $\mathcal{T}(W_1) = A$ and $\mathcal{T}(W_2) = B$, which implies that $\mathcal{T}(W_1 \cup W_2) = \mathcal{T}(W_1) \cup \mathcal{T}(W_2)$, a contradiction. Therefore, $\mathcal{T}(W_1 \cup W_2)$ is connected. □

2.1.50 Corollary. *Let X be a continuum. If p and q are in X such that $\mathcal{T}(\{p, q\}) \neq \mathcal{T}(\{p\}) \cup \mathcal{T}(\{q\})$, then $\mathcal{T}(\{p, q\})$ is a continuum.*

The next theorem gives relationships between the images of maps and the images of $\mathcal{T}$. We subscript $\mathcal{T}$ to differentiate the continua on which $\mathcal{T}$ is defined.

2.1.51 Theorem. *Let X and Y be continua, and let $f\colon X \twoheadrightarrow Y$ be a surjective map. If A is a subset of X and B is a subset of Y, then the following hold.*

(a) $\mathcal{T}_Y(B) \subset f\mathcal{T}_X f^{-1}(B)$.
(b) If f is monotone, then $f\mathcal{T}_X(A) \subset \mathcal{T}_Y f(A)$ and $\mathcal{T}_X f^{-1}(B) \subset f^{-1}\mathcal{T}_Y(B)$.
(c) If f is monotone, then $\mathcal{T}_Y(B) = f\mathcal{T}_X f^{-1}(B)$.
(d) If f is open, then $f^{-1}\mathcal{T}_Y(B) \subset \mathcal{T}_X f^{-1}(B)$.
(e) If f is monotone and open, then $f^{-1}\mathcal{T}_Y(B) = \mathcal{T}_X f^{-1}(B)$.

Proof. We show (a). Let $y \in Y \setminus f\mathcal{T}_X f^{-1}(B)$. Then $f^{-1}(y) \cap \mathcal{T}_X f^{-1}(B) = \emptyset$. Thus, for each $x \in f^{-1}(y)$, there exists a subcontinuum W_x of X such that $x \in \text{Int}(W_x) \subset W_x \subset X \setminus f^{-1}(B)$. Since $f^{-1}(y)$ is compact, there exist $x_1, \ldots, x_n$ in $f^{-1}(y)$ such that $f^{-1}(y) \subset \bigcup_{j=1}^n \text{Int}(W_{x_j})$. Note that $\left(\bigcup_{j=1}^n W_{x_j}\right) \cap f^{-1}(B) = \emptyset$ and, for each $j \in \{1, \ldots, n\}$, $f^{-1}(y) \cap W_{x_j} \neq \emptyset$. Hence, $f\left(\bigcup_{j=1}^n W_{x_j}\right) \cap B = \emptyset$, and $f\left(\bigcup_{j=1}^n W_{x_j}\right)$ is a continuum ($y \in f(W_{x_j}) \cap f(W_{x_k})$ for every j and k in $\{1, \ldots, n\}$).

Observe that $Y \setminus f\left(X \setminus \bigcup_{j=1}^n \text{Int}(W_{x_j})\right)$ is an open subset of Y, and it is contained in $f\left(\bigcup_{j=1}^n W_{x_j}\right)$. To see this, note that, since $\bigcup_{j=1}^n \text{Int}(W_{x_j}) \subset \bigcup_{j=1}^n W_{x_j}$, $X \setminus \bigcup_{j=1}^n W_{x_j} \subset X \setminus \bigcup_{j=1}^n \text{Int}(W_{x_j})$. Then $f\left(X \setminus \bigcup_{j=1}^n W_{x_j}\right) \subset f\left(X \setminus \bigcup_{j=1}^n \text{Int}(W_{x_j})\right)$ and, consequently, $Y \setminus f\left(X \setminus \bigcup_{j=1}^n \text{Int}(W_{x_j})\right) \subset Y \setminus f\left(X \setminus \bigcup_{j=1}^n W_{x_j}\right)$. Since f is onto, $Y \setminus f\left(X \setminus \bigcup_{j=1}^n \text{Int}(W_{x_j})\right) \subset f\left(X \setminus \left(X \setminus \bigcup_{j=1}^n W_{x_j}\right)\right) = f\left(\bigcup_{j=1}^n W_{x_j}\right)$. Therefore, we obtain $Y \setminus f\left(X \setminus \bigcup_{j=1}^n \text{Int}(W_{x_j})\right) \subset f\left(\bigcup_{j=1}^n W_{x_j}\right)$.

Also, since $f^{-1}(y) \subset \bigcup_{j=1}^n \text{Int}(W_{x_j})$, we have that $\{y\} \cap f\left(X \setminus \bigcup_{j=1}^n \text{Int}(W_{x_j})\right) = \emptyset$. Hence, $y \in Y \setminus f\left(X \setminus \bigcup_{j=1}^n \text{Int}(W_{x_j})\right)$. Therefore, $y \in Y \setminus \mathcal{T}_Y(B)$.

We prove (b). First, we see $f\mathcal{T}_X(A) \subset \mathcal{T}_Y f(A)$. Let $y \in Y \setminus \mathcal{T}_Y f(A)$. Then there exists a subcontinuum W of Y such that $y \in \text{Int}(W) \subset W \subset Y \setminus f(A)$. It follows that $f^{-1}(y) \subset f^{-1}(\text{Int}(W)) \subset f^{-1}(W) \subset f^{-1}(Y) \setminus f^{-1}f(A) \subset X \setminus A$. Hence, $f^{-1}(y) \subset X \setminus A$. Since f is monotone, $f^{-1}(W)$ is a subcontinuum of X (Lemma 1.4.46). Thus, since $f^{-1}(y) \subset f^{-1}(\text{Int}(W)) \subset f^{-1}(W) \subset X \setminus A$, $f^{-1}(y) \cap \mathcal{T}_X(A) = \emptyset$. Therefore, $y \in Y \setminus f\mathcal{T}_X(A)$.

Now, we show $\mathcal{T}_X f^{-1}(B) \subset f^{-1}\mathcal{T}_Y(B)$. Let $x \in X \setminus f^{-1}\mathcal{T}_Y(B)$. Then $f(x) \in Y \setminus \mathcal{T}_Y(B)$. Hence, there exists a subcontinuum W of Y such that $f(x) \in \text{Int}(W) \subset W \subset Y \setminus B$. This implies that $x \in f^{-1}(f(x)) \subset f^{-1}(\text{Int}(W)) \subset f^{-1}(W) \subset X \setminus f^{-1}(B)$. Since f is monotone, $f^{-1}(W)$ is a subcontinuum of X. Therefore, $x \in X \setminus \mathcal{T}_X f^{-1}(B)$.

We prove (c). By (a), $\mathcal{T}_Y(B) \subset f\mathcal{T}_X f^{-1}(B)$. Since $\mathcal{T}_X f^{-1}(B) \subset f^{-1}\mathcal{T}_Y(B)$, by (b), $f\mathcal{T}_X f^{-1}(B) \subset ff^{-1}\mathcal{T}_Y(B) = \mathcal{T}_Y(B)$. Therefore, $f\mathcal{T}_X f^{-1}(B) = \mathcal{T}_Y(B)$.

We show (d). Let $x \in X \setminus \mathcal{T}_X f^{-1}(B)$. Then there exists a subcontinuum W of X such that $x \in \text{Int}(W) \subset W \subset X \setminus f^{-1}(B)$. Hence, $f(x) \in f(\text{Int}(W)) \subset f(W) \subset Y \setminus B$. Since f is open, $f(\text{Int}(W))$ is an open subset of Y. Therefore, $f(x) \in Y \setminus \mathcal{T}_Y(B)$.

Note that (e) follows directly from (b) and (d). □

The following two theorems are applications of Theorem 2.1.51.

2.1.52 Theorem. *Let X and Y be continua, and let $f\colon X \twoheadrightarrow Y$ be a surjective map. If X is locally connected, then Y is locally connected.*

Proof. By Theorem 2.1.37, it suffices to show that $\mathcal{T}_Y(B) = B$ for each closed subset B of Y. Let B be a closed subset of Y. By Remark 2.1.5, $B \subset \mathcal{T}_Y(B)$. By Theorem 2.1.51 (a), $\mathcal{T}_Y(B) \subseteq f\mathcal{T}_X f^{-1}(B)$. Since X is locally connected, $\mathcal{T}_X f^{-1}(B) = f^{-1}(B)$. Hence, $\mathcal{T}_Y(B) \subseteq ff^{-1}(B) = B$. Therefore, $\mathcal{T}_Y(B) = B$, and Y is locally connected. □

2.1.53 Theorem. *The monotone image of an indecomposable continuum is an indecomposable continuum.*

Proof. Let $f\colon X \twoheadrightarrow Y$ be a monotone map, where X is an indecomposable continuum. Let $y \in Y$. Then, by Theorem 2.1.51 (b), $\mathcal{T}_X f^{-1}(\{y\}) \subset f^{-1}\mathcal{T}_Y(\{y\})$. Since X is indecomposable, by Theorem 2.1.44, $\mathcal{T}_X f^{-1}(\{y\}) = X$. Hence, $Y = f(X) \subset ff^{-1}\mathcal{T}_Y(\{y\}) \subset Y$. Thus, $\mathcal{T}_Y(\{y\}) = Y$. Therefore, by Theorem 2.1.44, Y is indecomposable. □

2.1.54 Definition. Let X and Z be continua, and let $f\colon X \twoheadrightarrow Z$ be a map. Then f is an *atomic map* if for each subcontinuum K of X such that $f(K)$ is nondegenerate, then $K = f^{-1}(f(K))$.

2.1.55 Lemma. *Let X and Y be continua, and let $f\colon X \twoheadrightarrow Y$ be an atomic map. Then for every $x \in X$, $f^{-1}(f(x)) \subset \mathcal{T}_X(\{x\})$.*

Proof. Let $x \in X$, and let $z \in X \setminus \mathcal{T}_X(\{x\})$. Then there exists a subcontinuum W of X such that $z \in \text{Int}_X(W) \subset W \subset X \setminus \{x\}$. Since $f^{-1}(f(x))$ is a terminal subcontinuum of X [92, Theorem 8.1.25], by Lemma 1.4.51, we have that $f^{-1}(f(x)) \cap W = \emptyset$. In particular, $z \in X \setminus f^{-1}(f(x))$. Therefore, $f^{-1}(f(x)) \subset \mathcal{T}_X(\{x\})$. □

2.1.56 Lemma. *Let X and Y be continua, where Y is aposyndetic. If $f\colon X \twoheadrightarrow Y$ is a monotone map, then $\mathcal{T}_X(\{x\}) \subset f^{-1}f(x)$ for each $x \in X$.*

Proof. Let $x \in X$, and let $x' \in X \setminus f^{-1}(f(x))$. This implies that $f(x) \neq f(x')$. Hence, since Y is aposyndetic, there exists a subcontinuum W of Y such that $f(x') \in \mathrm{Int}_Y(W) \subset W \subset Y \setminus \{f(x)\}$. Thus, $x' \in f^{-1}(f(x')) \subset \mathrm{Int}_X(f^{-1}(W)) \subset f^{-1}(W) \subset X \setminus f^{-1}(f(x)) \subset X \setminus \{x\}$. Since f is monotone, $f^{-1}(W)$ is a subcontinuum of X, Lemma 1.4.46. It follows that $x' \in X \setminus \mathcal{T}_X(\{x\})$. Therefore, $\mathcal{T}_X(\{x\}) \subset f^{-1}(f(x))$. □

Since atomic maps are monotone [92, Theorem 8.1.24], as a consequence of Lemmas 2.1.55 and 2.1.56, we have the following.

2.1.57 Corollary. *Let X and Y be continua, where Y is aposyndetic, and let $f\colon X \twoheadrightarrow Y$ be an atomic map. Then for every $x \in X$, $\mathcal{T}_X(\{x\}) = f^{-1}(f(x))$.*

2.1.58 Definition. Let X and Y be continua. A surjective map $f\colon X \twoheadrightarrow Y$ is *quasi-monotone*, provided that for every subcontinuum Q of Y, with nonempty interior, $f^{-1}(Q)$ only has finitely many components, and if C is a component of $f^{-1}(Q)$, then $f(C) = Q$.

2.1.59 Theorem. *Let X and Y be continua, and let $f\colon X \twoheadrightarrow Y$ be a quasi-monotone map. If A is a subset of X and B is a subset of Y, then*

(1) *$f\mathcal{T}_X(A) \subset \mathcal{T}_Y f(A)$ and $\mathcal{T}_X f^{-1}(B) \subset f^{-1}\mathcal{T}_Y(B)$;*
(2) *$\mathcal{T}_Y(B) = f\mathcal{T}_X f^{-1}(B)$;*
(3) *if f is also open, then $f^{-1}\mathcal{T}_Y(B) = \mathcal{T}_X f^{-1}(B)$.*

Proof. First, we show that $f\mathcal{T}_X(A) \subset \mathcal{T}_Y f(A)$. To this end, let $y \in Y \setminus \mathcal{T}_Y f(A)$. Then there exists a subcontinuum W of Y such that $y \in \mathrm{Int}(W) \subset W \subset Y \setminus \mathcal{T}_Y f(A)$. Let $x \in f^{-1}(y)$ and let C_x be the component of $f^{-1}(W)$ containing x. Note that $x \in \mathrm{Int}(f^{-1}(W))$. Since f is quasi-monotone, $f^{-1}(W)$ only has a finite number of components. Hence, by Lemma 1.4.1, $x \in \mathrm{Int}(C_x)$. Also, $C_x \subset X \setminus f^{-1}f(A) \subset X \setminus A$. Thus, $x \in X \setminus \mathcal{T}_X(A)$. Since x is an arbitrary point of $f^{-1}(y)$, $f^{-1}(y) \subset X \setminus \mathcal{T}_X(A)$. Therefore, $y \in Y \setminus f\mathcal{T}_X(A)$.

Next, we prove $\mathcal{T}_X f^{-1}(B) \subset f^{-1}\mathcal{T}_Y(B)$. Let $x \in X \setminus f^{-1}\mathcal{T}_Y(B)$. Then $f(x) \in Y \setminus \mathcal{T}_Y(B)$. Hence, there exists a subcontinuum W of Y such that $f(x) \in \mathrm{Int}(W) \subset W \subset Y \setminus B$. Let C be the component of $f^{-1}(W)$ that contains x. Observe that $x \in \mathrm{Int}(f^{-1}(W))$. Since f is quasi-monotone, $f^{-1}(W)$ only has a finite number of components. Thus, by Lemma 1.4.1, $x \in \mathrm{Int}(C)$. Also, $C \subset f^{-1}(Y \setminus B) \subset X \setminus f^{-1}(B)$. As a consequence of this, $x \in X \setminus \mathcal{T}_X f^{-1}(B)$.

Now, we see that $\mathcal{T}_Y(B) = f\mathcal{T}_X f^{-1}(B)$. By Theorem 2.1.51 part (a), we have that $\mathcal{T}_Y(B) \subset f\mathcal{T}_X f^{-1}(B)$. By part (1) of this theorem, $\mathcal{T}_X f^{-1}(B) \subset f^{-1}\mathcal{T}_Y(B)$. Hence, $f\mathcal{T}_X f^{-1}(B) \subset ff^{-1}\mathcal{T}_Y(B)$. Since f is surjective, $ff^{-1}\mathcal{T}_Y(B) = \mathcal{T}_Y(B)$. Therefore, $\mathcal{T}_Y(B) = f\mathcal{T}_X f^{-1}(B)$.

Suppose f is also open. By Theorem 2.1.51 part (d), $f^{-1}\mathcal{T}_Y(B) \subset \mathcal{T}_X f^{-1}(B)$. Also, by part (1) of this theorem, $\mathcal{T}_X f^{-1}(B) \subset f^{-1}\mathcal{T}_Y(B)$. Therefore, $f^{-1}\mathcal{T}_Y(B) = \mathcal{T}_X f^{-1}(B)$. □

Now, we present some relations between $\mathcal{T}$ and inverse limits due to H. S. Davis [30].

2.1.60 Lemma. *Let $\{X_\lambda, f_\lambda^\alpha, \Lambda\}$ be an inverse system of continua whose inverse limit is $\mathcal{X}$. Let $\{p_\lambda\}_{\lambda\in\Lambda}$ be a point of $\mathcal{X}$, and let $\mathcal{A}$ be a nonempty closed subset of $\mathcal{X}$. If there exist $\lambda_0 \in \Lambda$, two open subsets U_{λ_0} and V_{λ_0} of X_{λ_0} and an inverse system $\{W_\lambda, f_\lambda^\alpha|_{W_\alpha}, \Lambda\}$ of subcontinua such that $p_{\lambda_0} \in U_{\lambda_0}$, $f_{\lambda_0}(\mathcal{A}) \subset V_{\lambda_0}$ and for each $\alpha \geq \lambda_0$,*

$$(f_{\lambda_0}^\alpha)^{-1}(U_{\lambda_0}) \subset W_\alpha \subset X_\alpha \setminus (f_{\lambda_0}^\alpha)^{-1}(V_{\lambda_0}),$$

then $\{p_\lambda\}_{\lambda\in\Lambda} \in \mathcal{X} \setminus \mathcal{T}(\mathcal{A})$.

Proof. Let $\mathcal{W} = \varprojlim\{W_\lambda, f_\lambda^\alpha|_{W_\alpha}, \Lambda\}$. Then $\mathcal{W}$ is a subcontinuum of $\mathcal{X}$, Theorem 1.2.20. Note that $f_{\lambda_0}^{-1}(U_{\lambda_0}) \subset \mathcal{W}$ since, for each $\alpha \geq \lambda_0$, $(f_{\lambda_0}^\alpha)^{-1}(U_{\lambda_0}) \subset \mathrm{Cl}((f_{\lambda_0}^\alpha)^{-1}(U_{\lambda_0})) \subset \mathrm{Cl}(f_\alpha^{-1}(f_{\lambda_0}^\alpha)^{-1}(U_{\lambda_0})) \subset W_\alpha$ and $\mathcal{W} = \bigcap_{\alpha\geq\lambda_0} f_\alpha^{-1}(W_\alpha)$ (Corollary 1.2.19). Since $\{p_\lambda\}_{\lambda\in\Lambda} \in f_{\lambda_0}^{-1}(U_{\lambda_0})$, $\{p_\lambda\}_{\lambda\in\Lambda} \in \mathrm{Int}(W)$.

Since, for every $\alpha \geq \lambda_0$, $f_\alpha^{-1}(W_\alpha) \subset \mathcal{X} \setminus f_\alpha^{-1}(f_{\lambda_0}^\alpha)^{-1}(V_{\lambda_0}) = \mathcal{X} \setminus f_{\lambda_0}^{-1}(V_{\lambda_0}) = f_{\lambda_0}^{-1}(X_{\lambda_0} \setminus V_{\lambda_0}) \subset \mathcal{X} \setminus \mathcal{A}$, $\bigcap_{\alpha\geq\lambda_0} f_\alpha^{-1}(W_\alpha) \subset \mathcal{X} \setminus \mathcal{A}$. Thus, $\mathcal{W} \cap \mathcal{A} = \emptyset$. Therefore, $\{p_\lambda\}_{\lambda\in\Lambda} \in \mathcal{X} \setminus \mathcal{T}(\mathcal{A})$. □

2.1.61 Lemma. *Let $\{X_\lambda, f_\lambda^\alpha, \Lambda\}$ be an inverse system of continua whose inverse limit is $\mathcal{X}$. Let $\{p_\lambda\}_{\lambda\in\Lambda}$ be a point of $\mathcal{X}$, and let $\mathcal{A}$ be a nonempty closed subset of $\mathcal{X}$. If $\{p_\lambda\}_{\lambda\in\Lambda} \in \mathcal{X} \setminus \mathcal{T}(\mathcal{A})$, then there exist $\lambda_0 \in \Lambda$, two open subsets U_{λ_0} and V_{λ_0} of X_{λ_0} and an inverse system $\{W_\lambda, f_\lambda^\alpha|_{W_\alpha}, \Lambda\}$ of subcontinua such that $p_{\lambda_0} \in U_{\lambda_0}$ $f_{\lambda_0}(\mathcal{A}) \subset V_{\lambda_0}$ and, for every $\alpha \geq \lambda_0$,*

$$(f_{\lambda_0}^\alpha)^{-1}(U_{\lambda_0}) \cap f_\alpha(\mathcal{X}) \subset W_\alpha \subset f_\alpha(\mathcal{X}) \setminus (f_{\lambda_0}^\alpha)^{-1}(V_{\lambda_0}).$$

Proof. Since $\{p_\lambda\}_{\lambda\in\Lambda} \in \mathcal{X} \setminus \mathcal{T}(\mathcal{A})$, there exists a subcontinuum $\mathcal{W}$ of $\mathcal{X}$ such that $\{p_\lambda\}_{\lambda\in\Lambda} \in \mathrm{Int}(\mathcal{W}) \subset \mathcal{W} \subset \mathcal{X} \setminus \mathcal{A}$. By Proposition 1.2.13, there exist $\lambda(p) \in \Lambda$ and an open subset $U_{\lambda(p)}$ of $X_{\lambda(p)}$ such that $\{p_\lambda\}_{\lambda\in\Lambda} \in f_{\lambda(p)}^{-1}(U_{\lambda(p)}) \subset \mathrm{Int}(\mathcal{W})$. Similarly, for each $\{a_\lambda\}_{\lambda\in\Lambda} \in \mathcal{A}$, there exist $\lambda(a) \in \Lambda$ and an open subset $U_{\lambda(a)}$ of $X_{\lambda(a)}$ such that $\{a_\lambda\}_{\lambda\in\Lambda} \in f_{\lambda(a)}^{-1}(U_{\lambda(a)}) \subset \mathcal{X} \setminus \mathcal{W}$. Since $\mathcal{A}$ is compact, there exist $\{a_\lambda^1\}_{\lambda\in\Lambda}, \ldots, \{a_\lambda^m\}_{\lambda\in\Lambda}$ in $\mathcal{A}$ such that $\mathcal{A} \subset \bigcup_{j=1}^m f_{\lambda(a^j)}^{-1}(U_{\lambda(a^j)})$. Let $\lambda_0 = \max\{\lambda(p), \lambda(a^1), \ldots, \lambda(a^m)\}$. Define $U_{\lambda_0} = (f_{\lambda(p)}^{\lambda_0})^{-1}(U_{\lambda(p)})$ and $V_{\lambda_0} = \bigcup_{j=1}^m (f_{\lambda(a^j)}^{\lambda_0})^{-1}(U_{\lambda(a^j)})$.

For each $\lambda \in \Lambda$, let $W_\lambda = f_\lambda(\mathcal{W})$. Let $\alpha \geq \lambda_0$, and let $z_\alpha \in (f_{\lambda_0}^\alpha)^{-1}(U_{\lambda_0}) \cap f_\alpha(\mathcal{X})$. Since $z_\alpha \in f_\alpha(\mathcal{X})$, there exists $\{z_\lambda\}_{\lambda\in\Lambda} \in X$ such that $f_\alpha(\{z_\lambda\}_{\lambda\in\Lambda}) = z_\alpha$. Since $f_{\lambda(p)}^\alpha(z_\alpha) \in U_{\lambda(p)}$, $f_{\lambda(p)}(\{z_\lambda\}_{\lambda\in\Lambda}) \in U_{\lambda(p)}$. Therefore,

$$\{z_\lambda\}_{\lambda\in\Lambda} \in f_{\lambda(p)}^{-1}(U_{\lambda(p)}) \subset \mathrm{Int}(\mathcal{W}) \subset \mathcal{W}.$$

Thus, $z_\alpha = f_\alpha(\{z_\lambda\}_{\lambda\in\Lambda}) \in f_\alpha(\mathcal{W}) = W_\alpha$. Hence, $(f^\alpha_{\lambda_0})^{-1}(U_{\lambda_0}) \cap f_\alpha(\mathcal{X}) \subset W_\alpha$.

Now, let $w_\alpha \in W_\alpha = f_\alpha(\mathcal{W})$. Then there exists $\{w_\lambda\}_{\lambda\in\Lambda} \in \mathcal{W}$ such that $f_\alpha(\{w_\lambda\}_{\lambda\in\Lambda}) = w_\alpha$. Clearly, $w_\alpha \in f_\alpha(\mathcal{X})$. Suppose $w_\alpha \in (f^\alpha_{\lambda_0})^{-1}(V_{\lambda_0})$. Then there exists $j \in \{1,\dots,m\}$ such that $f^\alpha_{\lambda(a^j)}(w_\alpha) \in U_{\lambda(a^j)}$. Hence, $f_{\lambda(a^j)}(\{w_\lambda\}_{\lambda\in\Lambda}) \in U_{\lambda(a^j)}$. Therefore, $\{w_\lambda\}_{\lambda\in\Lambda} \in \mathcal{X} \setminus \mathcal{W}$. This contradiction establishes that $W_\alpha \subset f_\alpha(\mathcal{X}) \setminus (f^\alpha_{\lambda_0})^{-1}(V_{\lambda_0})$. □

2.1.62 Theorem. *Let $\{X_\lambda, f^\alpha_\lambda, \Lambda\}$ be an inverse system of continua whose inverse limit is $\mathcal{X}$. Let $\{p_\lambda\}_{\lambda\in\Lambda}$ be a point of $\mathcal{X}$, and let $\mathcal{A}$ be a nonempty closed subset of $\mathcal{X}$. Then the following are equivalent:*

(a) $\{p_\lambda\}_{\lambda\in\Lambda} \in \mathcal{X} \setminus \mathcal{T}(\mathcal{A})$*;*
(b) there exist $\lambda_0 \in \Lambda$, two open subsets U_{λ_0} and V_{λ_0} of X_{λ_0} and an inverse system $\{W_\lambda, f^\alpha_\lambda|_{W_\alpha}, \Lambda\}$ of subcontinua such that $p_{\lambda_0} \in U_{\lambda_0}$, $f_{\lambda_0}(\mathcal{A}) \subset V_{\lambda_0}$ and, for every $\alpha \geq \lambda_0$,

$$(f^\alpha_{\lambda_0})^{-1}(U_{\lambda_0}) \cap f_\alpha(\mathcal{X}) \subset W_\alpha \subset f_\alpha(\mathcal{X}) \setminus (f^\alpha_{\lambda_0})^{-1}(V_{\lambda_0}).$$

Proof. Suppose (a). Then Lemma 2.1.61 shows (b).

Now, suppose (b). Note that if $\mathcal{W} = \varprojlim\{W_\lambda, f^\alpha_\lambda|_{W_\alpha}, \Lambda\}$, then for each $\lambda \in \Lambda$, $f_\lambda(\mathcal{W}) = f_\lambda(\mathcal{X}) \cap W_\lambda$, $f_\lambda(\mathcal{W})$ is a continuum and $\mathcal{W} = \varprojlim\{f_\lambda(\mathcal{W}), f^\alpha_\lambda|_{f_\alpha(\mathcal{W})}, \Lambda\}$ (Proposition 1.2.17).

The relations in Lemma 2.1.60 imply that $(f^\alpha_{\lambda_0})^{-1}(U_{\lambda_0}) \cap f_\alpha(\mathcal{X}) \subset W_\alpha \subset f_\alpha(\mathcal{X}) \setminus (f^\alpha_{\lambda_0})^{-1}(V_{\lambda_0})$. □

2.1.63 Theorem. *Let $\{X_\lambda, f^\alpha_\lambda, \Lambda\}$ be an inverse system of continua, with surjective bonding maps, whose inverse limit is $\mathcal{X}$. Let $\{p_\lambda\}_{\lambda\in\Lambda}$ be a point of $\mathcal{X}$, and let $\mathcal{A}$ be a nonempty closed subset of $\mathcal{X}$. Then the following are equivalent:*

(a) $\{p_\lambda\}_{\lambda\in\Lambda} \in \mathcal{X} \setminus \mathcal{T}(\mathcal{A})$*;*
(b) there exist $\lambda_0 \in \Lambda$, two open subsets U_{λ_0} and V_{λ_0} of X_{λ_0} and an inverse system $\{W_\lambda, f^\alpha_\lambda|_{W_\alpha}, \Lambda\}$ of subcontinua such that $p_{\lambda_0} \in U_{\lambda_0}$, $f_{\lambda_0}(\mathcal{A}) \subset V_{\lambda_0}$ and, for every $\alpha \geq \lambda_0$,

$$(f^\alpha_{\lambda_0})^{-1}(U_{\lambda_0}) \subset W_\alpha \subset X_\alpha \setminus (f^\alpha_{\lambda_0})^{-1}(V_{\lambda_0}).$$

Proof. Note that, since the bonding maps are surjective, the projection maps are surjective also (Theorem 1.2.16). Thus, the theorem now follows directly from Theorem 2.1.62. □

2.1.64 Corollary. *Let $\{X_\lambda, f^\alpha_\lambda, \Lambda\}$ be an inverse system of continua, with surjective bonding maps, whose inverse limit is $\mathcal{X}$. Let $\mathcal{A}$ be a nonempty closed subset of $\mathcal{X}$, and let Λ_0 be a cofinal subset of Λ. Then,*

$$\bigcap_{\lambda\in\Lambda_0} f_\lambda^{-1}\mathcal{T}_{X_\lambda} f_\lambda(\mathcal{A}) \subset \mathcal{T}_{\mathcal{X}}(\mathcal{A}).$$

Proof. Let $\{p_\lambda\}_{\lambda\in\Lambda} \in \mathcal{X} \setminus \mathcal{T}_{\mathcal{X}}(\mathcal{A})$. By Theorem 2.1.63, there exist $\lambda_0 \in \Lambda$, two open subsets U_{λ_0} and V_{λ_0} of X_{λ_0} and an inverse system $\{W_\lambda, f_\lambda^\alpha |_{W_\alpha}, \Lambda\}$ of subcontinua such that $p_{\lambda_0} \in U_{\lambda_0}$, $f_{\lambda_0}(\mathcal{A}) \subset V_{\lambda_0}$ and, for every $\alpha \geq \lambda_0$,

$$(f_{\lambda_0}^\alpha)^{-1}(U_{\lambda_0}) \subset W_\alpha \subset X_\alpha \setminus (f_{\lambda_0}^\alpha)^{-1}(V_{\lambda_0}).$$

Since Λ_0 is a cofinal subset of Λ, there exists $\alpha \in \Lambda_0$ such that $\lambda_0 \leq \alpha$. Then $p_\alpha \in (f_{\lambda_0}^\alpha)^{-1}(U_{\lambda_0}) \subset W_\alpha \subset X_\alpha \setminus (f_{\lambda_0}^\alpha)^{-1}(V_{\lambda_0}) \subset X_\alpha \setminus f_\alpha(\mathcal{A})$. Thus, $p_\alpha \in X_\alpha \setminus \mathcal{T}_{X_\alpha} f_\alpha(\mathcal{A})$. Hence, $\{p_\lambda\}_{\lambda\in\Lambda} \in \mathcal{X} \setminus f_\alpha^{-1}\mathcal{T}_{X_\alpha} f_\alpha(\mathcal{A})$. Therefore, $\{p_\lambda\}_{\lambda\in\Lambda} \in \mathcal{X} \setminus \bigcap_{\alpha\in\Lambda_0} f_\alpha^{-1}\mathcal{T}_{X_\alpha} f_\alpha(\mathcal{A})$. □

2.1.65 Corollary. *Let $\{X_\lambda, f_\lambda^\alpha, \Lambda\}$ be an inverse system of continua, with surjective bonding maps, whose inverse limit is $\mathcal{X}$. Let $\mathcal{A}$ be a nonempty closed subset of $\mathcal{X}$, and let Λ_0 be a cofinal subset of Λ. If the bonding maps are monotone, then*

$$\bigcap_{\lambda\in\Lambda_0} f_\lambda^{-1}\mathcal{T}_{X_\lambda} f_\lambda(\mathcal{A}) = \mathcal{T}_{\mathcal{X}}(\mathcal{A}).$$

Proof. Let $\{p_\lambda\}_{\lambda\in\Lambda} \in \mathcal{X} \setminus \bigcap_{\lambda\in\Lambda_0} f_\lambda^{-1}\mathcal{T}_{X_\lambda} f_\lambda(\mathcal{A})$. Then there exist $\lambda_0 \in \Lambda_0$ such that $p_{\lambda_0} \in X_{\lambda_0} \setminus \mathcal{T}_{X_{\lambda_0}} f_{\lambda_0}(\mathcal{A})$. Thus, there exist a subcontinuum W_{λ_0} of X_{λ_0} and two open subsets U_{λ_0} and V_{λ_0} of X_{λ_0} such that $p_{\lambda_0} \in U_{\lambda_0} \subset W_{\lambda_0} \subset X_{\lambda_0} \setminus V_{\lambda_0} \subset X_{\lambda_0} \setminus f_{\lambda_0}(\mathcal{A})$.

Let $\alpha \in \Lambda_0$ be such that $\lambda_0 \leq \alpha$, and let $W_\alpha = (f_{\lambda_0}^\alpha)^{-1}(W_{\lambda_0})$. Since the bonding maps are monotone, W_α is a subcontinuum of X_α (Lemma 1.4.46). Clearly, $\{W_\lambda, f_\lambda^\gamma|_{W_\gamma}, \lambda, \alpha \in \Lambda_0 \text{ and } \lambda_0 \leq \lambda \leq \gamma\}$ is an inverse system of subcontinua and $(f_{\lambda_0}^\alpha)^{-1}(U_{\lambda_0}) \subset W_\alpha \subset X_\alpha \setminus (f_{\lambda_0}^\alpha)^{-1}(V_{\lambda_0})$. Hence, by Theorem 2.1.63, $\{p_\lambda\}_{\lambda\in\Lambda} \in \mathcal{X} \setminus \mathcal{T}_{\mathcal{X}}(\mathcal{A})$. The other inclusion is given by Corollary 2.1.64. □

The next corollary is a consequence of Davis' work.

2.1.66 Corollary. *Let $\{X_\lambda, f_\lambda^\alpha, \Lambda\}$ be an inverse system of aposyndetic continua, with surjective bonding maps, whose inverse limit is $\mathcal{X}$. If the bonding maps are monotone, then $\mathcal{X}$ is aposyndetic.*

Proof. Let Λ_0 be a cofinal subset of Λ. Let $\{p_\lambda\}_{\lambda\in\Lambda} \in \mathcal{X}$. By Corollary 2.1.65, we have that $\mathcal{T}_{\mathcal{X}}(\{\{p_\lambda\}_{\lambda\in\Lambda}\}) = \bigcap_{\lambda\in\Lambda_0} f_\lambda^{-1}\mathcal{T}_{X_\lambda}(\{p_\lambda\})$. Since each factor space is aposyndetic, by Theorem 2.1.34,

$$\bigcap_{\lambda\in\Lambda_0} f_\lambda^{-1}\mathcal{T}_{X_\lambda}(\{p_\lambda\}) = \bigcap_{\lambda\in\Lambda_0} f_\lambda^{-1}(\{p_\lambda\}) = \{\{p_\lambda\}_{\lambda\in\Lambda}\}$$

(Corollary 1.2.19). Therefore, $\mathcal{T}_{\mathcal{X}}(\{\{p_\lambda\}_{\lambda\in\Lambda}\}) = \{\{p_\lambda\}_{\lambda\in\Lambda}\}$. Since $\{p_\lambda\}_{\lambda\in\Lambda}$ is arbitrary, by Theorem 2.1.34, $\mathcal{X}$ is aposyndetic. □

2.1.67 Theorem. *Let X be a continuum with the property of Kelley, and let W_1 and W_2 be disjoint subcontinua with nonempty interior. Then $\mathcal{T}(W_1) \cap \mathcal{T}(W_2) = \emptyset$.*

Proof. By Corollary 1.6.21, there exists a subcontinuum M_1 such that $W_1 \subset \mathrm{Int}(M_1)$ and $M_1 \cap W_2 = \emptyset$. This implies that $W_1 \cap \mathcal{T}(W_2) = \emptyset$. By Corollary 1.6.21, there exists a subcontinuum M_2 such that $\mathcal{T}(W_2) \subset \mathrm{Int}(M_2)$ and $M_2 \cap W_1 = \emptyset$. Hence, $\mathcal{T}(W_2) \cap \mathcal{T}(W_1) = \emptyset$. □

2.2 $\mathcal{T}$-Symmetry and $\mathcal{T}$-Additivity

We study the properties of continua for which the set function $\mathcal{T}$ is symmetric and additive.

2.2.1 Definition. A continuum X *is* $\mathcal{T}$*-symmetric* if for each pair, A and B, of closed subsets of X, $A \cap \mathcal{T}(B) = \emptyset$ if and only if $B \cap \mathcal{T}(A) = \emptyset$. We say that X *is point* $\mathcal{T}$*-symmetric* if for each pair, p and q, of points of X, $p \notin \mathcal{T}(\{q\})$ if and only if $q \notin \mathcal{T}(\{p\})$.

2.2.2 Theorem. *If X is a weakly irreducible continuum, then X is $\mathcal{T}$-symmetric.*

Proof. Let A and B be two closed subsets of X. Suppose $A \cap \mathcal{T}(B) = \emptyset$. Hence, for each $a \in A$, there exists a subcontinuum W_a of X such that $a \in \mathrm{Int}(W_a) \subset W_a \subset X \setminus B$. Since A is compact, there exist $a_1, \ldots, a_n$ in A such that $A \subset \bigcup_{j=1}^n \mathrm{Int}(W_{a_j}) \subset \bigcup_{j=1}^n W_{a_j} \subset X \setminus B$. This implies that $B \subset X \setminus \left(\bigcup_{j=1}^n W_{a_j}\right) \subset X \setminus \left(\bigcup_{j=1}^n \mathrm{Int}(W_{a_j})\right) \subset X \setminus A$.

Let $b \in B$. Since X is weakly irreducible, $X \setminus \left(\bigcup_{j=1}^n W_{a_j}\right)$ only has a finite number of components that are open subsets of X. Let C be the component of $X \setminus \left(\bigcup_{j=1}^n W_{a_j}\right)$ such that $b \in C$. Then $b \in C \subset \mathrm{Cl}(C) \subset X \setminus \left(\bigcup_{j=1}^n \mathrm{Int}(W_{a_j})\right) \subset X \setminus A$. Thus, $b \in X \setminus \mathcal{T}(A)$. Therefore, $B \cap \mathcal{T}(A) = \emptyset$.

Similarly, if $B \cap \mathcal{T}(A) = \emptyset$, then $A \cap \mathcal{T}(B) = \emptyset$. Therefore, X is $\mathcal{T}$-symmetric. □

2.2.3 Corollary. *If X is an irreducible continuum, then X is $\mathcal{T}$-symmetric.*

Proof. Let X be an irreducible continuum. By Theorem 1.4.43, X is weakly irreducible. Hence, by Theorem 2.2.2, X is $\mathcal{T}$-symmetric. □

As a consequence of Theorem 1.4.70 and Theorem 2.2.2, we have the following.

2.2.4 Corollary. *If X is a θ-continuum, then X is $\mathcal{T}$-symmetric.*

2.2.5 Theorem. *Let X be a $\mathcal{T}$-symmetric continuum, and let p be a point of X. Then X is connected im kleinen at p if and only if X is semi-locally connected at p.*

Proof. Suppose X is connected im kleinen at p. Let $q \in \mathcal{T}(\{p\})$. Since X is $\mathcal{T}$-symmetric, $p \in \mathcal{T}(\{q\})$. Thus, $p \in \{q\}$ (Corollary 2.1.31). Hence, $p = q$ and $\mathcal{T}(\{p\}) = \{p\}$. Therefore, by Theorem 2.1.35, X is semi-locally connected at p.

Now, suppose X is semi-locally connected at p. Then $\mathcal{T}(\{p\}) = \{p\}$ (Theorem 2.1.35). Let A be a closed subset of X such that $p \in \mathcal{T}(A)$. Hence, since X is $\mathcal{T}$-symmetric, $A \cap \mathcal{T}(\{p\}) \neq \emptyset$. Thus, $p \in A$. Therefore, by Corollary 2.1.31, X is connected im kleinen at p. □

2.2.6 Definition. A continuum X *is $\mathcal{T}$-additive* if for each pair, A and B, of closed subsets of X, $\mathcal{T}(A \cup B) = \mathcal{T}(A) \cup \mathcal{T}(B)$.

The next theorem gives us a sufficient condition for a continuum X to be $\mathcal{T}$-additive.

2.2.7 Theorem. *Let X be a continuum. If for each point $x \in X$ and any two subcontinua W_1 and W_2 of X such that $x \in \operatorname{Int}(W_j)$, $j \in \{1, 2\}$, there exists a subcontinuum W_3 of X such that $x \in \operatorname{Int}(W_3)$ and $W_3 \subset W_1 \cap W_2$, then X is $\mathcal{T}$-additive.*

Proof. Let A_1 and A_2 be closed subsets of X, and let $x \in X \setminus \mathcal{T}(A_1) \cup \mathcal{T}(A_2)$. Hence, there exist two subcontinua W_1 and W_2 of X such that $x \in \operatorname{Int}(W_j) \subset W_j \subset X \setminus A_j$, $j \in \{1, 2\}$. By hypothesis, there exists a subcontinuum W_3 of X such that $x \in \operatorname{Int}(W_3)$ and $W_3 \subset W_1 \cap W_2$. Thus, $x \in \operatorname{Int}(W_3) \subset W_3 \subset X \setminus (A_1 \cup A_2)$. Hence, $x \in X \setminus \mathcal{T}(A_1 \cup A_2)$. Therefore, by Corollary 2.1.8, $\mathcal{T}(A_1 \cup A_2) = \mathcal{T}(A_1) \cup \mathcal{T}(A_2)$. □

We need the following definition to give a characterization of $\mathcal{T}$-additive continua.

2.2.8 Definition. Let X be a compactum. A *filterbase* Γ *in* X is a family $\Gamma = \{A_\omega\}_{\omega \in \Omega}$ of subsets of X having two properties:

(a) for each $\omega \in \Omega$, $A_\omega \neq \emptyset$ and
(b) for each $\omega_1, \omega_2 \in \Omega$, there exists $\omega_3 \in \Omega$ such that $A_{\omega_3} \subset A_{\omega_1} \cap A_{\omega_2}$.

2.2.9 Lemma. *Let X be a compactum. If Γ is a filterbase of closed subsets of X, then $\mathcal{T}\left(\bigcap\{G \mid G \in \Gamma\}\right) = \bigcap\{\mathcal{T}(G) \mid G \in \Gamma\}$.*

Proof. By Proposition 2.1.7, $\mathcal{T}\left(\bigcap\{G \mid G \in \Gamma\}\right) \subset \bigcap\{\mathcal{T}(G) \mid G \in \Gamma\}$.

Let $p \in X \setminus \mathcal{T}\left(\bigcap\{G \mid G \in \Gamma\}\right)$. Then there exists a subcontinuum W of X such that $p \in \operatorname{Int}(W) \subset W \subset X \setminus \left(\bigcap\{G \mid G \in \Gamma\}\right)$. Since W is compact, there exist $G_1, \ldots, G_n$ in Γ such that $W \cap \left(\bigcap_{j=1}^n G_j\right) = \emptyset$. Since Γ is a filterbase, there exists $G \in \Gamma$ such that $G \subset \bigcap_{j=1}^n G_j$. Note that $W \cap G = \emptyset$. Thus, $p \in X \setminus \mathcal{T}(G)$. Hence, $p \in X \setminus \bigcap\{\mathcal{T}(G) \mid G \in \Gamma\}$. Therefore, $\mathcal{T}\left(\bigcap\{G \mid G \in \Gamma\}\right) = \bigcap\{\mathcal{T}(G) \mid G \in \Gamma\}$. □

The next theorem gives us a characterization of $\mathcal{T}$-additivity on continua.

2.2.10 Theorem. *Let X be a continuum. Then X is $\mathcal{T}$-additive if and only if for each family Λ of closed subsets of X whose union is closed, $\mathcal{T}\left(\bigcup\{L \mid L \in \Lambda\}\right) = \bigcup\{\mathcal{T}(L) \mid L \in \Lambda\}$.*

Proof. Suppose X is $\mathcal{T}$-additive. By Proposition 2.1.7,

$$\bigcup\{\mathcal{T}(L) \mid L \in \Lambda\} \subset \mathcal{T}\left(\bigcup\{L \mid L \in \Lambda\}\right).$$

Now, suppose $x \in X \setminus \bigcup\{\mathcal{T}(L) \mid L \in \Lambda\}$. Then for each $L \in \Lambda$, let $F(L) = \{A \subset X \mid A \text{ is closed and } L \subset \operatorname{Int}(A)\}$. If $L = \emptyset$, then $\mathcal{T}(L) = \bigcap\{\mathcal{T}(A) \mid A \in F(L)\}$. (Clearly $\mathcal{T}(L) \subset \bigcap\{\mathcal{T}(A) \mid A \in F(L)\}$. Let $z \in X \setminus \mathcal{T}(L)$. Then there exists a subcontinuum W of X such that $z \in \operatorname{Int}(W)$. Let A be any proper closed subset of $X \setminus W$. Then $A \in F(L)$ and $z \in X \setminus \mathcal{T}(A)$. Hence, $z \in X \setminus \bigcap\{\mathcal{T}(A) \mid A \in F(L)\}$.) If $L \neq \emptyset$, then $F(L)$ is a filterbase of closed subsets of X. Since $\bigcap\{A \mid A \in F(L)\} = L$, $\mathcal{T}(L) = \bigcap\{\mathcal{T}(A) \mid A \in F(L)\}$, by Lemma 2.2.9.

Hence, for each $L \in \Lambda$, $x \in X \setminus \bigcap\{\mathcal{T}(A) \mid A \in F(L)\}$; and thus, there exists, for each $L \in \Lambda$, $A_L \in F(L)$ such that $x \in X \setminus \mathcal{T}(A_L)$. Note that $\{\operatorname{Int}(A_L) \mid L \in \Lambda\}$ is an open cover of $\bigcup\{L \mid L \in \Lambda\}$. Since this set is compact, there exist $L_1, \ldots, L_m$ in Λ such that

$$\bigcup\{L \mid L \in \Lambda\} \subset \bigcup_{j=1}^{m} \operatorname{Int}(A_{L_j}).$$

Since, by hypothesis and mathematical induction, $\mathcal{T}\left(\bigcup_{j=1}^{m} A_{L_j}\right) = \bigcup_{j=1}^{m} \mathcal{T}(A_{L_j})$, $\mathcal{T}\left(\bigcup\{L \mid L \in \Lambda\}\right) \subset \bigcup_{j=1}^{m} \mathcal{T}(A_{L_j})$. Now, since for every $j \in \{1, \ldots, m\}$, $x \in X \setminus \mathcal{T}(A_{L_j})$, it follows that $x \in X \setminus \mathcal{T}\left(\bigcup\{L \mid L \in \Lambda\}\right)$. Thus, we have that $\mathcal{T}\left(\bigcup\{L \mid L \in \Lambda\}\right) \subset \bigcup\{\mathcal{T}(L) \mid L \in \Lambda\}$. □

2.2.11 Theorem. *Each $\mathcal{T}$-symmetric continuum is $\mathcal{T}$-additive.*

Proof. Let X be a $\mathcal{T}$-symmetric continuum, and let A and B be two closed subsets of X. By Corollary 2.1.8, $\mathcal{T}(A) \cup \mathcal{T}(B) \subset \mathcal{T}(A \cup B)$.

Let $x \in \mathcal{T}(A \cup B)$. Then $\{x\} \cap \mathcal{T}(A \cup B) \neq \emptyset$. Since X is $\mathcal{T}$-symmetric, $\mathcal{T}(\{x\}) \cap (A \cup B) \neq \emptyset$. Hence, either $\mathcal{T}(\{x\}) \cap A \neq \emptyset$ or $\mathcal{T}(\{x\}) \cap B \neq \emptyset$. Thus, since X is $\mathcal{T}$-symmetric, $\{x\} \cap \mathcal{T}(A) \neq \emptyset$ or $\{x\} \cap \mathcal{T}(B) \neq \emptyset$; that is, $x \in \mathcal{T}(A)$ or $x \in \mathcal{T}(B)$. Then $x \in \mathcal{T}(A) \cup \mathcal{T}(B)$. Therefore, X is $\mathcal{T}$-additive. □

2.2.12 Theorem. *If X is a hereditarily unicoherent continuum, then X is $\mathcal{T}$-additive.*

Proof. Let X be a hereditarily unicoherent continuum, and let A and B be two closed subsets of X. By Corollary 2.1.8, $\mathcal{T}(A) \cup \mathcal{T}(B) \subset \mathcal{T}(A \cup B)$.

Let $x \in X \setminus (\mathcal{T}(A) \cup \mathcal{T}(B))$. Then $x \notin \mathcal{T}(A)$ and $x \notin \mathcal{T}(B)$. Hence, there exist subcontinua W_A and W_B of X such that $x \in \operatorname{Int}(W_A) \cap \operatorname{Int}(W_B) \subset W_A \cap W_B \subset X \setminus (A \cup B)$. Since X is hereditarily unicoherent, $W_A \cap W_B$ is a subcontinuum of X. Therefore, $x \in X \setminus \mathcal{T}(A \cup B)$. □

The following corollary is a consequence of Theorem 2.2.10; however, we present a different proof based on Corollary 2.1.25.

2.2.13 Corollary. *If X is a $\mathcal{T}$-additive continuum and if A is a closed subset of X, then $\mathcal{T}(A) = \bigcup_{a\in A} \mathcal{T}(\{a\})$.*

Proof. Let A be a closed subset of the $\mathcal{T}$-additive continuum X. If $a \in A$, then $\{a\} \subset A$ and $\mathcal{T}(\{a\}) \subset \mathcal{T}(A)$, by Proposition 2.1.7. Hence, $\bigcup_{a\in A} \mathcal{T}(\{a\}) \subseteq \mathcal{T}(A)$.

Now, let $x \in X$ be such that $x \in X \setminus \mathcal{T}(\{a\})$ for each $a \in A$. Then, by Corollary 2.1.25, for every $a \in A$, there exists an open subset U_a of X such that $a \in U_a$ and $x \in X \setminus \mathcal{T}(\mathrm{Cl}(U_a))$. Since A is compact and $\{U_a\}_{a\in A}$ is an open cover of A, there exist $a_1, \ldots, a_n$ in A such that $A \subset \bigcup_{j=1}^n U_{a_j}$. Since for each $j \in \{1, \ldots, n\}$, $x \in X \setminus \mathcal{T}(\mathrm{Cl}(U_{a_j}))$, $x \in X \setminus \bigcup_{j=1}^n \mathcal{T}(\mathrm{Cl}(U_{a_j}))$. Thus, $x \in X \setminus \mathcal{T}\left(\bigcup_{j=1}^n \mathrm{Cl}(U_{a_j})\right)$ (since X is $\mathcal{T}$-additive). Hence, $x \in X \setminus \mathcal{T}(A)$. □

2.2.14 Theorem. *A continuum X is locally connected if and only if X is aposyndetic and $\mathcal{T}$-additive.*

Proof. Suppose X is a locally connected continuum. Then, by Theorem 2.1.37, $\mathcal{T}(A) = A$ for each closed subset A of X. Hence, $\mathcal{T}(\{p\}) = \{p\}$ for each point $p \in X$ and $\mathcal{T}(A \cup B) = A \cup B = \mathcal{T}(A) \cup \mathcal{T}(B)$ for each pair of closed subsets A and B of X. Therefore, X is aposyndetic (Theorem 2.1.34) and $\mathcal{T}$-additive.

Now, suppose X is aposyndetic and $\mathcal{T}$-additive. Let A be a closed subset of X. Since X is $\mathcal{T}$-additive, by Corollary 2.2.13, $\mathcal{T}(A) = \bigcup_{a\in A} \mathcal{T}(\{a\})$. Since X is aposyndetic, by Theorem 2.1.34, $\mathcal{T}(\{a\}) = \{a\}$ for each $a \in A$. Thus, $\mathcal{T}(A) = \bigcup_{a\in A} \mathcal{T}(\{a\}) = \bigcup_{a\in A}\{a\} = A$. Therefore, by Theorem 2.1.37, X is locally connected. □

As a consequence of Theorems 2.2.2, 2.2.11 and 2.2.14, we have the following.

2.2.15 Corollary. *If X is an aposyndetic weakly irreducible continuum, then X is locally connected.*

2.2.16 Corollary. *If X is an aposyndetic hereditarily unicoherent continuum, then X is locally connected.*

Proof. Let X be an aposyndetic hereditarily unicoherent continuum. Since X is hereditarily unicoherent, by Theorem 2.2.12, X is $\mathcal{T}$-additive. Then, since X is an aposyndetic $\mathcal{T}$-additive continuum, by Theorem 2.2.14, X is locally connected.

□

2.2.17 Theorem. *A continuum X is $\mathcal{T}$-symmetric if and only if X is point $\mathcal{T}$-symmetric and $\mathcal{T}$-additive.*

Proof. Suppose X is $\mathcal{T}$-symmetric. By Theorem 2.2.11, X is $\mathcal{T}$-additive. Clearly, X is point $\mathcal{T}$-symmetric.

Now, suppose X is point $\mathcal{T}$-symmetric and $\mathcal{T}$-additive. Let A and B be two closed subsets of X. Suppose that $A \cap \mathcal{T}(B) = \emptyset$ and $B \cap \mathcal{T}(A) \neq \emptyset$. Let $x \in B \cap \mathcal{T}(A)$. Then $x \in B$ and $x \in \mathcal{T}(A)$. Note that $x \in X \setminus A$ (if $x \in A$, then $x \in A \cap B \subset A \cap \mathcal{T}(B)$, a contradiction). Since X is $\mathcal{T}$-additive, $\mathcal{T}(A) = \bigcup_{a\in A} \mathcal{T}(\{a\})$, Corollary 2.2.13. Thus, there exists $a \in A$ such that $x \in \mathcal{T}(\{a\})$. This implies that $a \in \mathcal{T}(\{x\})$ (since X is point $\mathcal{T}$-symmetric). Consequently, since $x \in B$, $a \in \mathcal{T}(\{x\}) \subset \mathcal{T}(B)$ (Proposition 2.1.7). Hence, $a \in A \cap \mathcal{T}(B)$, a contradiction to our assumption. Therefore, $B \cap \mathcal{T}(A) = \emptyset$, and X is $\mathcal{T}$-symmetric. □

2.2.18 Theorem. *Let X be a semi-aposyndetic irreducible metric continuum. Then X is homeomorphic to $[0, 1]$.*

Proof. Since X is an irreducible metric continuum, by Corollary 2.2.3, X is $\mathcal{T}$-symmetric. Let x and y be points of X. Since X is semi-aposyndetic, by Theorem 2.1.32, either $x \in X \setminus \mathcal{T}(\{y\})$ or $y \in X \setminus \mathcal{T}(\{x\})$. This implies that $x \in X \setminus \mathcal{T}(\{y\})$ if and only if $y \in X \setminus \mathcal{T}(\{x\})$, by the $\mathcal{T}$-symmetry of X. Hence, $\mathcal{T}(\{x\}) = \{x\}$ for each $x \in X$. Therefore, X is aposyndetic (Theorem 2.1.34).

Since X is $\mathcal{T}$-symmetric, X is $\mathcal{T}$-additive (Theorem 2.2.11). Since X is an aposyndetic $\mathcal{T}$-additive continuum, X is locally connected (Theorem 2.2.14). Thus, X is an irreducible locally connected metric continuum. Consequently, X is an irreducible arcwise connected metric continuum [60, Theorem 3-15]. Therefore, X is homeomorphic to $[0, 1]$. □

2.2.19 Theorem. *The monotone image of a $\mathcal{T}$-symmetric continuum is $\mathcal{T}$-symmetric.*

Proof. Let X be a $\mathcal{T}$-symmetric continuum, and let $f\colon X \twoheadrightarrow Y$ be a monotone map. Let A and B be two closed subsets of Y such that $A \cap \mathcal{T}_Y(B) = \emptyset$. Then, by Theorem 2.1.51 (c), $A \cap f\mathcal{T}_X f^{-1}(B) = \emptyset$. Thus, $f^{-1}(A) \cap \mathcal{T}_X f^{-1}(B) = \emptyset$. Since X is $\mathcal{T}$-symmetric, $\mathcal{T}_X f^{-1}(A) \cap f^{-1}(B) = \emptyset$. Hence, $f(\mathcal{T}_X f^{-1}(A) \cap f^{-1}(B)) = f\mathcal{T}_X f^{-1}(A) \cap B = \mathcal{T}_Y(A) \cap B = \emptyset$. Therefore, Y is $\mathcal{T}$-symmetric. □

2.2.20 Theorem. *The monotone image of a $\mathcal{T}$-additive continuum is $\mathcal{T}$-additive.*

Proof. Let X be a $\mathcal{T}$-additive continuum, and let $f\colon X \twoheadrightarrow Y$ be a monotone map. Let A and B be two closed subsets of Y. Then, by Theorem 2.1.51 (c), $\mathcal{T}_Y(A \cup B) = f\mathcal{T}_X f^{-1}(A \cup B) = f\mathcal{T}_X(f^{-1}(A) \cup f^{-1}(B)) = f(\mathcal{T}_X f^{-1}(A) \cup \mathcal{T}_X f^{-1}(B)) = f\mathcal{T}_X f^{-1}(A) \cup f\mathcal{T}_X f^{-1}(B) = \mathcal{T}_Y(A) \cup \mathcal{T}_Y(B)$. □

As a consequence of Theorem 2.1.59, we have the next results.

2.2.21 Theorem. *Let X and Y be continua, and let $f : X \to Y$ be a quasi-monotone map. If X is $\mathcal{T}_X$-symmetric, then Y is $\mathcal{T}_Y$-symmetric.*

Proof. Let B_1 and B_2 be two closed subsets of Y such that $B_1 \cap \mathcal{T}_Y(B_2) = \emptyset$. By Theorem 2.1.59 part (2), $\mathcal{T}_Y(B_2) = f\mathcal{T}_X f^{-1}(B_2)$. Hence, $B_1 \cap f\mathcal{T}_X f^{-1}(B_2) = \emptyset$. Since

$$f^{-1}(B_1) \cap \mathcal{T}_X f^{-1}(B_2) \subset f^{-1}(B_1) \cap f^{-1}f\mathcal{T}_X f^{-1}(B_2) = \emptyset,$$

we have that $f^{-1}(B_1) \cap \mathcal{T}_X f^{-1}(B_2) = \emptyset$. Thus, by $\mathcal{T}_X$-symmetry of X, $\mathcal{T}_X f^{-1}(B_1) \cap f^{-1}(B_2) = \emptyset$. Since $f(\mathcal{T}_X f^{-1}(B_1) \cap f^{-1}(B_2)) = f\mathcal{T}_X f^{-1}(B_1) \cap B_2$ and, by Theorem 2.1.59 part (2), $\mathcal{T}_Y(B_1) = f\mathcal{T}_X f^{-1}(B_1)$, we obtain that $\mathcal{T}_Y(B_1) \cap B_2 = \emptyset$. Therefore, Y is $\mathcal{T}_Y$-symmetric. □

2.2.22 Theorem. *Let X and Y be continua, and let $f : X \to Y$ be a quasi-monotone map. If X is $\mathcal{T}_X$-additive, then Y is $\mathcal{T}_Y$-additive.*

Proof. Let B_1 and B_2 be closed subsets of Y. By Theorem 2.1.59 part (2), we have

$$\mathcal{T}_Y(B_1 \cup B_2) = f\mathcal{T}_X f^{-1}(B_1 \cup B_2) = f(\mathcal{T}_X(f^{-1}(B_1) \cup f^{-1}(B_2))).$$

Since X is $\mathcal{T}_X$-additive, $\mathcal{T}_X(f^{-1}(B_1) \cup f^{-1}(B_2)) = \mathcal{T}_X f^{-1}(B_1) \cup \mathcal{T}_X f^{-1}(B_2)$. Thus, $f(\mathcal{T}_X(f^{-1}(B_1) \cup f^{-1}(B_2))) = f(\mathcal{T}_X f^{-1}(B_1) \cup \mathcal{T}_X(f^{-1}(B_2))) = f\mathcal{T}_X f^{-1}(B_1) \cup f\mathcal{T}_X f^{-1}(B_2) = \mathcal{T}_Y(B_1) \cup \mathcal{T}_Y(B_2)$ (Theorem 2.1.59 part (2)). Therefore, $\mathcal{T}_Y(B_1 \cup B_2) = \mathcal{T}_Y(B_1) \cup \mathcal{T}_Y(B_2)$, and Y is $\mathcal{T}_Y$-additive. □

2.3 Idempotency of $\mathcal{T}$

We present a characterization of idempotency of $\mathcal{T}$ and consequences of this. We also give a sufficient condition for the idempotency of $\mathcal{T}$ on closed sets and give an example that shows this condition is not necessary.

In 1980, David P. Bellamy asked: *If X and Y are indecomposable continua, is $\mathcal{T}$ idempotent on $X \times Y$? Even for only the closed sets of $X \times Y$?*. We prove that the set function $\mathcal{T}$ is not idempotent on the product of two indecomposable continua. This gives a negative answer to the first question, and the second one remains open (Question 8.1.3). We also prove that $\mathcal{T}$ is never idempotent on the cone and suspension over an indecomposable continuum. We give an example of a decomposable metric continuum Z such that $\mathcal{T}$ is not idempotent on the family of closed sets of either $Z \times [0, 1]$ or $K(Z)$ or $\Sigma(Z)$. We consider continua for which $\mathcal{T}$ is idempotent on continua. We present a couple of classes of continua on which idempotency of $\mathcal{T}$ on continua implies idempotency of $\mathcal{T}$ on closed sets.

2.3.1 Definition. Let X be a continuum. We say that $\mathcal{T}$ *is idempotent on* X provided that $\mathcal{T}^2(A) = \mathcal{T}(A)$ for each subset A of X. We say that $\mathcal{T}$ *is idempotent on closed sets* (*on continua*) if $\mathcal{T}^2(A) = \mathcal{T}(A)$ for each closed subset (subcontinuum) A of X.

2.3.2 Proposition. *Let X be a continuum, and let Z be a nonempty closed subset of X such that $\mathcal{T}^2(Z) = \mathcal{T}(Z)$. If $\mathcal{T}(Z) = A \cup B$, where A and B are nonempty closed subsets of X, then $\mathcal{T}(A \cup B) = \mathcal{T}(A) \cup \mathcal{T}(B) = A \cup B$.*

Proof. Since $\mathcal{T}^2(Z) = \mathcal{T}(Z)$, by Remark 2.1.5 and Proposition 2.1.7, we have

$$\mathcal{T}(Z) = A \cup B \subset \mathcal{T}(A) \cup \mathcal{T}(B) \subset \mathcal{T}(A \cup B) = \mathcal{T}^2(Z) = \mathcal{T}(Z).$$

□

2.3.3 Proposition. *Let X be a continuum. If Z is a nonempty closed subset of X such that $\mathcal{T}^2(Z) = \mathcal{T}(Z)$, then*

$$\mathcal{T}(Z) = \bigcup\{\mathcal{T}(\{w\}) \mid w \in \mathcal{T}(Z)\}.$$

Proof. Let $x \in \bigcup\{\mathcal{T}(\{w\}) \mid w \in \mathcal{T}(Z)\}$. Then there exists $w \in \mathcal{T}(Z)$ such that $x \in \mathcal{T}(\{w\})$. Since $\mathcal{T}^2(Z) = \mathcal{T}(Z)$, $\mathcal{T}(\{w\}) \subset \mathcal{T}^2(Z) = \mathcal{T}(Z)$. Hence, $x \in \mathcal{T}(Z)$, and $\bigcup\{\mathcal{T}(\{w\}) \mid w \in \mathcal{T}(Z)\} \subset \mathcal{T}(Z)$. The other inclusion is clear. □

2.3.4 Theorem. *Let X be a continuum. Then $\mathcal{T}$ is idempotent on X if and only if for each subcontinuum W of X and each point $x \in \mathrm{Int}(W)$, there exists a subcontinuum M of X such that $x \in \mathrm{Int}(M) \subset M \subset \mathrm{Int}(W)$.*

Proof. Suppose $\mathcal{T}$ is idempotent on X. Let W be a subcontinuum of X, and let $x \in \mathrm{Int}(W)$. Hence, $x \in X \setminus \mathcal{T}(X \setminus W)$. Since $\mathcal{T}$ is idempotent, $x \in X \setminus \mathcal{T}^2(X \setminus W)$. Thus, there exists a subcontinuum M of X such that $x \in \mathrm{Int}(M) \subset M \subset X \setminus \mathcal{T}(X \setminus W) \subset X \setminus (X \setminus W) = W$. Therefore, $x \in \mathrm{Int}(M) \subset M \subset \mathrm{Int}(W)$.

Now, suppose the condition stated holds. We show $\mathcal{T}$ is idempotent. By Remark 2.1.5, for each subset A of X, $\mathcal{T}(A) \subset \mathcal{T}^2(A)$. Let B be a subset of X, and let $x \in X \setminus \mathcal{T}(B)$. Then there exists a subcontinuum W of X such that $x \in \mathrm{Int}(W) \subset W \subset X \setminus B$. By hypothesis, there exists a subcontinuum M such that $x \in \mathrm{Int}(M) \subset M \subset \mathrm{Int}(W)$. Since $\mathrm{Int}(W) \subset X \setminus \mathcal{T}(B)$, $x \in \mathrm{Int}(M) \subset M \subset X \setminus \mathcal{T}(B)$. Hence, $x \in X \setminus \mathcal{T}^2(B)$. Therefore, $\mathcal{T}$ is idempotent. □

2.3.5 Corollary. *Let X be a continuum. If $\mathcal{T}$ is idempotent on X, W is a subcontinuum of X and K is a component of $\mathrm{Int}(W)$, then K is open.*

Proof. Let W be a subcontinuum of X, and let K be a component of $\mathrm{Int}(W)$. If $x \in K$, then, by Theorem 2.3.4, there exists a subcontinuum M of X such that $x \in \mathrm{Int}(M) \subset M \subset \mathrm{Int}(W)$. Hence, $M \subset K$ and x is an interior point of K. Therefore, K is open. □

2.3.6 Definition. Let X be a continuum. A subcontinuum M of X is a *continuum domain* if $M = \mathrm{Cl}(\mathrm{Int}(M))$. A continuum domain M is a *strong continuum domain* provided that $\mathrm{Int}(M)$ is connected.

2.3.7 Corollary. *Let X be a continuum. If $\mathcal{T}$ is idempotent on X, $x \in X$ and W is a subcontinuum of X such that $x \in \mathrm{Int}(W)$, then there exists a strong continuum domain M of X such that $x \in \mathrm{Int}(M) \subset M \subset W$.*

Proof. Let W be a subcontinuum of X, and let $x \in \mathrm{Int}(W)$. By Corollary 2.3.5, the component, K, of W containing x is open. Let $M = \mathrm{Cl}(K)$. Then M is a strong continuum domain and $x \in \mathrm{Int}(M) \subset M \subset W$. □

2.3.8 Theorem. *Let X be a continuum, and let A be a subset of X. If $\mathcal{T}^2(A) = \mathcal{T}(A)$, then the components of $X \setminus \mathcal{T}(A)$ are open and continuumwise connected.*

Proof. If $A = \emptyset$ or $\mathcal{T}(A) = X$, then the result is clear. Let A be a nonempty subset of X such that $\mathcal{T}(A) \neq X$. Let L be a component of $X \setminus \mathcal{T}(A)$, and let $x \in L$. Since $\mathcal{T}^2(A) = \mathcal{T}(A)$, $x \in X \setminus \mathcal{T}^2(A)$. Hence, there exists a subcontinuum W of X such that $x \in \mathrm{Int}(W) \subset W \subset X \setminus \mathcal{T}(A)$. Since L is a component of $X \setminus \mathcal{T}(A)$ and $L \cap W \neq \emptyset$, $W \subset L$. Thus, x is an interior point of L. Therefore, L is open.

We show that L is continuumwise connected. Let ℓ' and ℓ'' be two points of L. For each $\ell \in L$, there exists a subcontinuum W_ℓ of X such that $\ell \in \text{Int}(W_\ell) \subset W_\ell \subset X \setminus \mathcal{T}(A)$. Observe that $W_\ell \subset L$ for all $\ell \in L$. Also note that $\{\text{Int}(W_\ell) \mid \ell \in L\}$ forms an open cover of L, since L is connected, there exist $\ell_1, \ldots, \ell_n$ in L such that $\{\text{Int}(W_{\ell_1}), \ldots, \text{Int}(W_{\ell_n})\}$ forms a chain such that $\ell' \in \text{Int}(W_{\ell_1})$ and $\ell'' \in \text{Int}(W_{\ell_n})$ [24, (2.F.2)]. Hence, $W = \bigcup_{j=1}^n W_{\ell_j}$ is a subcontinuum of X such that $\{\ell', \ell''\} \subset W \subset L$. Therefore, L is continuumwise connected. □

A similar result to Corollary 2.3.7 is true when $\mathcal{T}$ is idempotent on closed sets.

2.3.9 Theorem. *Let X be a continuum such that $\mathcal{T}$ is idempotent on closed sets. If A is a nonempty closed subset of X and $x \in X \setminus \mathcal{T}(A)$, then there exists a strong continuum domain W of X such that $x \in \text{Int}(W) \subset W \subset X \setminus \mathcal{T}(A) \subset X \setminus A$.*

Proof. Let A be a nonempty closed subset of X, and let $x \in X \setminus \mathcal{T}(A)$. Since $\mathcal{T}$ is idempotent on closed sets, we have that $x \in X \setminus \mathcal{T}^2(A)$. By Corollary 2.1.25, there exists an open subset U of X such that $\mathcal{T}(A) \subset U$ and $x \in X \setminus \mathcal{T}(\text{Cl}(U))$. Let L be the component of $X \setminus \mathcal{T}(\text{Cl}(U))$ containing x. Since $\mathcal{T}$ is idempotent on closed sets, by Theorem 2.3.8, L is open. Let $W = \text{Cl}(L)$. Then W is a strong continuum domain of X such that $x \in \text{Int}(W) \subset W \subset X \setminus U \subset X \setminus \mathcal{T}(A)$. □

2.3.10 Theorem. *Let X be an aposyndetic continuum such that $\mathcal{T}$ is idempotent on closed sets. If $\mathcal{T}(W) = W$ for each strong continuum domain W of X, then X is locally connected.*

Proof. Note that, by Theorem 2.1.38, it is enough to show that $\mathcal{T}(A) = A$ for each subcontinuum A of X. Clearly $\mathcal{T}(X) = X$. Let A be a proper subcontinuum of X. By Remark 2.1.5, $A \subset \mathcal{T}(A)$. Let $x \in X \setminus A$. Since X is aposyndetic, by Theorem 2.3.9, for each $a \in A$, there exists a strong continuum domain W_a of X such that $a \in \text{Int}(W_a) \subset W_a \subset X \setminus \{x\}$. Since A is compact, there exist $a_1, \ldots, a_n$ in A such that $A \subset \bigcup_{j=1}^n \text{Int}(W_{a_j}) \subset \bigcup_{j=1}^n W_{a_j} \subset X \setminus \{x\}$. Observe that $\bigcup_{j=1}^n \text{Int}(W_{a_j})$ is a connected open set. Let $W = \text{Cl}\left(\bigcup_{j=1}^n \text{Int}(W_{a_j})\right)$. Then W is a strong continuum domain such that $A \subset W \subset X \setminus \{x\}$. By hypothesis, $\mathcal{T}(W) = W$. Hence, by Proposition 2.1.7, $\mathcal{T}(A) \subset \mathcal{T}(W) = W \subset X \setminus \{x\}$. Thus, $x \in X \setminus \mathcal{T}(A)$. Therefore, $\mathcal{T}(A) = A$, and X is locally connected. □

The next theorem gives a sufficient condition to have $\mathcal{T}$ idempotent on closed sets.

2.3.11 Theorem. *Let X be a continuum. If for each nonempty subset A of X and every subcontinuum K of X such that $\text{Int}(K) \neq \emptyset$ and $A \cap K = \emptyset$, there exists a subcontinuum W of X such that $K \subset \text{Int}(W) \subset W \subset X \setminus A$, then $\mathcal{T}$ is idempotent on closed sets.*

Proof. Suppose that for each nonempty subset A of X and every subcontinuum K of X such that $\text{Int}(K) \neq \emptyset$ and $K \cap A = \emptyset$, there exists a subcontinuum W of X such that $K \subset \text{Int}(W) \subset W \subset X \setminus A$. Let A be a nonempty closed subset of X, and let $x \in X \setminus \mathcal{T}(A)$. Then there exists a subcontinuum K of X such that

$x \in \text{Int}(K) \subset K \subset X \setminus A$. By hypothesis, there exists a subcontinuum W of X such that $K \subset \text{Int}(W) \subset W \subset X \setminus A$. This implies that $x \in X \setminus \mathcal{T}^2(A)$. Hence, $\mathcal{T}^2(A) \subset \mathcal{T}(A)$. By Remark 2.1.5 and Proposition 2.1.7, $\mathcal{T}(A) \subset \mathcal{T}^2(A)$. Therefore, $\mathcal{T}$ is idempotent on closed sets. □

The next example shows that the converse of Theorem 2.3.11 is not true even for metric continua.

2.3.12 Example. Let

$$X = (\{0\} \times [-1, 2]) \cup \left\{ \left(x, \sin\left(\frac{1}{x}\right) \right) \mid x \in \left(0, \frac{\pi}{2}\right] \right\},$$

let $K = \{0\} \times \left[0, \frac{3}{2}\right]$ and let $A = \{0\} \times \left[-\frac{1}{3}, -\frac{1}{2}\right]$. Then $\mathcal{T}$ is idempotent on closed sets, also, $\text{Int}(K) = \{0\} \times \left(1, \frac{3}{2}\right)$, $K \cap A = \emptyset$, and if W is a subcontinuum of X such that $K \subset \text{Int}(W)$, then $A \subset W$.

2.3.13 Theorem. *Let X be a continuum. If $\mathcal{T}$ is idempotent on X and $\mathcal{T}(\{p, q\})$ is a continuum for all p and q in X, then X is indecomposable.*

Proof. Suppose X is decomposable. Then, by Corollary 2.3.7, there exists a strong continuum domain W of X. Let p_0 and q_0 be any two points in $X \setminus \text{Int}(W)$. Note that, by Theorem 2.3.4, $\mathcal{T}(\{p_0, q_0\}) \cap \text{Int}(W) = \emptyset$.

Since $\mathcal{T}(\{p_0, q_0\})$ is connected and $\mathcal{T}(\{p_0, q_0\}) \cap \text{Int}(W) = \emptyset$, p_0 and q_0 belong to the same component of $X \setminus \text{Int}(W)$. Thus, $X \setminus \text{Int}(W)$ is a continuum. By Theorem 2.3.4, there exists a subcontinuum M, with nonempty interior, such that $M \subset \text{Int}(X \setminus \text{Int}(W)) \subset X \setminus W$. Then let $K = X \setminus (\text{Int}(W) \cup \text{Int}(M))$, and let p_1 and q_1 be any two points of K. By Theorem 2.3.4, $\mathcal{T}(\{p_1, q_1\}) \subset K$. Hence, K is a continuum also. Now, let $p \in \text{Int}(M)$, and let $q \in \text{Int}(W)$. Observe that, by Theorem 2.3.4, $\mathcal{T}(\{p, q\}) \cap \text{Int}(K) = \emptyset$. Then $\mathcal{T}(\{p, q\}) \subset X \setminus \text{Int}(K) \subset W \cup M$. Hence, $\mathcal{T}(\{p, q\})$ is not a continuum, a contradiction. □

2.3.14 Theorem. *Let X be a point $\mathcal{T}$-symmetric continuum for which $\mathcal{T}$ is idempotent on singletons (Definition 3.1.1). Suppose there exists a point $p \in X$ such that $\mathcal{T}(\{p\}) = X$. Then X is indecomposable.*

Proof. Let p be a point of X such that $\mathcal{T}(\{p\}) = X$, and let $q \in X \setminus \{p\}$. Then either $p \in X \setminus \mathcal{T}(\{q\})$ or $p \in \mathcal{T}(\{q\})$. If $p \in X \setminus \mathcal{T}(\{q\})$, then, by point $\mathcal{T}$-symmetry, $q \in X \setminus \mathcal{T}(\{p\}) = \emptyset$, a contradiction. Hence, $p \in \mathcal{T}(\{q\})$. Thus, $\mathcal{T}(\{p\}) \subset \mathcal{T}^2(\{q\}) = \mathcal{T}(\{q\})$ ($\mathcal{T}$ is idempotent of singletons). Then $\mathcal{T}(\{q\}) = X$. Therefore, $\mathcal{T}(\{x\}) = X$ for each point x of X, and X is indecomposable, Theorem 2.1.44. □

Now, we give a characterization of locally connected continua.

2.3.15 Theorem. *Let X be a continuum such that $\mathcal{T}$ is idempotent on continua. Then X is locally connected if and only if $\mathcal{T}|_{\mathcal{C}_1(X)} : \mathcal{C}_1(X) \twoheadrightarrow \mathcal{C}_1(X)$ is surjective.*

Proof. First note that, by Remark 2.1.5, $A \subset \mathcal{T}(A)$ for all subsets A of X. Thus, $\mathcal{T}(\{x\}) = \{x\}$ for each $x \in X$. Hence, X is aposyndetic, Theorem 2.1.34.

Let $A \in \mathcal{C}_1(X)$. Since $\mathcal{T}|_{\mathcal{C}_1(X)}$ is surjective, there exists $A' \in \mathcal{C}_1(X)$ such that $\mathcal{T}(A') = A$. Thus, since $\mathcal{T}$ is idempotent on continua, $\mathcal{T}(A) = \mathcal{T}^2(A') = \mathcal{T}(A') = A$. Therefore, by Theorem 2.1.38, X is locally connected.

The converse implication is clear since for locally connected continua X, $\mathcal{T}$ is the identity map on 2^X, Theorem 2.1.37. In particular, $\mathcal{T}|_{\mathcal{C}_1(X)}$ is the identity map on $\mathcal{C}_1(X)$. □

2.3.16 Corollary. *Let X be a continuum such that $\mathcal{T}$ is idempotent on closed sets. Then X is locally connected if and only if $\mathcal{T}\colon 2^X \twoheadrightarrow 2^X$ is surjective.*

2.3.17 Theorem. *Let X and Y be continua such that $\mathcal{T}_X$ is idempotent on continua, and let $f\colon X \twoheadrightarrow Y$ be a quasi-monotone map. If for each subcontinuum Z of Y, there exists a subcontinuum W of X such that $\mathcal{T}_X(W) = f^{-1}(Z)$, then Y is locally connected.*

Proof. Let Z be a subcontinuum of Y. Since f is quasi-monotone, by Theorem 2.1.59 (2), we have that $\mathcal{T}_Y(Z) = f\mathcal{T}_X f^{-1}(Z)$. By hypothesis, there exists a subcontinuum W of X such that $\mathcal{T}_X(W) = f^{-1}(Z)$. Hence, since $\mathcal{T}_X^2(W) = \mathcal{T}_X(W)$, we have that $f\mathcal{T}_X f^{-1}(Z) = f\mathcal{T}_X\mathcal{T}_X(W) = f\mathcal{T}_X(W) = ff^{-1}(Z) = Z$. Thus, $\mathcal{T}_Y(Z) = Z$. Therefore, by Theorem 2.1.38, Y is locally connected. □

The next theorem is an application of Theorem 2.3.11.

2.3.18 Theorem. *If X is a continuum with the property of Kelley, then $\mathcal{T}$ is idempotent on closed sets.*

Proof. Let A be a nonempty closed set of X. We know that $\mathcal{T}(A) \subset \mathcal{T}^2(A)$. Let $x \in X \setminus \mathcal{T}(A)$. Then there exists a subcontinuum W of X such that $x \in \operatorname{Int}(W) \subset W \subset X \setminus A$. By Corollary 1.6.21, there exists a subcontinuum M of X such that $W \subset \operatorname{Int}(M)$ and $M \cap A = \emptyset$. By Theorem 2.3.11, $\mathcal{T}$ is idempotent on closed sets. □

2.3.19 Remark. Theorem 2.3.18 cannot be improved to obtain that the set function $\mathcal{T}$ is idempotent, even for metric continua. Let X be the product of a solenoid and a simple closed curve. Then X is a homogeneous metric continuum. By [92, Theorem 4.2.32], $\mathcal{T}_X$ is idempotent on closed sets and, by Theorem 2.3.22, $\mathcal{T}_X$ is not idempotent.

Next, we prove that the set function $\mathcal{T}$ is not idempotent on the product of two continua one of which is indecomposable. In particular, $\mathcal{T}$ is not idempotent on the product of two indecomposable continua. We need the following.

2.3.20 Lemma. *Let X be an indecomposable continuum, let $x_0 \in X$ and let Y be a continuum. If W is a subcontinuum of $X \times Y$ such that* $\operatorname{Int}_{X\times Y}(W) \neq \emptyset$, *then $W \cap (\{x_0\} \times Y) \neq \emptyset$.*

Proof. Let $\pi_X\colon X \times Y \twoheadrightarrow X$ be the projection map. Note that $\pi_X(W)$ is a subcontinuum of X such that $\operatorname{Int}_X(\pi_X(W)) \neq \emptyset$. Thus, since X is indecomposable, $\pi_X(W) = X$, Corollary 1.4.35. Let $w \in W$ be such that $\pi_X(w) = x_0$. Then $w \in W \cap (\{x_0\} \times Y)$. Therefore, $W \cap (\{x_0\} \times Y) \neq \emptyset$. □

2.3.21 Corollary. *If X is an indecomposable continuum, $x_0 \in X$ and Y is a continuum, then $\mathcal{T}_{X\times Y}(\{x_0\} \times Y) = X \times Y$.*

2.3.22 Theorem. *If X is an indecomposable continuum and Y is a continuum, then $\mathcal{T}_{X\times Y}$ is not idempotent on $X \times Y$.*

Proof. Let $(x_0, y_0) \in X \times Y$, let $A = (\{x_0\} \times Y) \setminus \{(x_0, y_0)\}$ and let $(x, y) \in (X \times Y) \setminus A$, where $y \neq y_0$. Let U be an open subset of X such that $x \in U \subset \mathrm{Cl}_X(U) \subset X \setminus \{x_0\}$. Let

$$W = (\mathrm{Cl}_X(U) \times Y) \cup (X \times \{y_0\}).$$

Then W is a subcontinuum of $X \times Y$ such that $(x, y) \in \mathrm{Int}_{X\times Y}(W) \subset W \subset (X \times Y) \setminus A$. Hence, $\mathcal{T}_{X\times Y}(A) \neq X \times Y$. Since $\mathcal{T}_{X\times Y}(A)$ is closed, $\mathrm{Cl}_{X\times Y}(A) \subset \mathcal{T}_{X\times Y}(A)$. Since A is dense in $\{x_0\} \times Y$, $\mathrm{Cl}_{X\times Y}(A) = \{x_0\} \times Y$. Hence, by Corollary 2.3.21, $X \times Y \subset \mathcal{T}_{X\times Y}(\{x_0\} \times Y) \subset \mathcal{T}^2_{X\times Y}(A)$. Therefore, $\mathcal{T}_{X\times Y}$ is not idempotent on $X \times Y$ (Figure 2.5). □

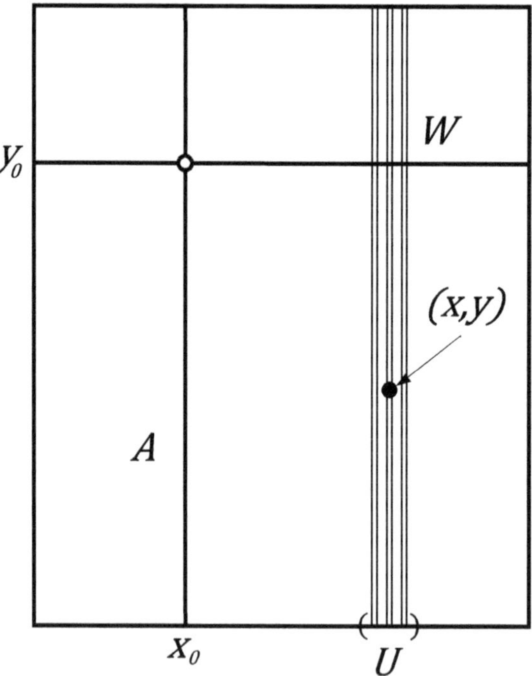

Fig. 2.5 $\mathcal{T}$ not idempotent on products

2.3.23 Corollary. *If X and Y are indecomposable continua, then $\mathcal{T}_{X\times Y}$ is not idempotent on $X \times Y$.*

2.3.24 Theorem. *Let X and Y be continua. If Z is a closed subset of $X \times Y$ such that $\pi_X(Z) \neq X$ and $\pi_Y(Z) \neq Y$, where π_X and π_Y are the projection maps, then $\mathcal{T}_{X\times Y}(Z) \subset \pi_X(Z) \times \pi_Y(Z)$.*

Proof. Let $(x, y) \in (X \times Y) \setminus (\pi_X(Z) \times \pi_Y(Z))$. Without loss of generality, we assume that $x \in X \setminus \pi_X(Z)$. Since $\pi_X(Z)$ is closed in X, there exists an open set U of X such that $x \in U \subset \mathrm{Cl}_X(U) \subset X \setminus \pi_X(Z)$. Let $y' \in Y \setminus \pi_Y(Z)$. Let $W = (\mathrm{Cl}_X(U) \times Y) \cup (X \times \{y'\})$. Then W is a subcontinuum of $X \times Y$ such that $(x, y) \in \mathrm{Int}_{X\times Y}(W) \subset W \subset (X \times Y) \setminus \pi_X(Z) \times \pi_Y(Z)$. Therefore, $\mathcal{T}_{X\times Y}(Z) \subset \pi_X(Z) \times \pi_Y(Z)$. □

2.3.25 Corollary. *Let X and Y be continua. If Z is a closed subset of $X \times Y$ such that $\mathcal{T}_{X\times Y}(Z) = X \times Y$, then either $\pi_X(Z) = X$ or $\pi_Y(Z) = Y$.*

2.3.26 Corollary. *Let X and Y be continua. If A and B are nonempty proper closed subsets of X and Y, respectively, then $\mathcal{T}_{X\times Y}(A \times B) = A \times B$.*

Regarding chainable metric continua [92, Definition 2.4.3], we have the next result.

2.3.27 Theorem. *Let X and Y be indecomposable chainable metric continua, and let Z be a subcontinuum of $X \times Y$. Then $\mathcal{T}_{X\times Y}(Z) = X \times Y$ if and only if either $\pi_X(Z) = X$ or $\pi_Y(Z) = Y$.*

Proof. If $\mathcal{T}_{X\times Y}(Z) = X \times Y$, then, by Corollary 2.3.25, either $\pi_X(Z) = X$ or $\pi_Y(Z) = Y$.

Now, suppose $\pi_X(Z) = X$. Let W be a subcontinuum of $X \times Y$ such that $\mathrm{Int}_{X\times Y}(W) \neq \emptyset$. We show that $W \cap Z \neq \emptyset$. Note that, since X is indecomposable, $\pi_X(W) = X$, Corollary 1.4.35.

Suppose that $W \cap Z = \emptyset$. Then there exists $\varepsilon > 0$ such that $\varepsilon < d(W, Z)$. Note that $\pi_Y(Z)$ is a chainable continuum, by [92, Theorem 2.4.10]. Let $f \colon X \twoheadrightarrow [0, 1]$ and $g \colon \pi_Y(Z) \twoheadrightarrow [0, 1]$ be ε-maps [92, Theorem 2.4.22]. Then $f \times g \colon X \times \pi_Y(Z) \twoheadrightarrow [0, 1] \times [0, 1]$ is an ε-map (we use the "max" metric). Hence, $(f \times g)(Z) \cap (f \times g)(W) = \emptyset$, otherwise $f \times g$ would not be an ε-map.

Since $\pi_X((f \times g)(W)) = f(\pi_X(W)) = [0, 1]$ and $\pi_Y((f \times g)(Z)) = g(\pi_Y((Z)) = [0, 1]$, we have that $(f \times g)(Z) \cap (f \times g)(W) \neq \emptyset$ [102, Theorem 130, p. 158], a contradiction. Therefore, $W \cap Z \neq \emptyset$, and $\mathcal{T}_{X\times Y}(Z) = X \times Y$. □

As a consequence of Theorems 2.3.24 and 2.3.27, we have the following.

2.3.28 Corollary. *Let X and Y be indecomposable chainable metric continua. If Z is a subcontinuum of $X \times Y$, then either $\mathcal{T}_{X\times Y}(Z) = X \times Y$ or $\mathcal{T}_{X\times Y}(Z) \subset \pi_X(Z) \times \pi_Y(Z)$.*

Now, we turn our attention to cones, $q \colon X \times [0, 1] \twoheadrightarrow K(X)$ denotes the quotient map.

2.3.29 Lemma. *Let X be an indecomposable continuum and let $x_0 \in X$. If W is a subcontinuum of $K(X)$ such that $\mathrm{Int}_{K(X)}(W) \neq \emptyset$, then $W \cap q(\{x_0\} \times [0, 1]) \neq \emptyset$.*

Proof. If $\nu_X \in W$, then it is clear that $W \cap q(\{x_0\} \times Y) \neq \emptyset$. Suppose that $\nu_X \notin W$. Then, since q is monotone, $q^{-1}(W)$ is a subcontinuum of $X \times [0,1]$ (Lemma 1.4.46) such that $\text{Int}_{X \times [0,1]}(q^{-1}(W)) \neq \emptyset$. By Lemma 2.3.20, $q^{-1}(W) \cap (\{x_0\} \times [0,1]) \neq \emptyset$. Therefore, $W \cap q(\{x_0\} \times [0,1]) \neq \emptyset$. □

2.3.30 Corollary. *If X is an indecomposable continuum and $x_0 \in X$, then $\mathcal{T}_{K(X)} q(\{x_0\} \times [0,1]) = K(X)$.*

2.3.31 Theorem. *If X is an indecomposable continuum, then $\mathcal{T}_{K(X)}$ is not idempotent on $K(X)$.*

Proof. Let $(x_0, t_0) \in X \times (0,1)$, let $A = (X \times [0,1]) \setminus \{(x_0, t_0)\}$ and let $(x,t) \in (X \times (0,1)) \setminus A$, where $t \neq t_0$. Construct W as in the proof of Theorem 2.3.22 in such a way that $W \cap (X \times \{1\}) = \emptyset$. Then $q((x,t)) \in \text{Int}_{K(X)}(q(W)) \subset q(W) \subset K(X) \setminus q(A)$. Thus, $\mathcal{T}_{K(X)} q(A) \neq K(X)$. Since $q(A)$ is dense in $q(\{x_0\} \times Y)$, $\text{Cl}_{K(X)}(q(A)) = q(\{x_0\} \times Y)$. Hence, by Corollary 2.3.30, $K(X) \subset \mathcal{T}_{K(X)} q(\{x_0\} \times Y) \subset \mathcal{T}^2_{K(X)} q(A)$. Therefore, $\mathcal{T}_{K(X)}$ is not idempotent on $K(X)$ (Figure 2.6). □

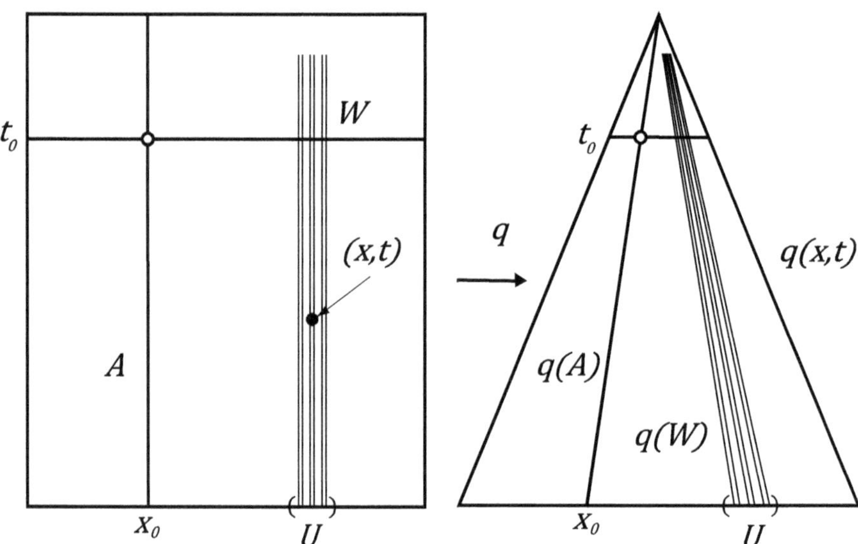

Fig. 2.6 $\mathcal{T}$ not idempotent on cones

Next, we consider suspensions. The results about suspensions are similar to the ones about cones. We state them and present the necessary changes. Here, $q: X \times [0,1] \twoheadrightarrow \Sigma(X)$ denotes the quotient map.

2.3.32 Lemma. *Let X be an indecomposable continuum and let $x_0 \in X$. If W is a subcontinuum of $\Sigma(X)$ such that $\text{Int}_{\Sigma(X)}(W) \neq \emptyset$, then $W \cap q(\{x_0\} \times [0,1]) \neq \emptyset$.*

Proof. If either $v^+ \in W$ or $v^- \in W$, then it is clear that $W \cap q(\{x_0\} \times [0,1]) \neq \emptyset$. Suppose that $\{v^+, v^-\} \cap W = \emptyset$. Then, since q is monotone, $q^{-1}(W)$ is a subcontinuum of $X \times [0,1]$ (Lemma 1.4.46) such that $\text{Int}_{X\times[0,1]}(q^{-1}(W)) \neq \emptyset$. By Lemma 2.3.20, $q^{-1}(W) \cap (\{x_0\} \times [0,1]) \neq \emptyset$. Therefore, $W \cap q(\{x_0\} \times [0,1]) \neq \emptyset$.
□

2.3.33 Corollary. *If X is an indecomposable continuum and $x_0 \in X$, then $\mathcal{T}_{\Sigma(X)} q(\{x_0\} \times [0,1]) = \Sigma(X)$.*

2.3.34 Theorem. *If X is an indecomposable continuum, then $\mathcal{T}_{\Sigma(X)}$ is not idempotent on $\Sigma(X)$.*

Proof. Let $(x_0, t_0) \in X \times (0,1)$, let $A = (X \times [0,1]) \setminus \{(x_0, t_0)\}$, and let $(x,t) \in (X \times (0,1)) \setminus A$, where $t \neq t_0$. Construct W as in the proof of Theorem 2.3.22 in such a way that $W \cap (X \times \{1\} \cup X \times \{0\}) = \emptyset$. Then $q((x,t)) \in \text{Int}_{\Sigma(X)}(q(W)) \subset q(W) \subset \Sigma(X) \setminus q(A)$. Thus, $\mathcal{T}_{\Sigma(X)} q(A) \neq \Sigma(X)$. Since $q(A)$ is dense in $q(\{x_0\} \times [0,1])$, $\text{Cl}_{\Sigma(X)}(q(A)) = q(\{x_0\} \times [0,1])$. Hence, by Corollary 2.3.33, $\Sigma(X) \subset \mathcal{T}_{\Sigma(X)} q(\{x_0\} \times Y) \subset \mathcal{T}^2_{\Sigma(X)} q(A)$. Therefore, $\mathcal{T}_{\Sigma(X)}$ is not idempotent on $\Sigma(X)$.
□

Now, we give a decomposable metric continuum Z such that $\mathcal{T}$ is idempotent on the family of closed sets of neither $Z \times [0,1]$ nor $K(Z)$ nor $\Sigma(Z)$. In particular, $\mathcal{T}$ is not idempotent on either $Z \times [0,1]$ or $K(Z)$ or $\Sigma(Z)$.

2.3.35 Example. Let us consider the Knaster continuum X, [92, Example 2.4.7], and let X' be the reflection of X in $\mathbb{R}^2$ with respect to the origin $p = (0,0)$. Let $Z = X \cup X'$, [92, Example 2.4.9] (Figure 2.7). It is well known that X is an indecomposable continuum, [92, Remark 2.4.8].

To see that $\mathcal{T}_{Z\times[0,1]}$ is not idempotent on the family of closed subsets of $Z \times [0,1]$, let us note that, using Corollary 2.3.21, we have that $\mathcal{T}_{Z\times[0,1]}(\{p\} \times [0,1]) = Z \times [0,1]$. Let $z \in X \setminus \{p\}$. Then, using again Corollary 2.3.21, we obtain that $\mathcal{T}_{Z\times[0,1]}(\{z\} \times [0,1]) = X \times [0,1]$. In particular, $\{p\} \times [0,1] \subset \mathcal{T}_{Z\times[0,1]}(\{z\} \times [0,1])$. Hence,

$$\mathcal{T}_{Z\times[0,1]}\mathcal{T}_{Z\times[0,1]}(\{z\} \times [0,1]) = \mathcal{T}_{Z\times[0,1]}(X \times [0,1]) = Z \times [0,1].$$

Therefore, $\mathcal{T}_{Z\times[0,1]}$ is not idempotent on the family of closed subsets of $Z \times [0,1]$.

To show that $\mathcal{T}_{K(Z)}$ is not idempotent on the family of closed subsets of $K(Z)$, let us observe that, using Corollary 2.3.30, we obtain that $\mathcal{T}_{K(Z)}(q(\{p\} \times [0,1])) = K(Z)$. Let $z \in X \setminus \{p\}$. Then, using Corollary 2.3.30, we have that $\mathcal{T}_{K(Z)}(q(\{z\} \times [0,1])) = K(X)$. Observe that $q(\{p\} \times [0,1]) \subset \mathcal{T}_{K(Z)}(q(\{z\} \times [0,1])$. Thus,

$$\mathcal{T}_{K(Z)}\mathcal{T}_{K(Z)}(q(\{z\} \times [0,1])) = \mathcal{T}_{K(Z)}(K(X)) = K(Z).$$

Therefore, $\mathcal{T}_{K(Z)}$ is not idempotent on the family of closed subsets of $K(Z)$.

To see that $\mathcal{T}_{\Sigma(Z)}$ is not idempotent on the family of closed subsets of $\Sigma(Z)$, let us note that, using Corollary 2.3.33, we obtain that $\mathcal{T}_{\Sigma(Z)}(q(\{p\} \times [0,1])) = \Sigma(Z)$. Let $z \in X \setminus \{p\}$. Then, by Corollary 2.3.33, we have that $\mathcal{T}_{\Sigma(Z)}(q(\{z\} \times [0,1])) = \Sigma(X)$. Hence, $q(\{p\} \times [0,1]) \subset \mathcal{T}_{\Sigma(Z)}(q(\{z\} \times [0,1])$. Thus,

$$\mathcal{T}_{\Sigma(Z)}\mathcal{T}_{\Sigma(Z)}(q(\{z\} \times [0,1])) = \mathcal{T}_{\Sigma(Z)}(\Sigma(X)) = \Sigma(Z).$$

Therefore, $\mathcal{T}_{\Sigma(Z)}$ is not idempotent on the family of closed subsets of $\Sigma(Z)$.

Next, we consider the idempotency of $\mathcal{T}$ on continua. Note that the idempotency of $\mathcal{T}$ on closed sets implies the idempotency of $\mathcal{T}$ on continua. The next example shows that the converse is not true for metric continua.

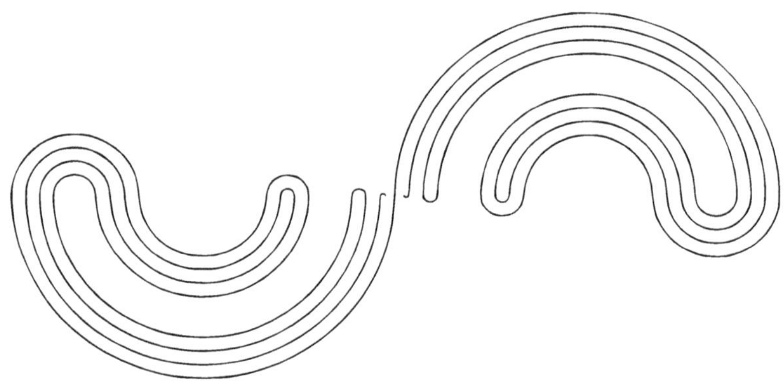

Fig. 2.7 Double Knaster

2.3.36 Example. There exists a metric continuum X such that $\mathcal{T}$ is idempotent on continua, and $\mathcal{T}$ is not idempotent on closed sets. Let X be the continuum defined by the closure of the union of a sequence $\{X_n\}_{n=1}^{\infty}$ of harmonic suspensions such that of the size of X_{n+1} is $\frac{1}{2}$ the size of X_n. Also, one of the vertexes of X_2 is the mid point of the limit segment of X_1, and one of the vertexes of X_3 is the mid point of the limit segment of X_2, etc., as shown in Figure 2.8.

Note that $\mathcal{T}$ is idempotent on continua. The limit segment in X_n is denoted by $\overline{x_{2n-1}x_{2n}}$, for each $n \in \mathbb{N}$. Also, observe that given $n \in \mathbb{N}$:

- $\mathcal{T}(\{x_1, x_2, x_4, x_6, \ldots, x_{2n}\}) = \overline{x_1x_2} \cup \{x_4, \ldots, x_{2n}\}$;
- $\mathcal{T}^2(\{x_1, x_2, x_4, x_6, \ldots, x_{2n}\}) = \overline{x_1x_2} \cup \overline{x_3x_4} \cup \{x_6, \ldots, x_{2n}\}$;

  $\vdots$

- $\mathcal{T}^n(\{x_1, x_2, x_4, x_6, \ldots, x_{2n}\}) = \overline{x_1x_2} \cup \overline{x_3x_4} \cup \ldots \cup \overline{x_{2n-1}x_{2n}}$; and
- $\mathcal{T}^{n+1}(\{x_1, x_2, x_4, x_6, \ldots, x_{2n}\}) = \mathcal{T}^n(\{x_1, x_2, x_4, x_6, \ldots, x_{2n}\})$.

Therefore, $\mathcal{T}$ is not idempotent on closed sets.

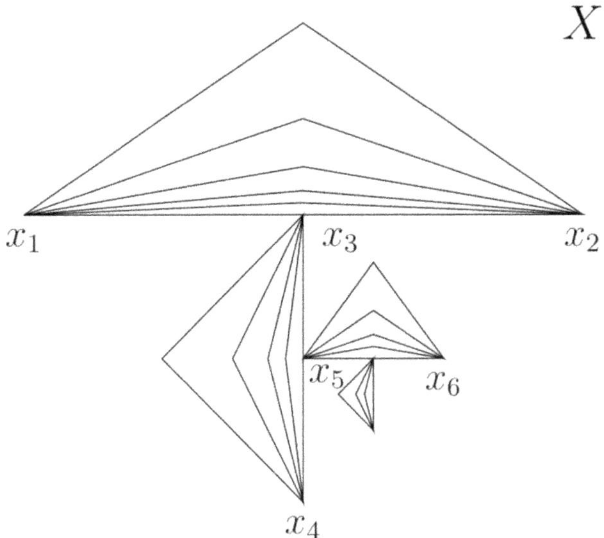

Fig. 2.8 Sequence of harmonic suspensions

2.3.37 Theorem. *Let X be a $\mathcal{T}$-additive continuum. If $\mathcal{T}$ is idempotent on continua, then $\mathcal{T}$ is idempotent on closed sets.*

Proof. Let A be a nonempty closed subset of X. Since X is $\mathcal{T}$-additive, by Corollary 2.2.13, $\mathcal{T}(A) = \bigcup\{\mathcal{T}(\{a\}) \mid a \in A\}$. Note that $\mathcal{T}(\{x\})$ is a continuum, Theorem 2.1.27, for all $x \in X$. Hence, $\mathcal{T}^2(A) = \mathcal{T}(\bigcup\{\mathcal{T}(\{a\}) \mid a \in A\}) = \bigcup\{\mathcal{T}^2(\{a\}) \mid a \in A\} = \bigcup\{\mathcal{T}(\{a\}) \mid a \in A\} = \mathcal{T}(A)$. The second equality is by $\mathcal{T}$-additivity (Corollary 2.2.13), and the third equality is by the idempotency of $\mathcal{T}$ on continua. □

2.3.38 Corollary. *If X is either a weakly irreducible or a hereditarily unicoherent continuum, then the idempotency on continua of $\mathcal{T}$ implies the idempotency on closed sets of $\mathcal{T}$.*

Proof. Suppose X is weakly irreducible. Then X is $\mathcal{T}$-symmetric, Theorem 2.2.2 Hence, X is $\mathcal{T}$-additive, Theorem 2.2.11. If X is hereditarily unicoherent, then X is $\mathcal{T}$-additive, Theorem 2.2.12. In both cases, the corollary now follows from Theorem 2.3.37. □

As a consequence of Theorem 1.4.43 and Corollary 2.3.38, we obtain that the following.

2.3.39 Corollary. *If X is an irreducible continuum, then the idempotency on continua of $\mathcal{T}$ implies the idempotency of closed sets of $\mathcal{T}$.*

2.3.40 Proposition. *Let X be a continuum and let $n \in \mathbb{N}$. If $A = A_1 \cup \cdots \cup A_n \in \mathcal{C}_n(X)$, where $A_1, \ldots, A_n$ are the components of A, is such that $\mathcal{T}(A) \in \mathcal{C}_n(X) \setminus \mathcal{C}_{n-1}(X)$, then $\mathcal{T}(A) = \mathcal{T}(A_1) \cup \cdots \cup \mathcal{T}(A_n)$.*

Proof. Suppose that $\mathcal{T}(A) = D_1 \cup \cdots \cup D_n$, where $D_1, \ldots, D_n$ are the components of $\mathcal{T}(A)$. Since $A \subseteq D_1 \cup \cdots \cup D_n$ and $D_j \cap A \neq \emptyset$ for each $j \in \{1, \ldots, n\}$, Corollary 2.1.20, we may assume that $A_j \subseteq D_j$ for every $j \in \{1, \ldots, n\}$. Observe that $\mathcal{T}(A_j) = \mathcal{T}(A \cap D_j) = D_j$, Corollary 2.1.48, for all $j \in \{1, \ldots, n\}$. Therefore, $\mathcal{T}(A) = \mathcal{T}(A_1) \cup \cdots \cup \mathcal{T}(A_n)$. □

2.3.41 Theorem. *Let X be a continuum, and let $n \in \mathbb{N}$. If $\mathcal{T}$ is idempotent on continua, then $\mathcal{T}^{2n}(A) = \mathcal{T}^{2n-1}(A)$ for each $A \in \mathcal{C}_n(X)$.*

Proof. We do the proof by induction over n. Since $\mathcal{T}$ is idempotent on continua, $\mathcal{T}^2(A) = \mathcal{T}(A)$ for each $A \in \mathcal{C}_1(X)$. Suppose that $\mathcal{T}^{2k-2}(A) = \mathcal{T}^{2k-3}(A)$ for every $A \in \mathcal{C}_{k-1}(A)$. Let $B \in \mathcal{C}_k(X)$. We show that $\mathcal{T}^{2k}(B) = \mathcal{T}^{2k-1}(B)$.

Note that $\mathcal{T}(B) \in \mathcal{C}_k(X)$ (Corollary 2.1.21). If $\mathcal{T}(B) \in \mathcal{C}_{k-1}(X)$, then $\mathcal{T}^{2k-2}\mathcal{T}(B) = \mathcal{T}^{2k-3}\mathcal{T}(B)$, and $\mathcal{T}^{2k-1}(B) = \mathcal{T}^{2k-2}(B)$. Hence, suppose that $\mathcal{T}(B) \in \mathcal{C}_k(X) \setminus \mathcal{C}_{k-1}(X)$. By Proposition 2.3.40, $\mathcal{T}(B) = \mathcal{T}(B_1) \cup \cdots \cup \mathcal{T}(B_k)$, where $B_1, \ldots, B_k$ are the components of B. We consider two cases.

Case 1. $\mathcal{T}^2(B) \in \mathcal{C}_{k-1}(X)$.

Then $\mathcal{T}^{2k-2}\mathcal{T}^2(B) = \mathcal{T}^{2k-3}\mathcal{T}^2(B)$. Hence, $\mathcal{T}^{2k}(B) = \mathcal{T}^{2k-1}(B)$.

Case 2. $\mathcal{T}^2(B) \in \mathcal{C}_k(X) \setminus \mathcal{C}_{k-1}(X)$.

By Proposition 2.3.40, $\mathcal{T}^2(B) = \mathcal{T}(\mathcal{T}(B_1) \cup \cdots \cup \mathcal{T}(B_k)) = \mathcal{T}^2(B_1) \cup \cdots \cup \mathcal{T}^2(B_k)$. Since $\mathcal{T}$ is idempotent on continua, $\mathcal{T}^2(B) = \mathcal{T}(B)$. Therefore, $\mathcal{T}^{2k}(B) = \mathcal{T}^{2k-1}(B)$ for every $B \in \mathcal{C}_k(X)$. □

2.3.42 Remark. Observe that Example 2.3.36 shows that Theorem 2.3.41 cannot be improved to obtain a global bound for all the elements of $\mathcal{C}_\infty(X) = \bigcup_{n=1}^{\infty} \mathcal{C}_n(X)$. Also in the same example, note that if $A = \{x_1, p\} \cup \{x_{2n}\}_{n=1}^{\infty}$, where $\lim_{n\to\infty} x_{2n} = p$, then $\mathcal{T}^k(A) \neq \mathcal{T}^{k+1}(A)$ for any $k \in \mathbb{N}$.

2.4 Finite Powers of $\mathcal{T}$

We study the properties of continua using the set functions $\mathcal{T}^n$, $n \in \mathbb{N}$.

2.4.1 Definition. Let X be a compactum. If A is a subset of X, then let $\mathcal{T}^0(A) = A$. If A is a subset of X and $n \in \mathbb{N}$, then $\mathcal{T}^n(A) = \mathcal{T}\mathcal{T}^{n-1}(A)$.

2.4.2 Remark. Let X be a compactum. Observe that, by Remark 2.1.5, for each $n \in \mathbb{N}$ and every subset A of X, $\mathcal{T}^n(A)$ is a closed subset of X. Hence, $\mathcal{T}^n(A)$ is compact, for all $n \in \mathbb{N}$. Also, by Proposition 2.1.7, if B is a subset of X and $A \subset B$, then $\mathcal{T}^n(A) \subset \mathcal{T}^n(B)$ for each $n \in \mathbb{N}$. In addition, by Theorem 2.1.27, if A is a subcontinuum of X, then $\mathcal{T}^n(A)$ is a subcontinuum of X for all $n \in \mathbb{N}$.

2.4.3 Definition. Let X be a compactum, and let $n \in \mathbb{N}$. Then X is *point $\mathcal{T}$-n-symmetric* provided that for each pair of points p and q of X, $p \in \mathcal{T}^n(\{q\})$ if and only if $q \in \mathcal{T}^n(\{p\})$.

2.4.4 Lemma. *Let X be a θ-continuum. Let x_1 and x_2 be two points of X, and let j and k be in $\mathbb{N} \cup \{0\}$. If $j < k$ and $\mathcal{T}^j(\{x_1\}) \cap \mathcal{T}^{k-j}(\{x_2\}) = \emptyset$, then $\mathcal{T}^{j+1}(\{x_1\})) \cap \mathcal{T}^{k-j-1}(\{x_2\}) = \emptyset$.*

Proof. Let $z \in \mathcal{T}^j(\{x_1\})$. Then $z \in X \setminus \mathcal{T}^{k-j}(\{x_2\})$. Hence, there exists a subcontinum W_z of X such that $z \in \mathrm{Int}(W_z) \subset W_z \subset X \setminus \mathcal{T}^{k-j-1}(\{x_2\})$. Note that $\{\mathrm{Int}(W_z) \mid z \in \mathcal{T}^j(\{x_1\})\}$ is an open cover of $\mathcal{T}^j(\{x_1\})$. Since $\mathcal{T}^j(\{x_1\})$ is compact (Remark 2.4.2), there exist $z_1, \ldots, z_m$ in $\mathcal{T}^j(\{x_1\})$ such that $\mathcal{T}^j(\{x_1\}) \subset \bigcup_{l=1}^m \mathrm{Int}(W_{z_l}) \subset \bigcup_{l=1}^m W_{z_l} \subset X \setminus \mathcal{T}^{k-j-1}(\{x_2\})$. Thus, since $\mathcal{T}^j(\{x_1\})$ is a continuum (Remark 2.4.2), $\bigcup_{l=1}^m W_{z_l}$ is a continuum. Hence, if $z' \in \mathcal{T}^{k-j-1}(\{x_2\})$, then $z' \in X \setminus \bigcup_{l=1}^m W_{z_l} \subset X \setminus \bigcup_{l=1}^m \mathrm{Int}(W_{z_l}) \subset X \setminus \mathcal{T}^j(\{x_1\})$. Since X is a θ-continuum and $\bigcup_{l=1}^m W_{z_l}$ is a subcontinuum of X, $X \setminus \bigcup_{l=1}^m W_{z_l}$ only has finitely many open components (Theorem 1.4.70). Let A be the component of $X \setminus \bigcup_{l=1}^m W_{z_l}$ that contains z'. Then A is an open subset of X. Hence, $z' \in A \subset \mathrm{Cl}(A) \subset \bigcup_{l=1}^m \mathrm{Int}(W_{z_l}) \subset X \setminus \mathcal{T}^j(\{x_1\})$. Thus, $z' \in X \setminus \mathcal{T}^{j+1}(\{x_1\})$. Therefore, $\mathcal{T}^{j+1}(\{x_1\})) \cap \mathcal{T}^{k-j-1}(\{x_2\}) = \emptyset$. □

As a consequence of Theorems 1.4.43 and 1.4.70 and Lemma 2.4.4, we have the following.

2.4.5 Corollary. *Let X be an irreducible continuum between a and b. Let x_1 and x_2 be two points of X, and let j and k in $\mathbb{N} \cup \{0\}$. If $j < k$ and $\mathcal{T}^j(\{x_1\}) \cap \mathcal{T}^{k-j}(\{x_2\}) = \emptyset$, then $\mathcal{T}^{j+1}(\{x_1\})) \cap \mathcal{T}^{k-j-1}(\{x_2\}) = \emptyset$.*

2.4.6 Theorem. *If X is a θ-continuum and $n \in \mathbb{N}$, then X is point $\mathcal{T}$-n-symmetric.*

Proof. We apply Lemma 2.4.4 several times for $k = n$. Let x_1 and x_2 be two points of X and suppose $j = 0$. Then $\{x_1\} \cap \mathcal{T}^n(\{x_2\}) = \mathcal{T}^0(\{x_1\}) \cap \mathcal{T}^n(\{x_2\}) = \emptyset$ implies that $\mathcal{T}(\{x_1\}) \cap \mathcal{T}^{n-1}(\{x_2\}) = \emptyset$. The equality $\mathcal{T}(\{x_1\}) \cap \mathcal{T}^{n-1}(\{x_2\}) = \emptyset$ implies that $\mathcal{T}^2(\{x_1\}) \cap \mathcal{T}^{n-2}(\{x_2\}) = \emptyset$. Continuing with this process, we obtain that $\emptyset = \mathcal{T}^n(\{x_1\}) \cap \mathcal{T}^0(\{x_2\}) = \mathcal{T}^n(\{x_1\}) \cap \{x_2\}$. Therefore, X is point $\mathcal{T}$-n-symmetric. □

Also, from Theorems 1.4.43, 1.4.70 and 2.4.6, we obtain the following.

2.4.7 Corollary. *If X is an irreducible continuum between a and b, then X is point $\mathcal{T}$-n-symmetric.*

2.4.8 Theorem. *Let X be a θ-continuum. If there exists an $n \in \mathbb{N}$ such that $\mathcal{T}^{n+1}(\{x\}) = \mathcal{T}^n(\{x\})$ for each x in X, then $\mathcal{G}_n = \{\mathcal{T}^n(\{x\}) \mid x \in X\}$ is a decomposition of X.*

Proof. Let p and q be two points of X, and suppose that there exists $x \in \mathcal{T}^n(\{p\}) \cap \mathcal{T}^n(\{q\})$. By Remark 2.4.2, we have that $\mathcal{T}^n(\{x\}) \subset \mathcal{T}^n(\mathcal{T}^n(\{p\})) = \mathcal{T}^n(\{p\})$. By Theorem 2.4.6, $p \in \mathcal{T}^n(\{x\})$ and, by Remark 2.4.2, $\mathcal{T}^n(\{p\}) \subset \mathcal{T}^n(\mathcal{T}^n(\{x\})) = \mathcal{T}^n(\{x\})$. Hence, $\mathcal{T}^n(\{p\}) = \mathcal{T}^n(\{x\})$. Similarly, $\mathcal{T}^n(\{q\}) = \mathcal{T}^n(\{x\})$. Therefore, $\mathcal{T}^n(\{p\}) = \mathcal{T}^n(\{q\})$, and $\mathcal{G}_n$ is a decomposition. □

2.4.9 Definition. A metric continuum X is said to be n*-indecomposable* provided that X is the union of n continua such that no one of them is a subset of the union of the others and X is not the union of $n + 1$ such continua.

In the next result, we present characterizations of n-indecomposable metric continua given by C. E. Burgess [16, Theorem 1], [17, Theorems 1, 4 and 5] and [18, Theorem 7].

2.4.10 Theorem. *Let X be a metric continuum. Then the following are equivalent:*

(1) *X is n-indecomposable, for some $n \geq 2$;*
(2) *X is the union of n indecomposable continua such that no one of them is a subset of the union of the others and it is irreducible about some n points;*
(3) *X is the union of n indecomposable continua such that no one of them is a subset of the union of the others and it is irreducible about some finite set;*
(4) *X is the union of n indecomposable continua $M_1, \ldots, M_n$ such that for each $k \in \{1, \ldots, n\}$, there exists a composant C_k of M_k, where $C_k \cap M_j = \emptyset$, for each $j \in \{1, \ldots, n\}$, $j \neq k$;*
(5) *$\mathcal{T}(\mathcal{F}_1(X))$ is finite and $|\mathcal{T}(\mathcal{F}_1(X))| \geq 2$.*

2.4.11 Theorem. *Let X be an irreducible metric continuum between a and b, and suppose that $\mathcal{T}^n(\{a\}) = X$ and $\mathcal{T}^m(\{a\}) \neq X$ for any $m < n$. If $\mathcal{T}^k(\{a\})$ is a k-indecomposable continuum and if $\mathcal{T}^{k+1}(\{a\}) \neq \mathcal{T}^k(\{a\})$, then $\mathcal{T}^{k+1}(\{a\})$ is a $(k+1)$-indecomposable continuum.*

Proof. Since X is an irreducible continuum between a and b, by Theorem 1.4.40, $\mathcal{T}^{k+1}(\{a\}) \setminus \mathcal{T}^k(\{a\})$ is connected for $k+1 < n$. Hence, $\mathrm{Cl}(\mathcal{T}^{k+1}(\{a\}) \setminus \mathcal{T}^k(\{a\}))$ is an indecomposable continuum, otherwise, $\mathcal{T}^m(\{a\}) = X$ for some $m < n$. □

2.4.12 Theorem. *A necessary and sufficient condition for a metric irreducible continuum X, between a and b, to be n-indecomposable is that $\mathcal{T}(\{a\})$ is indecomposable and $\mathcal{T}^n(\{a\}) = X$ and $\mathcal{T}^m(\{a\}) \neq X$ for any $m < n$.*

Proof. We show the condition is necessary. Note that for each $k \in \{1, \ldots, n-1\}$, $\mathcal{T}^{k+1}(\{a\}) \neq \mathcal{T}^k(\{a\})$. Thus, if $\mathcal{T}(\{a\})$ is indecomposable, applying Theorem 2.4.11 several times, we obtain that $\mathcal{T}^n(\{a\}) = X$ is an n-indecomposable continuum.

By Theorem 2.4.10, if X is n-indecomposable, then X is the union of n-indecomposable continua such that no one of them is a subset of the union of the others. Hence, the condition is sufficient. □

2.4.13 Theorem. *Let X be a weakly irreducible continuum. If Λ is a family of closed subsets of X whose union is closed, then $\mathcal{T}^n\left(\bigcup\{L \mid L \in \Lambda\}\right) = \bigcup\{\mathcal{T}^n(L) \mid L \in \Lambda\}$, for each $n \in \mathbb{N}$.*

Proof. For $n = 1$, by Theorem 2.2.2, X is $\mathcal{T}$-symmetric and, by Theorem 2.2.11, X is $\mathcal{T}$-additive. Hence, by Theorem 2.2.10, we have that if Λ is a family of closed subsets of X whose union is closed, then $\mathcal{T}\left(\bigcup\{L \mid L \in \Lambda\}\right) = \bigcup\{\mathcal{T}(L) \mid L \in \Lambda\}$. Now, the theorem follows from mathematical induction. □

As a consequence of Theorems 2.4.13, 1.4.43 and 1.4.70, we obtain the following.

2.4.14 Theorem. *Let X be an irreducible continuum. If Λ is a family of closed subsets of X whose union is closed, then $\mathcal{T}^n\left(\bigcup\{L \mid L \in \Lambda\}\right) = \bigcup\{\mathcal{T}^n(L) \mid L \in \Lambda\}$, for every $n \in \mathbb{N}$.*

2.4.15 Theorem. *Let X be a hereditarily unicoherent continuum. If Λ is a family of closed subsets of X whose union is closed, then $\mathcal{T}^n\left(\bigcup\{L \mid L \in \Lambda\}\right) = \bigcup\{\mathcal{T}^n(L) \mid L \in \Lambda\}$, for all $n \in \mathbb{N}$.*

Proof. By Theorem 2.2.12, X is $\mathcal{T}$-additive. Thus, by Theorem 2.2.10, we have that if Λ is a family of closed subsets of X whose union is closed, then $\mathcal{T}\left(\bigcup\{L \mid L \in \Lambda\}\right) = \bigcup\{\mathcal{T}(L) \mid L \in \Lambda\}$. Now, the theorem follows from mathematical induction. □

2.4.16 Theorem. *If X is a hereditarily unicoherent continuum and K and L are subcontinua of X such that $K \cap L \neq \emptyset$, then $\mathcal{T}^n(K) \cap \mathcal{T}^n(L) = \mathcal{T}^n(K \cap L)$, for each $n \in \mathbb{N}$.*

Proof. For $n = 1$, by Proposition 2.1.7, $\mathcal{T}(K \cap L) \subset \mathcal{T}(K) \cap \mathcal{T}(L)$. Suppose there exists $x \in \mathcal{T}(K) \cap \mathcal{T}(L) \setminus \mathcal{T}(K \cap L)$. Then there exists a subcontinuum W of X such that $x \in \text{Int}(W) \subset W \subset X \setminus (K \cap L)$. Since X is hereditarily unicoherent, $W \cap K$ and $W \cap L$ are connected. Note that $W \cap K \neq \emptyset$, otherwise $x \in \text{Int}(W) \subset W \subset X \setminus K$ and $x \in X \setminus \mathcal{T}(K)$. Similarly, $W \cap L \neq \emptyset$. Since $W \cap (K \cap L) = \emptyset$, $W \cap K$ and $W \cap L$ are disjoint subcontinua. Since $K \cap L \neq \emptyset$, $K \cup L$ is a subcontinuum of X. Since X is hereditarily unicoherent, $W \cap (K \cup L)$ is a subcontinuum of X and $W \cap (K \cup L) = (W \cap K) \cup (W \cap L)$, a contradiction. Therefore, $\mathcal{T}(K) \cap \mathcal{T}(L) \subset \mathcal{T}(K \cap L)$ and the theorem is true for $n = 1$.

Let $n \in \mathbb{N}$, by Remark 2.4.2, $\mathcal{T}^n(K)$ and $\mathcal{T}^n(L)$ are subcontinua of X. By Remark 2.1.5 and Proposition 2.1.7, $K \subset \mathcal{T}(K) \subset \mathcal{T}^n(K)$ and $L \subset \mathcal{T}(L) \subset \mathcal{T}^n(L)$. Since $K \cap L \neq \emptyset$, $\mathcal{T}^n(K) \cap \mathcal{T}^n(L) \neq \emptyset$. Therefore, the theorem now follows by mathematical induction. □

2.4.17 Theorem. *Let X be a continuum, let $K_1, \ldots, K_h$ be subcontinua of X and let $K = \bigcup_{j=1}^{h} K_j$. Then, for each $n \in \mathbb{N} \cup \{0\}$, either $\mathcal{T}^n(K)$ is a subcontinuum of X or $\mathcal{T}^n(K) = M_1 \cup M_2$, where M_1 and M_2 are two disjoint closed subsets of X, and for $L_j = M_j \cap K$ $(j \in \{1, 2\})$, $K = L_1 \cup L_2$ and $\mathcal{T}^n(K) = \mathcal{T}^n(L_1) \cup \mathcal{T}^n(L_2)$.*

Proof. The theorem is true for $n = 0$ because $\mathcal{T}^0(L_1 \cup L_2) = L_1 \cup L_2$. Suppose the result is true for $n = k - 1$, but it is false for $n = k$. Note that, if $\mathcal{T}^{k-1}(K)$ is a continuum, by Theorem 2.1.27, $\mathcal{T}^k(K)$ is a continuum. Hence, if $\mathcal{T}^k(K)$ is disconnected, then $\mathcal{T}^{k-1}(K)$ is disconnected. Since the theorem is not true for $n = k$, $\mathcal{T}^k(K) = M_1 \cup M_2$, where M_1 and M_2 are disjoint closed subsets of X. Let $M_j' = \mathcal{T}^{k-1}(K) \cap M_j$ $(j \in \{1, 2\})$. Then M_1' and M_2' are disjoint closed subsets of

X and $\mathcal{T}^{k-1}(K) = M_1' \cup M_2'$. By Corollary 2.1.20, we obtain that both M_1' and M_2' are nonempty. Hence, $L_j = M_j \cap K = M_j' \cap K$ ($j \in \{1, 2\}$). Since the theorem is true for $n = k - 1$, $\mathcal{T}^{k-1}(K) = \mathcal{T}^{k-1}(L_1) \cup \mathcal{T}^{k-1}(L_2)$.

Since X is normal, there exist open subsets U_1 and U_2 of X such that $M_j \subset U_j$ ($j \in \{1, 2\}$) and $\mathrm{Cl}(U_1) \cap \mathrm{Cl}(U_2) = \emptyset$. Suppose there exists $x \in M_1 \setminus \mathcal{T}^k(L_1)$. Then there exists a subcontinuum W_x of X such that $x \in \mathrm{Int}(W_x) \subset W_x \subset X \setminus \mathcal{T}^{k-1}(L_1)$. For each $w \in W_x \cap \mathrm{Bd}(U_1)$, we have that $w \in X \setminus \mathcal{T}^k(K)$. Thus, there exists a subcontinuum W_w of X such that $w \in \mathrm{Int}(W_w) \subset W_w \subset X \setminus \mathcal{T}^{k-1}(K)$. Since $W_x \cap \mathrm{Bd}(U_1)$ is compact, there exist $w_1, \ldots, w_l$ in $W_x \cap \mathrm{Bd}(U_1)$ such that $W_x \cap \mathrm{Bd}(U_1) \subset \bigcup_{j=1}^{l} \mathrm{Int}(W_{w_j})$. Observe that $\bigcup_{j=1}^{l} W_{w_j} \cup (U_1 \cap W_x)$ only has a finite number of components (Theorem 1.4.36). Let S be the component of $\bigcup_{j=1}^{l} W_{w_j} \cup (U_1 \cap W_x)$ that contains x. Since $W_x \cap \mathcal{T}^{k-1}(L_1) = \emptyset$ and $\mathcal{T}^{k-1}(L_2) \subset M_2 \subset U_2$, we obtain that $\mathrm{Cl}(U_1) \cap \mathcal{T}^{k-1}(L_2) = \emptyset$. Thus, $(W_x \cap \mathrm{Cl}(U_1)) \cap \mathcal{T}^{k-1}(K) = \emptyset$. Since $\mathcal{T}^{k-1}(K) = \mathcal{T}^{k-1}(L_1) \cup \mathcal{T}^{k-1}(L_2)$ and $W_w \cap \mathcal{T}^{k-1}(K) = \emptyset$, for all $w \in W_x \cap \mathrm{Bd}(U_1)$, we have that $S \cap \mathcal{T}^{k-1}(K) = \emptyset$. Hence, $x \in \mathrm{Int}(W_x) \cap U_1 \subset S \subset X \setminus \mathcal{T}^{k-1}(K)$. This implies that $x \in X \setminus \mathcal{T}^k(K)$, a contradiction to the fact that $x \in M_1 \subset \mathcal{T}^k(L_1)$. Thus, $M_1 \subset \mathcal{T}^k(L_1)$. Similarly, $M_2 \subset \mathcal{T}^k(L_2)$. As a consequence of the above, $\mathcal{T}^k(K) = M_1 \cup M_2 \subset \mathcal{T}^k(L_1) \cup \mathcal{T}^k(L_2) \subset \mathcal{T}^k(K)$. Hence, $\mathcal{T}^k(K) = \mathcal{T}^k(L_1) \cup \mathcal{T}^k(L_2)$. Therefore, the theorem is true by mathematical induction. □

2.4.18 Corollary. *Let X be a continuum, let $K_1, \ldots, K_h$ be subcontinua of X and let $K = \bigcup_{j=1}^{h} K_j$. Then, for each $n \in \mathbb{N}$, either $\mathcal{T}^n(K)$ is a continuum or $\mathcal{T}^n(K) = \bigcup_{j=1}^{l} \mathcal{T}^n(L_j)$, where $l \leq h$ and $\mathcal{T}^n(L_1), \ldots, \mathcal{T}^n(L_l)$ are pairwise disjoint continua.*

The next result extends Corollary 2.1.49.

2.4.19 Corollary. *Let X be a continuum and let K_1 and K_2 be subcontinua of X. Then, for each $n \in \mathbb{N}$, either $\mathcal{T}^n(K_1 \cup K_2)$ is a continuum or $\mathcal{T}^n(K_1 \cup K_2) = \mathcal{T}^n(K_1) \cup \mathcal{T}^n(K_2)$.*

2.4.20 Theorem. *Let X be a continuum, let $p_1, \ldots, p_k$ be points of X and let $n \in \mathbb{N}$. If $X = \bigcup_{j=1}^{k} \mathcal{T}^n(\{p_j\})$, then $X = \mathcal{T}^n(\{p_1, \ldots, p_k\})$.*

Proof. By Remark 2.4.2, we have that for each $j \in \{1, \ldots, k\}$, $\mathcal{T}^n(\{p_j\}) \subset \mathcal{T}^n(\{p_1, \ldots, p_k\})$. Thus,

$$X = \bigcup_{j=1}^{k} \mathcal{T}^n(\{p_j\}) \subset \mathcal{T}^n(\{p_1, \ldots, p_k\}).$$

Therefore, $X = \mathcal{T}^n(\{p_1, \ldots, p_k\})$. □

2.4.21 Definition. Let X be a continuum, let A be a nonempty subset of X and let $n \in \mathbb{N}$. Then A is a $\mathcal{T}^n$*-set* if and only if

(1) if $a \in A$, then $\mathcal{T}^n(\{a\}) \subset A$;

and

(2) if $p \in X \setminus A$, then $\mathcal{T}^n(\{p\}) \cap A = \emptyset$.

A $\mathcal{T}^n$-set A is *total* if and only if for each $\mathcal{T}^n$-set B contained in A, we have that $B = A$.

2.4.22 Proposition. *If X is a $\mathcal{T}$-symmetric continuum, then each $\mathcal{T}$-closed subset of X (Definition 4.1.1) is a $\mathcal{T}^n$-set for each $n \in \mathbb{N}$.*

Proof. Let A be a $\mathcal{T}$-closed set and let $a \in A$. Then, by Proposition 2.1.7, $\mathcal{T}(\{a\}) \subset \mathcal{T}(A) = A$. Hence, for each $n \in \mathbb{N}$, $\mathcal{T}^n(\{a\}) \subset A$.

Now, let $p \in X \setminus A$. We show, by mathematical induction, that $\mathcal{T}^n(\{p\}) \cap A = \emptyset$. For $n = 1$, suppose that $\mathcal{T}(\{p\}) \cap A \neq \emptyset$. Let $a \in \mathcal{T}(\{p\}) \cap A$. Since X is point $\mathcal{T}$-symmetric, $p \in \mathcal{T}(\{a\}) \subset A$ (by the previous paragraph), a contradiction. Thus, $\mathcal{T}(\{p\}) \cap A = \emptyset$. Suppose $\mathcal{T}^n(\{p\}) \cap A = \emptyset$ and assume that $\mathcal{T}^{n+1}(\{p\}) \cap A \neq \emptyset$. Let $a \in \mathcal{T}^{n+1}(\{p\}) \cap A$. Since X is $\mathcal{T}$-symmetric, we have that $\emptyset \neq \mathcal{T}(\{a\}) \cap \mathcal{T}^n(\{p\}) \subset A \cap \mathcal{T}^n(\{p\})$, a contradiction. Hence, $\mathcal{T}^{n+1}(\{p\}) \cap A = \emptyset$. Therefore, A is a $\mathcal{T}^n$-set for all $n \in \mathbb{N}$. □

2.4.23 Theorem. *Let X be a continuum, and let $n \in \mathbb{N}$. If A is $\mathcal{T}^n$-set and A_1 is a component of A, then A_1 is a $\mathcal{T}^n$-set.*

Proof. Let A be a $\mathcal{T}^n$-set, and let A_1 be a component of A. Let $a \in A_1$. Then, since A is a $\mathcal{T}^n$-set, $\mathcal{T}^n(\{a\}) \subset A$. Since $\mathcal{T}^n(\{a\})$ is a continuum (Remark 2.4.2), $\mathcal{T}(\{a\}) \cap A_1 \neq \emptyset$ and A_1 is a component of A, we have that $\mathcal{T}^n(\{a\}) \subset A_1$. Next, let $p \in X \setminus A_1$ and assume that there exists $q \in \mathcal{T}^n(\{p\}) \cap A_1$. If $p \in X \setminus A$, then, since A is a $\mathcal{T}^n$-set, $\mathcal{T}^n(\{p\}) \cap A = \emptyset$. Thus, $\mathcal{T}^n(\{p\}) \cap A_1 = \emptyset$, a contradiction. Thus, $p \in A$. This implies that $\mathcal{T}^n(\{p\}) \subset A$. Since $\mathcal{T}^n(\{p\})$ is a continuum (Remark 2.4.2), $\mathcal{T}(\{p\}) \cap A_1 \neq \emptyset$ and A_1 is a component of A, we have that $\mathcal{T}^n(\{p\}) \subset A_1$. In particular, $p \in A_1$, a contradiction. Hence, $\mathcal{T}^n(\{p\}) \cap A_1 = \emptyset$. Therefore, A_1 is a $\mathcal{T}^n$-set. □

2.4.24 Corollary. *Let X be a continuum, and let $n \in \mathbb{N}$. If A is a total $\mathcal{T}^n$ set, then A is connected.*

Proof. Let A be a total $\mathcal{T}^n$-set, and let A_1 be a component of A. By Theorem 2.4.23, A_1 is a $\mathcal{T}^n$-set and $A_1 \subset A$. Hence, by Definition 2.4.21, we have that $A_1 = A$. Therefore, A is connected. □

2.4.25 Theorem. *Let X be a continuum, let $\{A_m\}_{m=1}^{\infty}$ be a sequence of nonempty subsets of X and let $n \in \mathbb{N}$. If for each $m \in \mathbb{N}$, A_m is a $\mathcal{T}^n$-set, then both $\bigcup_{m=1}^{\infty} A_m$ and $\bigcap_{m=1}^{\infty} A_m$ are $\mathcal{T}^n$-sets.*

Proof. Let $a \in \bigcap_{m=1}^{\infty} A_m$. Then, for every $m \in \mathbb{N}$, $a \in A_m$. Let $m \in \mathbb{N}$. Since A_m is a $\mathcal{T}^n$-set, $\mathcal{T}^n(\{a\}) \subset A_m$. Hence, since $m \in \mathbb{N}$ is arbitrary, $\mathcal{T}^n(\{a\}) \subset \bigcap_{m=1}^{\infty} A_m$. Let $p \in X \setminus \bigcap_{m=1}^{\infty} A_m$. Then there exists $m_0 \in \mathbb{N}$ such that $p \in X \setminus$

A_{m_0}. Since A_{m_0} is $\mathcal{T}^n$-set, $\mathcal{T}^n(\{p\}) \cap A_{m_0} = \emptyset$. Hence, $\mathcal{T}^n(\{p\}) \cap \bigcap_{m=1}^{\infty} A_m = \emptyset$. Therefore, $\bigcap_{m=1}^{\infty} A_m$ is $\mathcal{T}^n$-set.

Let $a \in \bigcup_{m=1}^{\infty} A_m$. Then there exists $m_0 \in \mathbb{N}$ such that $a \in A_{m_0}$. Since A_{m_0} is a $\mathcal{T}^n$-set, $\mathcal{T}^n(\{a\}) \subset A_{m_0} \subset \bigcup_{m=1}^{\infty} A_m$. Let $p \in X \setminus \bigcup_{m=1}^{\infty} A_m$. Then, for each $m \in \mathbb{N}$, $p \in X \setminus A_m$. Since every A_m is a $\mathcal{T}^n$-set, $\mathcal{T}^n(\{p\}) \cap A_m = \emptyset$ for all $m \in \mathbb{N}$. Hence, $\mathcal{T}^n(\{p\}) \cap \bigcup_{m=1}^{\infty} A_m = \emptyset$. Therefore, $\bigcup_{m=1}^{\infty} A_m$ is a $\mathcal{T}^n$-set. □

2.4.26 Theorem. *Let X be an irreducible continuum between q and s, and let A be a subcontinuum of X. Then A is a $\mathcal{T}$-set if and only if for each $n \in \mathbb{N}$, A is a $\mathcal{T}^n$-set.*

Proof. Suppose A is $\mathcal{T}$-set. Let $a_0 \in A$. Then, by hypothesis, $\mathcal{T}(\{a\}) \subset A$. Hence, by Proposition 2.1.7, $\mathcal{T}^2(\{a_0\}) \subset \mathcal{T}(A)$. Thus, by mathematical induction, $\mathcal{T}^n(\{a_0\}) \subset \mathcal{T}^{n-1}(A)$. We prove that A is a $\mathcal{T}$-closed set. Let $x \in X \setminus A$. Since A is a $\mathcal{T}$-set, $\mathcal{T}(\{x\}) \cap A = \emptyset$. Thus, for each $a \in A$, there exists a subcontinuum W_a of X such that $a \in \operatorname{Int}(W_a) \subset W_a \subset X \setminus \{x\}$. Since A is compact, there exist $a_1, \ldots, a_k$ in A such that $A \subset \bigcup_{j=1}^{k} \operatorname{Int}(W_{a_j}) \subset \bigcup_{j=1}^{k} W_{a_j} \subset X \setminus \{x\}$. Since A a continuum, $\bigcup_{j=1}^{k} W_{a_j}$ is a subcontinuum of X. Since X is irreducible, by Theorem 1.4.40, $X \setminus \bigcup_{j=1}^{k} W_{a_j}$ has at most two open components. Let K be the component of $X \setminus \bigcup_{j=1}^{k} W_{a_j}$ that contains x. Hence, $\operatorname{Cl}(K)$ is a subcontinuum of X such that $x \in K \subset \operatorname{Cl}(K) \subset X \setminus A$. Thus, $x \in X \setminus \mathcal{T}(A)$, and $\mathcal{T}(A) = A$. It now follows that $\mathcal{T}^{n-1}(A) \subset A$. Therefore, $\mathcal{T}^n(\{a_0\}) \subset \mathcal{T}^{n-1}(A) \subset A$.

Next, we show that if $p \in X \setminus A$, then $\mathcal{T}^n(\{p\}) \cap A = \emptyset$. Suppose there exists $x \in \mathcal{T}^n(\{p\}) \cap A$. Since X is point $\mathcal{T}$-n-symmetric (Corollary 2.4.7), we have that $p \in \mathcal{T}^n(\{x\})$ and, by the previous paragraph, $\mathcal{T}^n(\{x\}) \subset A$. Thus, $p \in A$, a contradiction. Hence, $\mathcal{T}^n(\{p\}) \cap A = \emptyset$. Therefore, A is a $\mathcal{T}^n$-set for all $n \in \mathbb{N}$. □

2.4.27 Definition. Let X be a continuum, let $p \in X$ and let $n \in \mathbb{N}$. A subset V_p of X is a *$\mathcal{T}^n$-chain composant from p* of X if and only if

$$V_p = \{x \mid \text{there exist } \mathcal{T}^n(\{x_0\}), \ldots, \mathcal{T}^n(\{x_m\}) \text{ such that } x_0 = p,$$

$$x_m = x \text{ and } \mathcal{T}^n(\{x_j\}) \cap \mathcal{T}^n(\{x_k\}) \neq \emptyset \text{ if and only if } |j - k| \leq 1\}.$$

Note that $\{\mathcal{T}^n(\{x_0\}), \ldots, \mathcal{T}^n(\{x_m\})\}$ forms a chain contained in V_p. This composant V is an *essential $\mathcal{T}^n$-chain composant from p* if and only if each link in these finite chain has an essential part in the chain union; that is, each link is not contained in the union of the other links.

2.4.28 Theorem. *Let X be a continuum, and let $n \in \mathbb{N}$. If V_p is a $\mathcal{T}^n$-chain composant from p of X, then V_p is a connected $\mathcal{T}^n$-set.*

Proof. Observe that, for each $x \in V_p$, if $\{\mathcal{T}^n(\{x_0\}), \ldots, \mathcal{T}^n(\{x_m\})\}$ is the chain joining p and x, then $\bigcup_{j=0}^{m} \mathcal{T}^n(\{x_j\})$ is a connected set (Remark 2.4.2). Hence, V_p is connected.

To see that V_p is a $\mathcal{T}^n$-set, let $x \in V_p$. Then, by Definition 2.4.27, $\mathcal{T}^n(\{x\}) \subset V_p$. Now, let $x \in X \setminus V_p$ and assume there exists $x' \in \mathcal{T}^n(\{x\}) \cap V_p$. Then there exists a chain $\{\mathcal{T}^n(\{x_0\}), \ldots, \mathcal{T}^n(\{x_m\})\}$ such that $x_0 = p$ and $x_m = x'$. Then $\{\mathcal{T}^n(\{x_0\}), \ldots, \mathcal{T}^n(\{x_m\}), \mathcal{T}^n(\{x\})\}$ contains a chain from p to x. Hence, $x \in V_p$, a contradiction. Thus, $\mathcal{T}^n(\{x\}) \cap V_p = \emptyset$. Therefore, V_p is a $\mathcal{T}^n$-set. □

2.4.29 Theorem. *Let X be a continuum, and let $n \in \mathbb{N}$. Then $\mathfrak{V} = \{V_x \mid x \in X\}$, the family of $\mathcal{T}^n$-chain composants of X is a monotone decomposition of X.*

Proof. By Theorem 2.4.28, the elements of $\mathfrak{V}$ are connected. Let p and q be elements of X, and assume there exists $x \in V_p \cap V_q$. Since $x \in V_p$, there exists a chain $\{\mathcal{T}^n(\{x_0\}), \ldots, \mathcal{T}^n(\{x_m\})\}$ such that $x_0 = p$ and $x_m = x$. Since $x \in V_q$, there exists a chain $\{\mathcal{T}^n(\{z_0\}), \ldots, \mathcal{T}^n(\{z_l\})\}$ such that $z_0 = q$ and $z_l = x$. Then $\{\mathcal{T}^n(\{x_0\}), \ldots, \mathcal{T}^n(\{x_m\}), \mathcal{T}^n(\{z_0\}), \ldots, \mathcal{T}^n(\{z_l\})\}$ contains a chain from p to q. Therefore, $V_p = V_q$ and $\mathfrak{V}$ is a decomposition of X. □

2.4.30 Theorem. *Let X be a continuum, let A be a nonempty subset of X and let $n \in \mathbb{N}$. Then A is a total $\mathcal{T}^n$-set if and only if A is a $\mathcal{T}^n$-chain composant.*

Proof. Suppose A is a total $\mathcal{T}^n$-set. By Theorem 2.4.29, $\mathfrak{V} = \{V_x \mid x \in X\}$ is a monotone decomposition of X. Let $a \in A$. Then V_a consists of the union of all finite chains of $\mathcal{T}^n(\{x\})$'s from a. By condition (2) of Definition 2.4.21, we have that A contains the union of all these finite chains from a. Hence, $V_a \subset A$. By Theorem 2.4.28, V_a is a $\mathcal{T}^n$-set and, since A is a total $\mathcal{T}^n$-set, we obtain that $V_a = A$.

Now, assume that A is a $\mathcal{T}^n$-chain composant and let $a \in A$. Then, by Theorem 2.4.29, $A = V_a$. By Theorem 2.4.28, A is a $\mathcal{T}^n$-set. By condition (2) of Definition 2.4.21, V_a cannot contain a proper $\mathcal{T}^n$-set. Therefore, $A = V_a$ is a total $\mathcal{T}^n$-set. □

2.4.31 Definition. Let X be a continuum, and let L be a proper subset of X. Then L *is a C-set* if for each subcontinuum K of X such that $K \cap L \neq \emptyset$, either $K \subset L$ or $L \subset K$.

2.4.32 Remark. Let X be a continuum. Then C-sets are connected with empty interior. Note that each proper terminal subcontinuum X is a C-set. If X is an indecomposable continuum with more than one composant, then the composants of X are C-sets.

2.4.33 Theorem. *Let X be a continuum, and let L be a C-set in X. If K is a subset of X such that $K \subset \mathrm{Cl}(L)$, then $\mathrm{Cl}(L) \subset \mathcal{T}(K)$.*

Proof. Suppose there exists $x \in \mathrm{Cl}(L) \setminus \mathcal{T}(K)$. Then there exists a subcontinuum W of X such that $x \in \mathrm{Int}(W) \subset W \subset X \setminus K$. Note that $L \cap \mathrm{Int}(W) \neq \emptyset$. If $W \cap (X \setminus L) \neq \emptyset$, then, since L is a C-set, $L \subset W$. Hence, $K \subset \mathrm{Cl}(L) \subset W$, a contradiction. Thus, $W \cap (X \setminus L) = \emptyset$. Then $W \subset L$, a contradiction to the fact that C-sets have empty interior (Remark 2.4.32). Therefore, $\mathrm{Cl}(L) \subset \mathcal{T}(K)$. □

2.4.34 Theorem. *Let X be a continuum, let $n \in \mathbb{N}$ and let P be a subset of X such that $\mathcal{T}^n(P)$ is subcontinuum of X that is a C-set. If K is a subset of X such that $K \subset \mathcal{T}^n(P)$, then $\mathcal{T}^n(P) = \mathcal{T}^n(K)$.*

Proof. Let K be a subset of X such that $K \subset \mathcal{T}^n(P)$. By Theorem 2.4.33, $\mathcal{T}^n(P) \subset \mathcal{T}(K) \subset \mathcal{T}^n(K)$, Remark 2.4.2. By hypothesis, $K = \mathcal{T}^0(K) \subset \mathcal{T}^n(P)$. Suppose that $\mathcal{T}^n(K) \cap (X \setminus \mathcal{T}^n(P)) \neq \emptyset$. Hence, there exists $h \in \mathbb{N}$ such that $\mathcal{T}^h(K) \subset \mathcal{T}^n(P)$ and $\mathcal{T}^{h+1}(K) \cap (X \setminus \mathcal{T}^n(P)) \neq \emptyset$. Let $x \in \mathcal{T}^{h+1}(K) \setminus \mathcal{T}^n(P)$. Then there exists a subcontinuum W of X such that $x \in \operatorname{Int}(W) \subset W \subset X \setminus \mathcal{T}^{n-1}(P)$. Since $x \in \mathcal{T}^{h+1}(K)$, we have that $W \cap \mathcal{T}^h(K) \neq \emptyset$. But $\mathcal{T}^h(K) \subset \mathcal{T}^n(P)$, and there exists a point $w \in W \cap \mathcal{T}^h(K) \subset \mathcal{T}^n(P)$. Also, $x \in W \setminus \mathcal{T}^n(P)$. Since $\mathcal{T}^n(P)$ is a C-set, by Definition 2.4.31, $\mathcal{T}^{n-1}(P) \subset \mathcal{T}^n(P) \subset W$, a contradiction to the choice of W. Therefore, $\mathcal{T}^n(K) \subset \mathcal{T}^n(P)$ and $\mathcal{T}^n(K) = \mathcal{T}^n(P)$. □

2.4.35 Theorem. *Let X be a continuum, let $n \in \mathbb{N}$ and let P be a subcontinuum of X such that $\operatorname{Int}(\mathcal{T}^n(P)) \neq \emptyset$. Assume also that P has the property: if $p \in \mathcal{T}^n(P)$, then $\mathcal{T}^n(P) \subset \mathcal{T}^n(\{p\})$. If $n = 1$, then $\mathcal{T}(P)$ is an indecomposable continuum. If $n \geq 2$, then $\mathcal{T}^n(P)$ may contain uncountably many indecomposable continua.*

Proof. By Theorem 2.1.27, $\mathcal{T}(P)$ is a subcontinuum of X. Suppose that $\mathcal{T}(P)$ is the union of two proper subcontinua A and B. Since $\operatorname{Int}(\mathcal{T}(P)) \neq \emptyset$, there exists a point $x \in \operatorname{Int}(\mathcal{T}(P))$. Suppose there exists an open subset R of X such that $x \in R \subset A$. Let $b \in \mathcal{T}(P) \setminus A$. Then $x \in R \subset A \subset X \setminus \{b\}$. Thus, $x \in X \setminus \mathcal{T}(\{b\})$. By hypothesis, $\mathcal{T}(P) \subset \mathcal{T}(\{b\})$, a contradiction. Hence, $R \not\subset A$. Let $a \in \mathcal{T}(P) \subset B$ and let $z \in R \setminus A$, which is open. Then $z \in R \setminus A \subset B \subset X \setminus \{a\}$. Thus, $z \in X \setminus \mathcal{T}(\{a\})$. By hypothesis, $\mathcal{T}(P) \subset \mathcal{T}(\{a\})$, a contradiction. Therefore, $\mathcal{T}(P)$ is an indecomposable continuum.

For $n \geq 2$, consider the following example in $\mathbb{R}^3$. Let K be the Knaster continuum, [92, Example 2.4.7], in the (x, z)-plane tangent to the z-axis at $\overline{p} = (0, 0, 0)$ and lying to the right of the z-axis. Let X' be the set obtained by revolving K about the z-axis, and let C be a quarter of a circle in the (x, y)-plane traced by a point $q \in K \cap (x\text{-axis})$. Let C' be a copy of the Cantor ternary set on C, and let K_c be the indecomposable continuum in X' through $c \in C'$. Let $X = \bigcup\{K_c \mid c \in C'\}$. Then X is a continuum. Let $P = \{\overline{p}\}$. Note that $\mathcal{T}^2(\{\overline{p}\}) = X$ and, for each $\overline{x} \in \mathcal{T}^2(\{\overline{p}\})$, $\mathcal{T}^2(\{\overline{x}\}) = \mathcal{T}^2(\{\overline{p}\})$. Thus, the hypothesis is satisfied and X is the union of uncountably many metric indecomposable subcontinua. □

2.4.36 Theorem. *Let $n \in \mathbb{N}$, and let X be a point $\mathcal{T}$-n-symmetric continuum. Then the following are equivalent:*

(1) *if $x \in \mathcal{T}^n(\{p\})$, then $\mathcal{T}^n(\{x\}) \subset \mathcal{T}^n(\{p\})$;*
(2) *if $x \in X \setminus \mathcal{T}^n(\{p\})$, then $\mathcal{T}^n(\{x\}) \cap \mathcal{T}^n(\{p\}) = \emptyset$.*

Proof. Suppose (1) is true and (2) is false. Let $x \in X \setminus \mathcal{T}^n(\{p\})$, and let $z \in \mathcal{T}^n(\{p\}) \cap \mathcal{T}^n(\{x\})$. By (1), we have that $\mathcal{T}^n(\{z\}) \subset \mathcal{T}^n(\{p\})$. Since $z \in \mathcal{T}^n(\{x\})$ and X is point $\mathcal{T}$-n-symmetric, $x \in \mathcal{T}^n(\{z\}) \subset \mathcal{T}^n(\{p\})$, a contradiction. Therefore, (1) implies (2).

Now, assume (2) is true and (1) is false. Let $z \in \mathcal{T}^n(\{p\})$ be such that there exists $w \in \mathcal{T}^n(\{z\}) \setminus \mathcal{T}^n(\{p\})$. Since X is point $\mathcal{T}$-n-symmetric, we obtain that $z \in \mathcal{T}^n(\{w\})$. Hence, $z \in \mathcal{T}^n(\{w\}) \cap \mathcal{T}^n(\{p\})$. By (2), $\mathcal{T}^n(\{w\}) \cap \mathcal{T}^n(\{p\}) = \emptyset$, a contradiction. Therefore, (2) implies (1). □

As a consequence of Definition 2.4.21 and Theorem 2.4.36, we have the following.

2.4.37 Corollary. *Let $n \in \mathbb{N}$, and let X be a point $\mathcal{T}$-n-symmetric continuum. If for each $p \in X$, $x \in \mathcal{T}^n(\{p\})$ implies that $\mathcal{T}^n(\{x\}) \subset \mathcal{T}^n(\{p\})$, then $\mathcal{T}^n(\{p\})$ is a $\mathcal{T}^n$-set.*

References for Chapter 2

Section 2.1: [3–5, 29–31, 60, 91, 92, 96, 101, 120].
Section 2.2: [3–5, 8, 96].
Section 2.3: [3, 19, 24, 53, 86, 92, 94, 102].
Section 2.4: [16–18, 20, 32, 33, 118, 125].

Chapter 3
Decomposition Theorems

The set function $\mathcal{T}$ has also been used to prove several decomposition theorems. Here, first, we present three general decomposition theorems. Then we prove a Hausdorff version, for continua with the uniform property of Effros, of Jones' Aposyndetic Decomposition Theorem and Prajs' Mutual Decomposition Theorem. We present the way to show Rogers' Terminal Decomposition Theorem for homogeneous metric continua, and we end with other decomposition theorems.

3.1 Preliminaries

We present the necessary results to show the first three decomposition theorems of next section.

3.1.1 Definition. Let X be a continuum. We say that $\mathcal{T}$ *is idempotent on singletons* if $\mathcal{T}\mathcal{T}(\{x\}) = \mathcal{T}(\{x\})$ for all $x \in X$.

The next definition is based on a property obtained by Bellamy and Lum in [11, Lemma 5].

3.1.2 Definition. Let X be a continuum, and let $z \in X$. We say that $\mathcal{T}(\{z\})$ *has property* BL provided that $\mathcal{T}(\{z\}) \subset \mathcal{T}(\{x\})$ for each $x \in \mathcal{T}(\{z\})$.

3.1.3 Lemma. *Let X be a decomposable continuum for which $\mathcal{T}$ is idempotent on singletons, and let $z \in X$. If $\mathcal{T}(\{z\})$ has property BL, then $\mathcal{T}(\{w\}) = \mathcal{T}(\{z\})$ for every $w \in \mathcal{T}(\{z\})$.*

Proof. Let $z \in X$ be such that $\mathcal{T}(\{z\})$ has property BL, and let $w \in \mathcal{T}(\{z\})$. Since $w \in \mathcal{T}(\{z\})$ and $\mathcal{T}$ is idempotent on singletons, $\mathcal{T}(\{w\}) \subset \mathcal{T}\mathcal{T}(\{z\}) = \mathcal{T}(\{z\})$. Hence, $\mathcal{T}(\{w\}) \subset \mathcal{T}(\{z\})$. Since $\mathcal{T}(\{z\})$ has property BL, $\mathcal{T}(\{z\}) \subset \mathcal{T}(\{w\})$. Therefore, $\mathcal{T}(\{w\}) = \mathcal{T}(\{z\})$. □

S. Macías, *Set Function $\mathcal{T}$*, Developments in Mathematics 67,
https://doi.org/10.1007/978-3-030-65081-0_3

3.1.4 Corollary. *Let X be a decomposable continuum for which $\mathcal{T}$ is idempotent on singletons. If z_1 and z_2 are two points of X such that both $\mathcal{T}(\{z_1\})$ and $\mathcal{T}(\{z_2\})$ have property BL, then either $\mathcal{T}(\{z_1\}) = \mathcal{T}(\{z_2\})$ or $\mathcal{T}(\{z_1\}) \cap \mathcal{T}(\{z_2\}) = \emptyset$.*

Proof. Let z_1 and z_2 be points of X such that both $\mathcal{T}(\{z_1\})$ and $\mathcal{T}(\{z_2\})$ have property BL. Suppose $\mathcal{T}(\{z_1\}) \cap \mathcal{T}(\{z_2\}) \neq \emptyset$. Let $z_3 \in \mathcal{T}(\{z_1\}) \cap \mathcal{T}(\{z_2\})$. Then, by Lemma 3.1.3, $\mathcal{T}(\{z_1\}) = \mathcal{T}(\{z_3\}) = \mathcal{T}(\{z_2\})$. □

The proof following lemma is very similar to the proof of Proposition 2.3.3.

3.1.5 Lemma. *Let X be a decomposable continuum for which $\mathcal{T}$ is idempotent on singletons. If $z \in X$, then*

$$\mathcal{T}(\{z\}) = \bigcup\{\mathcal{T}(\{w\}) \mid w \in \mathcal{T}(\{z\})\}.$$

Proof. Let $z \in X$, and let $w_0 \in \mathcal{T}(\{z\})$. Since $\mathcal{T}$ is idempotent on singletons, $\mathcal{T}(\{w_0\}) \subset \mathcal{T}\mathcal{T}(\{z\}) = \mathcal{T}(\{z\})$. Hence, $\bigcup\{\mathcal{T}(\{w\}) \mid w \in \mathcal{T}(\{z\})\} \subset \mathcal{T}(\{z\})$. The other inclusion is clear. □

The proof of the next theorem is based on a technique of Bellamy and Lum [11, Lemma 5].

3.1.6 Theorem. *If X is a decomposable continuum for which $\mathcal{T}$ is idempotent on singletons, then, for each $x \in X$, there exists $z \in \mathcal{T}(\{x\})$ such that $\mathcal{T}(\{z\})$ has property BL.*

Proof. Let $x \in X$. Then by Lemma 3.1.5, we have that

$$\mathcal{T}(\{x\}) = \bigcup\{\mathcal{T}(\{w\}) \mid w \in \mathcal{T}(\{x\})\}.$$

Let $\mathcal{G}_x = \{\mathcal{T}(\{w\}) \mid w \in \mathcal{T}(\{x\})\}$. Partially order $\mathcal{G}_x$ by inclusion. Let $\mathcal{H} = \{\mathcal{T}(\{w_\lambda\})\}_{\lambda\in\Lambda}$ be a (set theoretic) chain of elements of $\mathcal{G}_x$. We show that $\mathcal{H}$ has a lower bound in $\mathcal{G}_x$.

Since $\mathcal{H}$ is a (set theoretic) chain of continua (Theorem 2.1.27), $\bigcap_{\lambda\in\Lambda} \mathcal{T}(\{w_\lambda\}) \neq \emptyset$. Let $w_0 \in \bigcap_{\lambda\in\Lambda} \mathcal{T}(\{w_\lambda\})$. Then,

$$\mathcal{T}(\{w_0\}) \subset \mathcal{T}\left(\bigcap_{\lambda\in\Lambda} \mathcal{T}(\{w_\lambda\})\right) \subset \bigcap_{\lambda\in\Lambda} \mathcal{T}\mathcal{T}(\{w_\lambda\}) =$$

$$\bigcap_{\lambda\in\Lambda} \mathcal{T}(\{w_\lambda\}) \subset \mathcal{T}(\{x\}).$$

Hence, by Kuratowski–Zorn Lemma, there exists $z \in \mathcal{T}(\{x\})$ such that $\mathcal{T}(\{z\})$ is a minimal element; that is, each $w \in \mathcal{T}(\{z\})$ satisfies that $\mathcal{T}(\{z\}) \subset \mathcal{T}(\{w\})$. Therefore, $\mathcal{T}(\{z\})$ has property BL. □

3.1.7 Lemma. *Let X be a continuum, let A and L be two nonempty closed subsets of X and let $\{A_\lambda\}_{\lambda\in\Lambda}$ be a net of nonempty closed subsets of X converging to A. If the net $\{\mathcal{T}(A_\lambda)\}_{\lambda\in\Lambda}$ converges to L, then $L \subset \mathcal{T}(A) \subset \mathcal{T}(L)$.*

Proof. We prove that $L \subset \mathcal{T}(A)$. Let $l \in L$, and let W be a subcontinuum of X such that $l \in \text{Int}(W)$. Since $\{\mathcal{T}(A_\lambda)\}_{\lambda\in\Lambda}$ converges to L, there exists $\lambda_0 \in \Lambda$ such that $\mathcal{T}(A_\lambda) \cap \text{Int}(W) \neq \emptyset$ for all $\lambda \geq \lambda_0$. This implies that $A_\lambda \cap W \neq \emptyset$ for every $\lambda \geq \lambda_0$. By [40, 3.1.23 and 1.6.1], there exist $\Lambda' \subset \Lambda$ and $z_\lambda \in A_\lambda \cap W$, for each $\lambda \in \Lambda'$, and a point z in X such that $\{z_\lambda\}_{\lambda\in\Lambda'}$ converges to z. Note that $z \in A \cap W$. Since W is an arbitrary subcontinuum of X containing l in its interior, $l \in \mathcal{T}(A)$. Hence, $L \subset \mathcal{T}(A)$.

Since $A_\lambda \subset \mathcal{T}(A_\lambda)$, $\{A_\lambda\}_{\lambda\in\Lambda}$ converges to A and $\{\mathcal{T}(A_\lambda)\}_{\lambda\in\Lambda}$ converges to L, we have that $A \subset L$. Thus, $\mathcal{T}(A) \subset \mathcal{T}(L)$. □

3.1.8 Definition. Let X be a continuum. Define $\xi\colon X \twoheadrightarrow \mathcal{F}_1(X)$ by $\xi(x) = \{x\}$.

3.1.9 Remark. Let X be a continuum. Since X is a Hausdorff space, the function ξ given in Definition 3.1.8 is a homeomorphism [98, p. 153].

3.1.10 Theorem. *Let X be a continuum with the property of Kelley. If $\mathcal{G} = \{\mathcal{T}(\{x\}) \mid x \in X\}$ is a decomposition of X, then $\mathcal{T}(\{x\})$ is a terminal subcontinuum of X for all $x \in X$.*

Proof. Suppose there exists a point x_0 in X such that $\mathcal{T}(\{x_0\})$ is not a terminal subcontinuum of X (by Theorem 2.1.27, $\mathcal{T}(\{x_0\})$ is a continuum). Then there exists a subcontinuum L of X such that $L\cap\mathcal{T}(\{x_0\}) \neq \emptyset$, $L\setminus\mathcal{T}(\{x_0\}) \neq \emptyset$ and $\mathcal{T}(\{x_0\})\setminus L \neq \emptyset$. Let $p \in L \setminus \mathcal{T}(\{x_0\})$ and $\ell \in L \cap \mathcal{T}(\{x_0\})$. Since $\mathcal{G}$ is a decomposition, without loss of generality, we assume that $x_0 \in \mathcal{T}(\{x_0\})\setminus L$. Since $p \in L\setminus\mathcal{T}(\{x_0\})$, there exists a subcontinuum W of X such that $p \in \text{Int}(W) \subset W \subset X \setminus \{x_0\}$. Let $K = L \cup W$. Then K is a subcontinuum of X, $p \in \text{Int}(K)$ and $K \subset X \setminus \{x_0\}$. By Theorem 1.6.20, there exists a subcontinuum M of X such that $\ell \in \text{Int}(M)$ and $M \subset X \setminus \{x_0\}$, a contradiction to the choice of ℓ. Therefore, $\mathcal{T}(\{x\})$ is a terminal subcontinuum of X for all $x \in X$. □

3.2 Three Decomposition Theorems

We present three decomposition theorems: one for the class of point $\mathcal{T}$-symmetric continua X for which $\mathcal{T}$ is idempotent on singletons; the second one is for the class of continua X for which $\mathcal{T}|_{\mathcal{F}_1(X)}$ is continuous and $\mathcal{T}$ is idempotent on singletons; and the third for the class of continua for which the set function $\mathcal{T}$ is continuous. We restrict ourselves to decomposable nonlocally connected continua; since it is known that $\mathcal{T}$ is a constant map on indecomposable continua, Theorem 2.1.44, and $\mathcal{T}$ is the identity map on locally connected continua, Theorem 2.1.37.

The proof of the following theorem is similar to the one given for Theorem 5.1.1.

3.2.1 Theorem. *Let X be a continuum. If*

$$\mathcal{G} = \{\mathcal{T}(\{x\}) \mid x \in X\}$$

is a decomposition of X, then $\mathcal{G}$ is upper semicontinuous. Hence, $X/\mathcal{G}$ is a continuum.

Proof. Let $x_0 \in X$, and let U be an open subset of X such that $\mathcal{T}(\{x_0\}) \subset U$. Let $K = \{x \in X \mid \mathcal{T}(\{x\}) \cap (X \setminus U) \neq \emptyset\}$. We show that K is closed in X. Let $x \in \text{Cl}(K)$. Then there exists a net $\{x_\lambda\}_{\lambda\in\Lambda}$ of points of K converging to x [40, 1.6.3]. Since each $x_\lambda \in K$, there exists $z_\lambda \in \mathcal{T}(\{x_\lambda\}) \setminus U$ for every element $\lambda \in \Lambda$. Without loss of generality, we assume that the net $\{z_\lambda\}_{\lambda\in\Lambda}$ converges to a point z [40, 3.1.23 and 1.6.1]. Note that $z \in X \setminus U$. We prove that $z \in \mathcal{T}(\{x\})$. To this end, suppose that $z \in X \setminus \mathcal{T}(\{x\})$. Then there exists a subcontinuum H of X such that $z \in \text{Int}(H) \subset H \subset X \setminus \{x\}$. Since $\{z_\lambda\}_{\lambda\in\Lambda}$ converges to z and $\{x_\lambda\}_{\lambda\in\Lambda}$ converges to x, there exists $\lambda_0 \in \Lambda$ such that $z_\lambda \in \text{Int}(H)$ and $x_\lambda \in X \setminus H$ for all $\lambda \geq \lambda_0$. This implies that $z_\lambda \in X \setminus \mathcal{T}(\{x_\lambda\})$ for each $\lambda \geq \lambda_0$, a contradiction. Thus, $z \in \mathcal{T}(\{x\}) \setminus U$, and K is closed. Let $V = X \setminus K$. Then $V = \{x \in X \mid \mathcal{T}(\{x\}) \subset U\}$, $\mathcal{T}(\{x_0\}) \subset V$ and V is open in X. Therefore, $\mathcal{G}$ is upper semicontinuous. By Theorem 1.1.22, $X/\mathcal{G}$ is a Hausdorff space. Therefore, $X/\mathcal{G}$ is a continuum. □

3.2.2 Corollary. *Let X be a continuum such that $\mathcal{T}_X$ is idempotent on singletons. If*

$$\mathcal{G} = \{\mathcal{T}_X(\{x\}) \mid x \in X\}$$

is a decomposition of X, then $X/\mathcal{G}$ is an aposyndetic continuum.

Proof. By Theorem 3.2.1, $X/\mathcal{G}$ is a continuum. Let $q\colon X \twoheadrightarrow X/\mathcal{G}$ be the quotient map. By Theorem 2.1.27, q is a monotone map. Let $\chi \in X/\mathcal{G}$, and let $x \in X$ be such that $\mathcal{T}_X(\{x\}) = q^{-1}(\chi)$. Since q is monotone, by Theorem 2.1.51 (c), $\mathcal{T}_{X/\mathcal{G}}(\{\chi\}) = q\mathcal{T}_X q^{-1}(\chi) = q\mathcal{T}_X\mathcal{T}_X(\{x\}) = q\mathcal{T}_X(\{x\}) = qq^{-1}(\chi) = \{\chi\}$, the third equality is true because $\mathcal{T}_X$ is idempotent on singletons. Therefore, $X/\mathcal{G}$ is an aposyndetic continuum, Theorem 2.1.34. □

3.2.3 Lemma. *Let X be a continuum, and let $z \in X$. If*

$$\mathcal{G} = \{\mathcal{T}(\{x\}) \mid x \in X\}$$

is a decomposition and W is a subcontinuum of X such that $\mathcal{T}(\{z\}) \cap \text{Int}(W) \neq \emptyset$, then $\mathcal{T}(\{z\}) \subset W$.

Proof. Let W be a subcontinuum of X, and let z be a point of X such that $\mathcal{T}(\{z\}) \cap \text{Int}(W) \neq \emptyset$. Let $x \in \mathcal{T}(\{z\}) \cap \text{Int}(W)$, and suppose there exists $y \in \mathcal{T}(\{z\}) \setminus W$. Thus, $x \in \text{Int}(W) \subset W \subset X \setminus \{y\}$; that is, $x \notin \mathcal{T}(\{y\})$. Since $\mathcal{G}$ is a decomposition, $\mathcal{T}(\{x\}) = \mathcal{T}(\{z\}) = \mathcal{T}(\{y\})$, a contradiction. Therefore, $\mathcal{T}(\{z\}) \subset W$. □

Now we are ready to prove our first decomposition theorem.

3.2.4 Theorem. *Let X be a decomposable point $\mathcal{T}_X$-symmetric continuum for which $\mathcal{T}_X$ is idempotent on singletons. Then,*

$$\mathcal{G} = \{\mathcal{T}_X(\{x\}) \mid x \in X\}$$

is an upper semicontinuous decomposition of X such that the quotient space $X/\mathcal{G}$ is an aposyndetic continuum.

Proof. Let $x \in X$. Then by Theorem 3.1.6, there exists $z \in \mathcal{T}_X(\{x\})$ such that $\mathcal{T}_X(\{z\})$ has property BL. Since X is point $\mathcal{T}_X$-symmetric and $z \in \mathcal{T}_X(\{x\})$, we have that $x \in \mathcal{T}_X(\{z\})$. Thus, since $\mathcal{T}_X(\{z\})$ has property BL, $\mathcal{T}_X(\{x\}) = \mathcal{T}_X(\{z\})$, Lemma 3.1.3. Hence, $\mathcal{T}_X(\{x\})$ has property BL. Therefore, by Corollary 3.1.4,

$$\mathcal{G} = \{\mathcal{T}_X(\{x\}) \mid x \in X\}$$

is a decomposition of X. Thus, by Theorem 3.2.1, $\mathcal{G}$ is upper semicontinuous and, by Corollary 3.2.2, $X/\mathcal{G}$ is an aposyndetic continuum. □

As an application of Theorem 3.2.4, we show the next result due to Javier Camargo.

3.2.5 Theorem. *Let X be an irreducible metric continuum. If $\mathcal{T}$ is idempotent on singletons, then $\mathcal{T}$ is idempotent on closed sets.*

Proof. If X is indecomposable, by Theorem 2.1.44, $\mathcal{T}$ is idempotent on closed sets. Hence, suppose X is decomposable. By Corollary 2.2.3, X is $\mathcal{T}$-symmetric; in particular, X is point $\mathcal{T}$-symmetric. By Theorem 3.2.4, $\mathcal{G} = \{\mathcal{T}(\{x\}) \mid x \in X\}$ is an upper semicontinuous decomposition of X such that $X/\mathcal{G}$ is an aposyndetic continuum. Let $q\colon X \twoheadrightarrow X/\mathcal{G}$ be the quotient map. Since for each $x \in X$, $q^{-1}q(x) = \mathcal{T}(\{x\})$ and $\mathcal{T}(\{x\})$ is a continuum (Theorem 2.1.27), we have that q is a monotone map. Thus, $X/\mathcal{G}$ is an irreducible metric continuum [75, Theorem 3, p. 192]. In fact, by Theorem 2.2.18, $X/\mathcal{G}$ is an arc.

Let A be a closed subset of X. Thus, $\mathcal{T}(\{a\}) = q^{-1}q(a) \subset \mathcal{T}(A)$, for each $a \in A$. Hence, $q^{-1}q(A) \subset \mathcal{T}(A)$. Let $z \in X \setminus q^{-1}q(A)$. Then $q(z) \in X/\mathcal{G} \setminus q(A)$. Since $X/\mathcal{G}$ is an arc, there exists a subcontinuum $\mathcal{W}$ of $X/\mathcal{G}$ such that $q(z) \in \text{Int}_{X/\mathcal{G}}(\mathcal{W}) \subset \mathcal{W} \subset X/\mathcal{G} \setminus q(A)$. Thus, $q^{-1}(\mathcal{W})$ is a subcontinuum of X (Lemma 1.4.46), $z \in \text{Int}_X(q^{-1}(\mathcal{W}))$ and $q^{-1}(\mathcal{W}) \cap A = \emptyset$. This implies that $z \in X \setminus \mathcal{T}(A)$. Hence, $q^{-1}q(A) = \mathcal{T}(A)$ and, by Theorem 4.1.25, $\mathcal{T}(A)$ is a $\mathcal{T}$-closed set. Therefore, $\mathcal{T}^2(A) = \mathcal{T}(A)$ for each closed subset A of X, and $\mathcal{T}$ is idempotent on closed sets. □

A proof of the following result may be found in [73, 2.1]. The theorem was originally proved by E. Dyer.

3.2.6 Theorem. *Let X and Y be nondegenerate metric continua. If $f\colon X \twoheadrightarrow Y$ is a surjective, monotone and open map, then there exists a dense G_δ subset W of Y having the following property: for every $y \in W$, for each subcontinuum B of*

$f^{-1}(y)$, for all $x \in \operatorname{Int}_{f^{-1}(y)}(B)$ and for every neighbourhood U of B in X, there exist a subcontinuum Z of X containing B and a neighbourhood V of y in Y such that $x \in \operatorname{Int}_X(Z)$, $(f|_Z)^{-1}(V) \subset U$ and $f|_Z\colon Z \twoheadrightarrow Y$ is a monotone surjective map.

Now, we are ready to prove our second decomposition theorem.

3.2.7 Theorem. *Let X be a decomposable continuum for which $\mathcal{T}_X|_{\mathcal{F}_1(X)}$ is continuous and for which $\mathcal{T}_X$ is idempotent on singletons. Then,*

$$\mathcal{G} = \{\mathcal{T}_X(\{x\}) \mid x \in X\}$$

is a continuous decomposition of X such that the quotient space $X/\mathcal{G}$ is an aposyndetic continuum. Moreover, all the elements of $\mathcal{G}$ are nowhere dense in X; and if X is metric, then there exists a dense G_δ subset $\mathcal{W}$ of $X/\mathcal{G}$ such that if $q(z) \in \mathcal{W}$, then $\mathcal{T}_X(\{z\})$ is an indecomposable metric continuum, where $q\colon X \twoheadrightarrow X/\mathcal{G}$ is the quotient map.

Proof. We show that $\mathcal{G}$ is a decomposition of X. To this end, let $M = \{x \in X \mid \mathcal{T}_X(\{x\})$ has property $BL\}$ (by Lemma 3.1.3, if $x \in M$, then $\mathcal{T}_X(\{z\}) = \mathcal{T}_X(\{x\})$ for all $z \in \mathcal{T}_X(\{x\})$). Note that for every $m \in M$, $\mathcal{T}_X(\{m\}) \subset M$. By Theorem 3.1.6, $M \neq \emptyset$, and $M \cap \mathcal{T}_X(\{x\}) \neq \emptyset$ for all $x \in X$. We show that $M = X$.

First, we prove that M is closed in X. Let $x \in \operatorname{Cl}(M)$. Then there exists a net $\{m_\lambda\}_{\lambda\in\Lambda}$ of elements of M converging to x [40, 1.6.3]. Let $z \in \mathcal{T}_X(\{x\})$. Since $\mathcal{T}_X$ is continuous on singletons, the net $\{\mathcal{T}_X(\{m_\lambda\})\}_{\lambda\in\Lambda}$ converges to $\mathcal{T}_X(\{x\})$ [97, Proposition 3.38]. Hence, for each $\lambda \in \Lambda$, there exists $z_\lambda \in \mathcal{T}_X(\{m_\lambda\})$ such that $\{z_\lambda\}_{\lambda\in\Lambda}$ converges to z [104, Theorems 2 and 4]. Since $\mathcal{T}_X$ is continuous on singletons, the net $\{\mathcal{T}_X(\{z_\lambda\})\}_{\lambda\in\Lambda}$ converges to $\mathcal{T}_X(\{z\})$. Since $\mathcal{T}_X(\{z_\lambda\}) = \mathcal{T}_X(\{m_\lambda\})$ for all $\lambda \in \Lambda$, we have that $\mathcal{T}_X(\{z\}) = \mathcal{T}_X(\{x\})$. Therefore, M is closed.

Now, we see that $M = X$. Suppose this is not true. Assume first that M is not connected. Then there exist two nonempty closed subsets M_1 and M_2 of X such that $M = M_1 \cup M_2$. Since X is normal, there exist two disjoint open subsets U_1 and U_2 of X such that $M_1 \subset U_1$ and $M_2 \subset U_2$. Let $j \in \{1, 2\}$. Observe that for each $m_j \in M_j$, $\mathcal{T}_X(\{m_j\}) \subset M_j \subset U_j$. Since $\mathcal{T}_X$ is upper semicontinuous (Theorem 5.1.1), for every $m_j \in M_j$, there exists an open subset V_{m_j} of X such that $m_j \in V_{m_j}$ and $\mathcal{T}_X(\{z\}) \subset U_j$ for each $z \in V_{m_j}$. Let $V_j = \bigcup\{V_{m_j} \mid m_j \in M_j\}$. Clearly, V_j is an open subset of X, $V_j \subset U_j$, $M_j \subset V_j$ and $\mathcal{T}_X(\{z\}) \subset U_j$ for every $z \in V_j$. Let

$$\begin{aligned} R_j &= \{x \in X \mid \mathcal{T}_X(\{x\}) \subset X \setminus M_j\} \\ &= \xi^{-1}((\mathcal{T}_X|_{\mathcal{F}_1(X)})^{-1}(\langle X \setminus M_j\rangle_1)), \end{aligned}$$

and let

$$R = \{x \in X \mid \mathcal{T}_X(\{x\}) \cap M_1 \neq \emptyset \text{ and } \mathcal{T}_X(\{x\}) \cap M_2 \neq \emptyset\}$$
$$= \xi^{-1}((\mathcal{T}_X|_{\mathcal{F}_1(X)})^{-1}(\langle X, M_1\rangle_1 \cap \langle X, M_2\rangle_1)).$$

Note that R_j is open in X and R is closed. Also, $V_k \subset R_j$, j and k in $\{1, 2\}$ and $k \neq j$, and $(V_1 \cup V_2) \cap R = \emptyset$. Thus, $R_j \neq \emptyset$, $j \in \{1, 2\}$. By Theorem 3.1.6, if $x \in X$, then $\mathcal{T}_X(\{x\}) \cap M \neq \emptyset$. Hence, $X \subset R \cup R_1 \cup R_2$, and $X = R \cup R_1 \cup R_2$. Observe that if $x \in R_1 \cap R_2$, then $\mathcal{T}_X(\{x\}) \subset X \setminus (M_1 \cup M_2)$, a contradiction to Theorem 3.1.6. This implies that $R_1 \cap R_2 = \emptyset$. As a consequence of this, since X is connected, $R \neq \emptyset$. Let $x \in R \cap (\mathrm{Cl}(R_1) \cup \mathrm{Cl}(R_2))$. Without loss of generality, we assume that $x \in \mathrm{Cl}(R_2)$. Then, by [40, 1.6.3], there exists a net $\{r_\lambda\}_{\lambda \in \Lambda}$ of elements of R_2 converging to x. Since $x \in R$, $\mathcal{T}_X(\{x\}) \cap M_2 \neq \emptyset$. Also, since $M_2 \subset V_2$, we have that $\mathcal{T}_X(\{x\}) \in \langle X, V_2\rangle_1$. Since $\mathcal{T}_X$ is continuous on singletons, there exists $\lambda_0 \in \Lambda$ such that $\mathcal{T}_X(\{r_{\lambda_0}\}) \in \langle X, V_2\rangle_1$. Let $z \in \mathcal{T}_X(\{r_{\lambda_0}\}) \cap V_2$. By the construction of V_2, $\mathcal{T}_X(\{z\}) \subset U_2$. By Theorem 3.1.6, $\mathcal{T}_X(\{z\}) \cap M \neq \emptyset$. Since $M_1 \cap U_2 = \emptyset$, we have that $\mathcal{T}_X(\{z\}) \cap M_2 \neq \emptyset$. Since $\mathcal{T}_X$ is idempotent on singletons, $\mathcal{T}_X(\{z\}) \subset \mathcal{T}_X^2(\{r_{\lambda_0}\}) = \mathcal{T}_X(\{r_{\lambda_0}\})$, and $\mathcal{T}_X(\{r_{\lambda_0}\}) \cap M_2 \neq \emptyset$, a contradiction to the fact that $r_{\lambda_0} \in R_2$. Thus, M cannot be disconnected.

Hence, M is connected. Let $x \in X \setminus M$. By Theorem 3.1.6, there exists $m_0 \in M$ such that $\mathcal{T}_X(\{m_0\}) \subset \mathcal{T}_X(\{x\})$. Since X is normal, there exist two disjoint open subsets U and V of X such that $M \subset U$ and $x \in V$. Since $\mathcal{T}_X$ is upper semicontinuous (Theorem 5.1.1), there exists an open subset W of X such that $m_0 \in W$ and $\mathcal{T}_X(\{z\}) \subset U$, for all $z \in W$. Let $K = \mathrm{Cl}\left(\bigcup\{\mathcal{T}_X(\{z\}) \mid z \in W\}\right) \cup M$. Since M is a subcontinuum of X and $\mathcal{T}_X(\{z\}) \cap M \neq \emptyset$ for each $z \in W$, we have that K is a subcontinuum of X. Note that $K \subset \mathrm{Cl}(U) \subset X \setminus V$ and $m_0 \in W \subset K$. Thus, $K \cap \{x\} = \emptyset$ and $m_0 \in X \setminus \mathcal{T}_X(\{x\})$, a contradiction to the fact that $\mathcal{T}_X(\{m_0\}) \subset \mathcal{T}_X(\{x\})$. Therefore, $M = X$, and $\mathcal{G} = \{\mathcal{T}_X(\{x\}) \mid x \in X\}$ is a decomposition of X. Hence, X is point $\mathcal{T}_X$-symmetric and, by Theorem 3.2.4, $X/\mathcal{G}$ is an aposyndetic continuum.

Since $\mathcal{T}_X|_{\mathcal{F}_1(X)}$ is continuous, in fact, $\mathcal{G}$ is a continuous decomposition of X. Note that, by Theorem 2.1.27, q is a monotone map. Also, since $\mathcal{G}$ is a continuous decomposition, by Corollary 1.1.24, q is an open map. Since q is an open map, all the elements of $\mathcal{G}$ are nowhere dense in X.

Suppose X is a metric continuum. Since the quotient map q is surjective, monotone and open, let $\mathcal{W}$ be the dense G_δ subset of $X/\mathcal{G}$ given by Theorem 3.2.6. Let $\omega \in \mathcal{W}$, and let $z \in X$ be such that $q(z) = \omega$. Suppose $\mathcal{T}_X(\{z\})$ is decomposable. Then there exist two subcontinua H and K of $\mathcal{T}_X(\{z\})$ such that $\mathcal{T}_X(\{z\}) = H \cup K$. Let $x \in H \setminus K$, and let U be an open subset of X such that $H \subset U$ and $K \setminus U \neq \emptyset$. By Theorem 3.2.6, there exist a subcontinuum Z of X containing H and a neighbourhood $\mathcal{V}$ of ω in $X/\mathcal{G}$ such that $x \in \mathrm{Int}_X(Z)$ and $(q|_Z)^{-1}(\mathcal{V}) \subset U$. Since $x \in \mathcal{T}_X(\{z\}) \cap \mathrm{Int}_X(Z)$, by Lemma 3.2.3, $\mathcal{T}_X(\{z\}) \subset Z$. Observe that this implies that $\mathcal{T}_X(\{z\}) \subset (q|_Z)^{-1}(\mathcal{V}) \subset U$, a contradiction. Therefore, $\mathcal{T}_X(\{z\})$ is an indecomposable metric continuum. □

As a consequence of Theorems 2.3.14 and 3.2.7, we obtain the following.

3.2.8 Corollary. *Let X be a continuum for which $\mathcal{T}|_{\mathcal{F}_1(X)}$ is continuous and for which $\mathcal{T}$ is idempotent on singletons. If there exists a point p of X such that $\mathcal{T}(\{p\}) = X$, then X is indecomposable.*

Next, we prove our third decomposition theorem.

3.2.9 Theorem. *Let X be a continuum for which $\mathcal{T}_X$ is continuous. Then,*

$$\mathcal{G} = \{\mathcal{T}_X(\{x\}) \mid x \in X\}$$

is a continuous decomposition of X, $X/\mathcal{G}$ is a locally connected continuum and $\mathcal{T}_X\left(2^X\right)$ is homeomorphic to $2^{X/\mathcal{G}}$. (In particular, if X is metric, then $\mathcal{T}_X\left(2^X\right)$ is homeomorphic to the Hilbert cube.) Moreover, all the elements of $\mathcal{G}$ are nowhere dense in X; and if X is metric, then there exists a dense G_δ subset $\mathcal{W}$ of $X/\mathcal{G}$ such that if $q(z) \in \mathcal{W}$, then $\mathcal{T}_X(\{z\})$ is an indecomposable metric continuum, where $q\colon X \twoheadrightarrow X/\mathcal{G}$ is the quotient map.

Proof. Since $\mathcal{T}$ is continuous, by Theorem 5.1.6, $\mathcal{T}$ is idempotent. Hence, by Theorem 3.2.7, $\mathcal{G}$ is a continuous decomposition such that $X/\mathcal{G}$ is an aposyndetic continuum.

Let $q\colon X \twoheadrightarrow X/\mathcal{G}$ be the quotient map. By Theorem 2.1.27, q is a monotone. By Theorem 1.1.24, q is an open and surjective map. Note that $q^{-1}\mathcal{T}_{X/\mathcal{G}}(\Gamma) = \mathcal{T}_X q^{-1}(\Gamma)$ for each subset Γ of $X/\mathcal{G}$ (Theorem 2.1.51 (e)). Hence, q is a $\mathcal{T}_{X X/\mathcal{G}}$-continuous surjective open map. Since $\mathcal{T}_X$ is continuous, $\mathcal{T}_{X/\mathcal{G}}$ is also continuous (Theorem 5.1.4). Since aposyndetic continua for which $\mathcal{T}$ is continuous are locally connected (Corollary 5.1.17), we have that $X/\mathcal{G}$ is locally connected.

To see that $\mathcal{T}_X\left(2^X\right)$ is homeomorphic to $2^{X/\mathcal{G}}$, let $\Im(q)\colon 2^{X/\mathcal{G}} \to 2^X$ be given by $\Im(q)(\Gamma) = q^{-1}(\Gamma)$. By Theorem 1.6.16, $\Im(q)$ is continuous. Note that $2^q \circ \Im(q) = 1_{2^{X/\mathcal{G}}}$. In particular, $\Im(q)\colon 2^{X/\mathcal{G}} \to \Im(q)\left(2^{X/\mathcal{G}}\right)$ is a homeomorphism. Thus, it is enough to show that $\mathcal{T}_X\left(2^X\right) = \Im(q)\left(2^{X/\mathcal{G}}\right)$.

Let $\Gamma \in 2^{X/\mathcal{G}}$. Then $\mathcal{T}_X\Im(q)(\Gamma) = \mathcal{T}_X q^{-1}(\Gamma) = q^{-1}\mathcal{T}_{X/\mathcal{G}}(\Gamma) = q^{-1}(\Gamma) = \Im(q)(\Gamma)$; the second equality is true by Theorem 2.1.51 (e), and the third equality is valid by Theorem 2.1.37. Thus, $\Im(q)(\Gamma) \in \mathcal{T}_X\left(2^X\right)$, and $\Im(q)\left(2^{X/\mathcal{G}}\right) \subset \mathcal{T}_X\left(2^X\right)$.

Now, let $K \in \mathcal{T}_X\left(2^X\right)$. Then there exists $A \in 2^X$ such that $\mathcal{T}_X(A) = K$. We prove that $K = \Im(q)q(A)$. Note that $q^{-1}q(A) = \bigcup\{q^{-1}q(a) \mid a \in A\} = \bigcup\{\mathcal{T}_X(\{a\}) \mid a \in A\}$. Since $\mathcal{G}$ is a decomposition, X is point $\mathcal{T}_X$-symmetric. Hence, X is $\mathcal{T}_X$-additive (Theorem 5.1.10). Since X is $\mathcal{T}_X$-additive, $\bigcup\{\mathcal{T}_X(\{a\}) \mid a \in A\} = \mathcal{T}_X(A)$ (Corollary 2.2.13). Thus, $\Im(q)q(A) = \mathcal{T}_X(A) = K$, $K \in \Im(q)\left(2^{X/\mathcal{G}}\right)$ and $\mathcal{T}_X\left(2^X\right) \subset \Im(q)\left(2^{X/\mathcal{G}}\right)$.

Therefore, $\mathcal{T}_X\left(2^X\right) = \Im(q)\left(2^{X/\mathcal{G}}\right)$. Since $\Im(q)\left(2^{X/\mathcal{G}}\right)$ is homeomorphic to $2^{X/\mathcal{G}}$, we see that $\mathcal{T}_X\left(2^X\right)$ is homeomorphic to $2^{X/\mathcal{G}}$. Suppose X is metric. Then $X/\mathcal{G}$ is a metric continuum [92, Theorem 1.3.7]. Since $2^{X/\mathcal{G}}$ is homeomorphic to the Hilbert cube [105, (1.97)], $\mathcal{T}_X\left(2^X\right)$ is homeomorphic to the Hilbert cube.

Now the theorem follows from Theorem 3.2.7. □

3.2.10 Corollary. *Let X be a hereditarily decomposable $\mathcal{T}$-additive metric continuum. Then $\mathcal{T}$ is continuous if and only if X is locally connected.*

Proof. If X is locally connected, then $\mathcal{T}$ is the identity map (Theorem 2.1.37). Hence, $\mathcal{T}$ is continuous.

Suppose X is a hereditarily decomposable $\mathcal{T}$-additive metric continuum. Note that if X is also aposyndetic, then, by Theorem 2.2.14, X is locally connected.

Assume that X is not aposyndetic. Since $\mathcal{T}$ is continuous, by Theorem 3.2.7, $\mathcal{G} = \{\mathcal{T}(\{x\}) \mid x \in X\}$ is a continuous decomposition of X. Since X is not aposyndetic, by Theorem 2.1.34, there exists $x_0 \in X$ such that $\mathcal{T}(\{x_0\})$ is nondegenerate. Then, by the continuity of $\mathcal{T}$, there exists an open subset U of X such that $x_0 \in U$, and for each $x \in U$, $\mathcal{T}(\{x\})$ is nondegenerate. Thus, by Theorem 3.2.9, there exists $x_1 \in U$ such that $\mathcal{T}(\{x_1\})$ is indecomposable, a contradiction. Hence, X is aposyndetic. Therefore, X is locally connected. □

3.2.11 Corollary. *Let X be a hereditarily decomposable and hereditarily unicoherent metric continuum. Then $\mathcal{T}$ is continuous if and only if X is locally connected.*

Proof. Since X is hereditarily unicoherent, X is $\mathcal{T}$-additive, Theorem 2.2.12. Now the corollary follows form Corollary 3.2.10. □

3.2.12 Theorem. *Let X be a decomposable continuum with the property of Kelley. If $\mathcal{T}_X$ is continuous on singletons and*

$$\mathcal{G} = \{\mathcal{T}_X(\{x\}) \mid x \in X\},$$

then $\mathcal{T}_X(Z) = q^{-1}\mathcal{T}_{X/\mathcal{G}}q(Z)$ for each nonempty closed subset Z of X, where $q\colon X \twoheadrightarrow X/\mathcal{G}$ is the quotient map.

Proof. Without loss of generality, we assume that X is not aposyndetic. Note that by Theorem 2.3.18 and Theorem 3.2.7, $\mathcal{G}$ is a decomposition of X. Let Z be a nonempty closed subset of X. We divide the proof in six steps.

Step 1. $q^{-1}q(Z) \subset \mathcal{T}_X(Z)$.
Let $x \in q^{-1}q(Z)$. Then $q(x) \in q(Z)$. Thus, there exists $z \in Z$ such that $q(z) = q(x)$. This implies that $\mathcal{T}_X(\{z\}) = \mathcal{T}_X(\{x\})$. Hence, since $\mathcal{T}_X$ is idempotent on closed sets (Theorem 2.3.18), $\mathcal{T}_X(\{x\}) = \mathcal{T}_X(\{z\}) \subset \mathcal{T}_X(Z)$. Therefore, $x \in \mathcal{T}_X(Z)$, and $q^{-1}q(Z) \subset \mathcal{T}_X(Z)$.

Step 2. $\mathcal{T}_X(Z) = q^{-1}q\mathcal{T}_X(Z)$.
Clearly, $\mathcal{T}_X(Z) \subset q^{-1}q\mathcal{T}_X(Z)$. Let $x \in q^{-1}q\mathcal{T}_X(Z)$. Then $q(x) \in q\mathcal{T}_X(Z)$. Hence, there exists $y \in \mathcal{T}_X(Z)$ such that $q(y) = q(x)$. This implies that $\mathcal{T}_X(\{y\}) = \mathcal{T}_X(\{x\})$. Thus, since $\mathcal{T}_X$ is idempotent on closed sets (Theorem 2.3.18), we have that $\mathcal{T}_X(\{y\}) \subset \mathcal{T}_X(Z)$. Hence, $x \in \mathcal{T}_X(Z)$, and $q^{-1}q\mathcal{T}_X(Z) \subset \mathcal{T}_X(Z)$. Therefore, $\mathcal{T}_X(Z) = q^{-1}q\mathcal{T}_X(Z)$.

Step 3. *If Z is connected and* $\mathrm{Int}_X(Z) \neq \emptyset$, *then* $Z = q^{-1}q(Z)$.
Clearly, $Z \subset q^{-1}q(Z)$. Let $x \in q^{-1}q(Z)$. Then $q(x) \in q(Z)$. Thus, there exists $z \in Z$ such that $q(z) = q(x)$. This implies that $\mathcal{T}_X(\{z\}) = \mathcal{T}_X(\{x\})$. Since $\mathcal{T}_X(\{x\})$ is a nowhere dense terminal subcontinuum of X

(Lemma 1.4.51 and Theorems 3.1.10 and 2.1.27), $\mathcal{T}_X(\{x\}) \cap Z \neq \emptyset$ and $\mathrm{Int}_X(Z) \neq \emptyset$, we have that $\mathcal{T}_X(\{x\}) \subset Z$. In particular, $x \in Z$. Therefore, $Z = q^{-1}q(Z)$.

Step 4. $q\mathcal{T}_X(Z) \subset \mathcal{T}_{X/\mathcal{G}}q(Z)$.
Let $\chi \in X/\mathcal{G} \setminus \mathcal{T}_{X/\mathcal{G}}q(Z)$. Then there exists a subcontinuum $\mathcal{W}$ of $X/\mathcal{G}$ such that $\chi \in \mathrm{Int}_{X/\mathcal{G}}(\mathcal{W}) \subset \mathcal{W} \subset X/\mathcal{G} \setminus q(Z)$. From these inclusions, we obtain that $q^{-1}(\chi) \subset \mathrm{Int}_X(q^{-1}(\mathcal{W})) \subset q^{-1}(\mathcal{W}) \subset X \setminus q^{-1}q(Z) \subset X \setminus Z$. Hence, since q is monotone (Theorem 2.1.27), $q^{-1}(\chi) \cap \mathcal{T}_X(Z) = \emptyset$. Thus, $qq^{-1}(\chi) \cap q\mathcal{T}_X(Z) = \emptyset$. Therefore, $\chi \in X/\mathcal{G} \setminus q\mathcal{T}_X(Z)$, and $q\mathcal{T}_X(Z) \subset \mathcal{T}_{X/\mathcal{G}}q(Z)$.

Step 5. $\mathcal{T}_{X/\mathcal{G}}q(Z) \subset q\mathcal{T}_X(Z)$.
Let $\chi \in X/\mathcal{G} \setminus q\mathcal{T}_X(Z)$. Then $\{\chi\} \cap q\mathcal{T}_X(Z) = \emptyset$. This implies that $q^{-1}(\chi) \cap q^{-1}q\mathcal{T}_X(Z) = \emptyset$. Hence, by Step 2, $q^{-1}(\chi) \cap \mathcal{T}_X(Z) = \emptyset$. Since $\mathcal{T}_X$ is idempotent on closed sets (Theorem 2.3.18), $q^{-1}(\chi) \cap \mathcal{T}_X^2(Z) = \emptyset$. Thus, there exists a subcontinuum W of X such that $q^{-1}(\chi) \subset \mathrm{Int}_X(W) \subset W \subset X \setminus \mathcal{T}_X(Z) \subset X \setminus Z$. From these inclusions, since q is an open map ($\mathcal{G}$ is a continuous decomposition), we obtain that $\{\chi\} = qq^{-1}(\chi) \subset \mathrm{Int}_{X/\mathcal{G}}(q(W)) \subset q(W) \subset q(X \setminus Z)$. To finish, we need to show that $q(W) \cap q(Z) = \emptyset$. Suppose there exists $\chi' \in q(W) \cap q(Z)$. Then, by Steps 3 and 1, $q^{-1}(\chi') \subset q^{-1}q(W) \cap q^{-1}q(Z) = W \cap q^{-1}q(Z) \subset W \cap \mathcal{T}_X(Z)$, a contradiction to the election of W. Hence, $q(W) \cap q(Z) = \emptyset$, and $\chi \in X/\mathcal{G} \setminus \mathcal{T}_{X/\mathcal{G}}q(Z)$. Therefore, $\mathcal{T}_{X/\mathcal{G}}q(Z) \subset q\mathcal{T}_X(Z)$.

Step 6. $q\mathcal{T}_X(Z) = \mathcal{T}_{X/\mathcal{G}}q(Z)$.
The equality follows from Steps 4 and 5.

From Steps 2 and 6, we have that

$$\mathcal{T}_X(Z) = q^{-1}q\mathcal{T}_X(Z) = q^{-1}\mathcal{T}_{X/\mathcal{G}}q(Z).$$

Therefore, $\mathcal{T}_X(Z) = q^{-1}\mathcal{T}_{X/\mathcal{G}}q(Z)$. □

3.3 Jones' Aposyndetic Decomposition

We prove a weak version of F. Burton Jones' Aposyndetic Decomposition Theorem only using the property of Kelley. Note that we obtain that the decomposition is only upper semicontinuous. Then we give a full version of the theorem using the uniform property of Effros.

We begin showing that the set function $\mathcal{T}$ commutes with homeomorphisms.

3.3.1 Lemma. *Let X and Y be continua, and let $h\colon X \twoheadrightarrow Y$ be a homeomorphism. If A is a subset of X, then $h(\mathcal{T}_X(A)) = \mathcal{T}_Y(h(A))$.*

Proof. Let A be a subset of X. Let $y \in Y \setminus h(\mathcal{T}_X(A))$. Then $h^{-1}(y) \in X \setminus \mathcal{T}_X(A)$. Hence, there exists a subcontinuum W of X such that $h^{-1}(y) \in \text{Int}(W) \subset W \subset X \setminus A$. This implies that $y \in \text{Int}(h(W)) \subset h(W) \subset Y \setminus h(A)$. Thus, $y \in Y \setminus \mathcal{T}_Y(h(A))$.

Now, let $y \in X \setminus \mathcal{T}_Y(h(A))$. Then there exists a subcontinuum K of Y such that $y \in \text{Int}(K) \subset K \subset Y \setminus h(A)$. This implies that $h^{-1}(y) \in \text{Int}(h^{-1}(K)) \subset h^{-1}(K) \subset X \setminus A$. Thus, $h^{-1}(y) \in X \setminus \mathcal{T}_X(A)$. Hence, $y \in Y \setminus h(\mathcal{T}_X(A))$. Therefore, $h(\mathcal{T}_X(A)) = \mathcal{T}_Y(h(A))$. □

The next theorem is the weak version of F. Burton Jones' Aposyndetic Decomposition Theorem.

3.3.2 Theorem. *Let X be a homogeneous continuum with the property of Kelley. If $\mathcal{G} = \{\mathcal{T}(\{x\}) \mid x \in X\}$, then*

(1) *$\mathcal{G}$ is an upper semicontinuous decomposition of X;*
(2) *the elements of $\mathcal{G}$ are homogeneous, terminal subcontinua of X which are mutually homeomorphic;*
(3) *the quotient space, $X/\mathcal{G}$, is an aposyndetic homogeneous continuum, and it does not contain nondegenerate proper terminal subcontinua;*
(4) *the quotient map $q\colon X \twoheadrightarrow X/\mathcal{G}$ is an atomic map.*

Proof. Let x be a point of X. We show that if $z \in \mathcal{T}(\{x\})$, then $\mathcal{T}(\{z\}) = \mathcal{T}(\{x\})$. Since X has the property of Kelley, by Theorem 2.3.18, $\mathcal{T}$ is idempotent on closed sets. Hence, $\mathcal{T}(\{z\}) \subset \mathcal{T}(\{x\})$. Also, by Theorem 3.1.6, there exists $x_0 \in \mathcal{T}(\{x\})$ such that $\mathcal{T}(\{x_0\})$ has property BL. Since X is homogeneous, there exists a homeomorphism $h\colon X \twoheadrightarrow X$ such that $h(x_0) = x$. Since $z \in \mathcal{T}(\{x\})$, $h^{-1}(z) \in \mathcal{T}(\{x_0\})$. Hence, by Lemma 3.1.3, $\mathcal{T}(\{h^{-1}(z)\}) = \mathcal{T}(\{x_0\})$. Thus, by Lemma 3.3.1,

$$\mathcal{T}(\{x\}) = h(\mathcal{T}(\{x_0\})) = h(\mathcal{T}(\{h^{-1}(z)\})) = \mathcal{T}(\{z\}).$$

Therefore, $\mathcal{G}$ is a decomposition of X. By Theorem 3.2.1, $\mathcal{G}$ is upper semicontinuous. The fact that each element of $\mathcal{G}$ is a terminal subcontinuum of X follows from Theorem 3.1.10. Also, by Corollary 3.2.2, $X/\mathcal{G}$ is an aposyndetic continuum. By Lemma 1.4.52, $X/\mathcal{G}$ does not contain nondegenerate proper terminal subcontinua. Also, by Lemma 3.3.1, we have that the elements of $\mathcal{G}$ are homogeneous and mutually homeomorphic.

We show that $X/\mathcal{G}$ is a homogeneous continuum. Since $\mathcal{G}$ is an upper semicontinuous decomposition of X, by Theorem 1.1.22, $X/\mathcal{G}$ is a Hausdorff space. Therefore, $X/\mathcal{G}$ is a continuum. Let $q\colon X \twoheadrightarrow X/\mathcal{G}$ be the quotient map. Let χ_1 and χ_2 be two elements of $X/\mathcal{G}$. We define $\zeta\colon X/\mathcal{G} \twoheadrightarrow X/\mathcal{G}$ as follows: let x_1 and x_2 be elements of X such that $q(x_1) = \chi_1$ and $q(x_2) = \chi_2$. Since X is homogeneous, there exists a homeomorphism $h\colon X \twoheadrightarrow X$ such that $h(x_1) = x_2$. Note that, by Lemma 3.3.1, $h(\mathcal{T}(\{x_1\})) = \mathcal{T}(\{x_2\})$; that is, $h(q^{-1}(\chi_1)) = q^{-1}(\chi_2)$. Define

$$\zeta(\chi) = q \circ h(q^{-1}(\chi)).$$

Then ζ is well defined, and $\zeta(\chi_1) = \chi_2$. To see that ζ is continuous, let $\mathcal{U}$ be an open subset $X/\mathcal{G}$. Then $q^{-1}(\mathcal{U})$ is a saturated open subset of X. Since h is a homeomorphism, $h^{-1}(q^{-1}(\mathcal{U}))$ is a saturated open subset of X (Lemma 3.3.1). Hence, $q(h^{-1}(q^{-1}(\mathcal{U}))) = \zeta^{-1}(\mathcal{U})$ is an open subset of $X/\mathcal{G}$. Thus, ζ is continuous. Similarly, ζ^{-1} is defined and is continuous. Therefore, $X/\mathcal{G}$ is homogeneous. To see that the quotient map is atomic, let K be a subcontinuum of X such that $q(K)$ is nondegenerate. Clearly, $K \subset q^{-1}q(K)$. Let $x \in q^{-1}q(K)$. Then there exists $k \in K$ such that $q(k) = q(x)$. Thus, $q^{-1}q(x) \cap K \neq \emptyset$. Since $q^{-1}q(x)$ is a terminal subcontinuum of X, Theorem 3.1.10, and $q(K)$ is nondegenerate, we have that $x \in q^{-1}q(x) \subset K$. Hence, $q^{-1}q(K) \subset K$. Therefore, q is an atomic map. □

3.3.3 Corollary. *Let X be a decomposable homogeneous continuum with the property of Kelley. If $x \in X$, then $\mathcal{T}(\{x\})$ is the maximal terminal proper subcontinuum of X containing x.*

Proof. If X is aposyndetic, then $\mathcal{T}(\{x\}) = \{x\}$ (Theorem 2.1.34). Since aposyndetic continua do not contain nondegenerate proper terminal subcontinua (Lemma 1.4.52), $\mathcal{T}(\{x\})$ is the maximal terminal proper subcontinuum of X containing x.

Suppose X is not aposyndetic, and let $x \in X$. Then, by Theorem 3.3.2, $\mathcal{T}(\{x\})$ is a terminal subcontinuum of X. Suppose K is a terminal proper subcontinuum of X such that $\mathcal{T}(\{x\}) \subsetneq K$. Let $q\colon X \twoheadrightarrow X/\mathcal{G}$ be the quotient map. By Lemma 1.4.53, $q(K)$ is a terminal subcontinuum of $X/\mathcal{G}$. Since $X/\mathcal{G}$ does not contain proper nondegenerate terminal subcontinua (Theorem 3.3.2), $q(K) = \{q(x)\}$. Hence, $K = q^{-1}q(x) = \mathcal{T}(\{x\})$, a contradiction. Therefore, $\mathcal{T}(\{x\})$ is the maximal terminal proper subcontinuum of X containing x. □

As a consequence of Theorem 3.3.2, we have the following:

3.3.4 Corollary. *If X is an arcwise connected homogeneous continuum with the property of Kelley, then X is aposyndetic.*

3.3.5 Definition. Let $\mathcal{G}$ be a decomposition of a compactum X, and let $\mathcal{H}$ be a family of homeomorphisms of X. We say that $\mathcal{H}$ *respects* $\mathcal{G}$ provided that for each pair, G and G', of elements of $\mathcal{G}$ and every $h \in \mathcal{H}$, either $h(G) = G'$ or $h(G) \cap G' = \emptyset$.

The following result is useful to prove Theorem 3.3.8.

3.3.6 Theorem. *Let X be a continuum with the uniform property of Effros, and let $\mathcal{G}$ be a decomposition of X such that the elements of $\mathcal{G}$ are nondegenerate closed sets of X. If the homeomorphism group, $\mathcal{H}(X)$, of X respects $\mathcal{G}$, then the following hold:*

(1) *$\mathcal{G}$ is a continuous decomposition of X.*
(2) *The elements of $\mathcal{G}$ are homogeneous and mutually homeomorphic.*

(3) *The quotient space $X/\mathcal{G}$ is homogeneous.*
(4) *If $q\colon X \twoheadrightarrow X/\mathcal{G}$ is the quotient map, then q is uniformly completely regular.*

Proof. By Theorem 1.4.59, X is a homogeneous continuum. First we show that $\mathcal{G}$ is upper semicontinuous. Let G be an element of $\mathcal{G}$, and let W be an open subset of X containing G. For each $g \in G$, there exists $U_g \in \mathfrak{U}_X$ such that $B(g, 2U_g) \subset W$. Since G is compact, there exist $g_1, \ldots, g_n$ in G such that $G \subset \bigcup_{j=1}^{n} \mathrm{Int}_X(B(g_j, U_{g_j}))$. Let $U = \bigcap_{j=1}^{n} U_{g_j}$. Then $U \in \mathfrak{U}_X$. Let V be an Effros entourage for U. Let $W' = \bigcup\{\mathrm{Int}_X(B(g, V)) \mid g \in G\}$, and let $G' \in \mathcal{G}$ be such that $G' \cap W' \neq \emptyset$. Let $x \in G' \cap W'$. Then there exists $g_0 \in G$ such that $x \in \mathrm{Int}_X(B(g_0, V))$. This implies that $\rho_X(x, g_0) < V$. Thus, there exists a U-homeomorphism $h\colon X \twoheadrightarrow X$ such that $h(g_0) = x$. Since $\mathcal{H}(X)$ respects $\mathcal{G}$ and $h(G) \cap G' \neq \emptyset$, we have that $h(G) = G'$. Let $g \in G$. Then there exists $j \in \{1, \ldots, n\}$ such that $g \in \mathrm{Int}_X(B(g_j, U_{g_j}))$. Since $U \subset U_{g_j}$, $\rho_X(h(g), g) < U_{g_j}$. Also, $\rho_X(g, g_j) < U_{g_j}$. Hence, $\rho_X(h(g), g_j) < 2U_{g_j}$. This implies that $h(g) \in W$. Therefore, $G' \subset W$, and $\mathcal{G}$ is an upper semicontinuous decomposition.

Next, we see that $\mathcal{G}$ is lower semicontinuous. Let $G \in \mathcal{G}$, let g_1 and g_2 be two elements of G and let W be an open subset of X containing g_1. Then there exists $U \in \mathfrak{U}_X$ such that $B(g_1, U) \subset W$. Let V be an Effros entourage for U. Let $G' \in \mathcal{G}$ be such that $G' \cap \mathrm{Int}_X(B(g_2, V)) \neq \emptyset$, and let $g' \in G' \cap \mathrm{Int}_X(B(g_2, V))$. Then $\rho_X(g', g_2) < V$. Thus, there exists a U-homeomorphism $h\colon X \twoheadrightarrow X$ such that $h(g_2) = g'$. Since $\mathcal{H}(X)$ respects $\mathcal{G}$ and $h(G) \cap G' \neq \emptyset$, we obtain that $h(G) = G'$. Then $h(g_1) \in G'$ and $\rho_X(g_1, h(g_1)) < U$. This implies that $h(g_1) \in G' \cap B(g_1, U) \subset G' \cap W$. Hence, $\mathcal{G}$ is lower semicontinuous. Therefore, $\mathcal{G}$ is a continuous decomposition of X.

Observe that, since X is homogeneous and $\mathcal{H}(X)$ respects $\mathcal{G}$, all the elements of $\mathcal{G}$ are homeomorphic and homogeneous.

Now, we prove that $X/\mathcal{G}$ is a homogeneous continuum. Since $\mathcal{G}$ is an upper semicontinuous decomposition of X, by Theorem 1.1.22, $X/\mathcal{G}$ is a Hausdorff space. Therefore, $X/\mathcal{G}$ is a continuum.

Let $q\colon X \twoheadrightarrow X/\mathcal{G}$ be the quotient map. Let χ_1 and χ_2 be two elements of $X/\mathcal{G}$. Let x_1 and x_2 be points of X such that $q(x_1) = \chi_1$ and $q(x_2) = \chi_2$. Since X is homogeneous, there exists a homeomorphism $h\colon X \twoheadrightarrow X$ such that $h(x_1) = x_2$. Since $\mathcal{H}(X)$ respects $\mathcal{G}$, we have that $h(q^{-1}(\chi_1)) = q^{-1}(\chi_2)$. Define $\zeta\colon X/\mathcal{G} \twoheadrightarrow X/\mathcal{G}$ by

$$\zeta(\chi) = q \circ h(q^{-1}(\chi)).$$

Then ζ is well defined ($\mathcal{H}(X)$ respects $\mathcal{G}$), and $\zeta(\chi_1) = \chi_2$. Since $\mathcal{G}$ is a continuous decomposition, q is an open map. Hence, since q is open and h is continuous, ζ is continuous. Note that

$$\zeta^{-1}\colon X/\mathcal{G} \twoheadrightarrow X/\mathcal{G}$$

is given by $\zeta^{-1}(\chi) = q \circ h^{-1}(q^{-1}(\chi))$, and it is continuous. Thus, ζ is a homeomorphism and, therefore, $X/\mathcal{G}$ is a homogeneous continuum.

Next, we demonstrate that the quotient map $q\colon X \twoheadrightarrow X/\mathcal{G}$ is uniformly completely regular. Let $U \in \mathfrak{U}_X$, and let $V \in \mathfrak{U}_X$ be an Effros entourage for U. Then, by Theorem 1.3.13, $\Omega = (q \times q)(V) \in \mathfrak{U}_{X/\mathcal{G}}$. Let χ and χ' be two elements of $X/\mathcal{G}$ such that $\rho_{X/\mathcal{G}}(\chi, \chi') < \Omega$. Then there exist x and x' in X such that $\rho_X(x, x') < V$, $q(x) = \chi$ and $q(x') = \chi'$. Hence, there exists a U-homeomorphism $h\colon X \twoheadrightarrow X$ such that $h(x) = x'$. Since $\mathcal{H}(X)$ respects $\mathcal{G}$ and $h(q^{-1}(\chi)) \cap q^{-1}(\chi') \neq \emptyset$, we have that $h(q^{-1}(\chi)) = q^{-1}(\chi')$. Thus, $h|_{q^{-1}(\chi)}\colon q^{-1}(\chi) \twoheadrightarrow q^{-1}(\chi')$ is a homeomorphism such that $\rho_X(z, h|_{q^{-1}(\chi)}(z)) < U$ for all $z \in q^{-1}(\chi)$. Therefore, q is a uniformly completely regular map. □

3.3.7 Definition. A continuum X is said to be *acyclic* provided that the first Čech cohomology group with integer coefficients is trivial; that is, $\check{H}^1(X, \mathbb{Z}) = \{0\}$.

We are ready to state and prove F. Burton Jones' Aposyndetic Decomposition Theorem.

3.3.8 Theorem. *Let X be a decomposable continuum with the uniform property of Effros. If $\mathcal{G} = \{\mathcal{T}_X(\{x\}) \mid x \in X\}$, then the following hold:*

(1) *$\mathcal{G}$ is a continuous, monotone and terminal decomposition of X.*
(2) *The elements of $\mathcal{G}$ are cell-like, acyclic, homogeneous and mutually homeomorphic continua.*
(3) *The quotient map $q\colon X \twoheadrightarrow X/\mathcal{G}$ is uniformly completely regular and atomic.*
(4) *The quotient space, $X/\mathcal{G}$, is an aposyndetic homogeneous continuum, and it does not contain nondegenerate terminal subcontinua.*
(5) *If X is metric, then the elements of $\mathcal{G}$ are indecomposable continua of the same dimension as X, and the quotient space is a one-dimensional metric continuum.*

Proof. By Theorem 1.4.59, X is a homogeneous continuum. By Theorem 2.1.27, the elements of $\mathcal{G}$ are continua. Since $\mathcal{H}(X)$ respects $\mathcal{G}$ (Lemma 3.3.1), by Theorem 3.3.6, $\mathcal{G}$ is a continuous decomposition of X, the elements of $\mathcal{G}$ are homogeneous and mutually homeomorphic and the quotient map is uniformly completely regular. By Theorem 3.3.2, the elements of $\mathcal{G}$ are terminal subcontinua, the quotient space is aposyndetic, it does not contain nondegenerate proper terminal subcontinua and the quotient map is atomic. By Theorem 1.5.8, q is a cell-like map. Hence, the elements of $\mathcal{G}$ are cell-like continua. Since the elements of $\mathcal{G}$ are cell-like continua, we have, in particular, that every map from an element of $\mathcal{G}$ into a simple closed curve is homotopic to a constant map. Hence, by [35, Theorem 8.1], the elements of $\mathcal{G}$ are acyclic. Part (5) follows from [92, Theorem 5.1.18]. □

3.3.9 Remark. Since every Effros homogeneous continuum has the uniform property of Effros, Theorem 1.4.60, Theorem 3.3.8 is true for Effros homogeneous continua.

Now we present some consequences of Theorem 3.3.8. We need the following:

3.3.10 Definition. A continuum X is *semi-indecomposable* provided that for every pair of subcontinua W_1 and W_2 of X such that $\mathrm{Int}_X(W_1) \neq \emptyset$ and $\mathrm{Int}_X(W_2) \neq \emptyset$, we have that $W_1 \cap W_2 \neq \emptyset$.

The next result shows that the semi-indecomposability of a decomposable continuum with the uniform property of Effros is equivalent to the semi-indecomposability of its quotient space with respect to Jones' decomposition.

3.3.11 Theorem. *Let X be a decomposable continuum with the uniform property of Effros, and let $\mathcal{G} = \{\mathcal{T}_X(\{x\}) \mid x \in X\}$. Then X is semi-indecomposable if and only if $X/\mathcal{G}$ is semi-indecomposable.*

Proof. By Theorem 1.4.59, X is a homogeneous continuum. By Theorem 3.3.8, $\mathcal{G}$ is a continuous monotone decomposition, and the elements of $\mathcal{G}$ are terminal subcontinua of X. Let $q\colon X \twoheadrightarrow X/\mathcal{G}$ be the quotient map. Then q is monotone and open.

If $X/\mathcal{G}$ is not semi-indecomposable, then there exist two disjoint subcontinua $\mathcal{W}_1$ and $\mathcal{W}_2$ of $X/\mathcal{G}$ such that $\mathrm{Int}_{X/\mathcal{G}}(\mathcal{W}_1) \neq \emptyset$ and $\mathrm{Int}_X(\mathcal{W}_2) \neq \emptyset$. Hence, $q^{-1}(\mathcal{W}_1)$ and $q^{-1}(\mathcal{W}_2)$ are two disjoint subcontinua of X (Lemma 1.4.46) such that $\mathrm{Int}_X(q^{-1}(\mathcal{W}_1)) \neq \emptyset$ and $\mathrm{Int}_X(q^{-1}(\mathcal{W}_1)) \neq \emptyset$. Therefore, X is not semi-indecomposable.

Suppose X is not semi-indecomposable. Then there exist two disjoint subcontinua W_1 and W_2 of X such that $\mathrm{Int}_X(W_1) \neq \emptyset$ and $\mathrm{Int}_X(W_2) \neq \emptyset$. Since the elements of $\mathcal{G}$ are terminal subcontinua of X, by Lemma 1.4.51, the elements of $\mathcal{G}$ are nowhere dense. Hence, since $\mathrm{Int}_X(W_1) \neq \emptyset$, if $w \in W_1$, then $\mathcal{T}_X(\{w\}) \subset W_1$. Similarly, the same is true for the elements of W_2. Let $j \in \{1,2\}$. Thus, $W_j = \bigcup\{\mathcal{T}_X(\{w\}) \mid w \in W_j\}$. This implies that $q(W_1)$ and $q(W_2)$ are disjoint subcontinua of $X/\mathcal{G}$ such that $\mathrm{Int}_X(q(W_1)) \neq \emptyset$ and $\mathrm{Int}_X(q(W_2)) \neq \emptyset$. Therefore, $X/\mathcal{G}$ is not semi-indecomposable. □

3.3.12 Theorem. *Let X be a continuum with the uniform property of Effros, and let $\mathcal{G} = \{\mathcal{T}_X(\{x\}) \mid x \in X\}$. Then $X/\mathcal{G}$ is locally connected if and only if $\mathcal{T}_X$ is continuous.*

Proof. By Theorem 1.4.59, X is homogeneous. Suppose $X/\mathcal{G}$ is locally connected. If X is an indecomposable continuum, then, by Theorem 2.1.44, $\mathcal{T}_X$ is a constant map, and $X/\mathcal{G}$ is a point. We assume that X is decomposable. By Theorem 3.3.8, $\mathcal{G}$ is a terminal continuous decomposition of X, and the quotient map is monotone and open. Let B be a proper subcontinuum of X, and let $x \in X \setminus B$. Since $\mathcal{T}_X(\{x\})$ is a terminal subcontinuum of X, either $\mathcal{T}_X(\{x\}) \cap B = \emptyset$ or $B \subset \mathcal{T}_X(\{x\})$. In any case, $q(B)$ is a proper subcontinuum of $X/\mathcal{G}$. Therefore, by Theorem 5.1.2, $\mathcal{T}_X$ is continuous. The converse implication follows directly from Theorem 3.2.9. □

3.3.13 Theorem. *Let X be a continuum with the uniform property of Effros. Let $\mathcal{G} = \{\mathcal{T}_X(\{x\}) \mid x \in X\}$, and let $q\colon X \twoheadrightarrow X/\mathcal{G}$ be the quotient map. If $\Im(q)\colon 2^{X/\mathcal{G}} \to 2^X$ is given by $\Im(q)(\Gamma) = q^{-1}(\Gamma)$, then $\mathcal{T}_X(2^X) \subset \Im(q)(2^{X/\mathcal{G}})$. Moreover, $\mathcal{T}_X(2^X) = \Im(q)(2^{X/\mathcal{G}})$ if and only if $\mathcal{T}_X$ is continuous.*

Proof. By Theorem 1.4.59, X is a homogeneous continuum. By Theorem 3.3.8, $\mathcal{G}$ is a monotone continuous decomposition of X. Hence, q is a monotone and open map. Thus, $\Im(q)$ is continuous, Theorem 1.6.16. To show that $\mathcal{T}_X(2^X) \subset \Im(q)(2^{X/\mathcal{G}})$, let $K \in \mathcal{T}_X(2^X)$. Then there exists $Z \in 2^X$ such that $\mathcal{T}_X(Z) = K$. By Theorems 1.6.22 and 3.2.12, $\mathcal{T}_X(Z) = q^{-1}\mathcal{T}_{X/\mathcal{G}}q(Z) = \Im(q)\mathcal{T}_{X/\mathcal{G}}q(Z)$. Since $K = \mathcal{T}_X(Z)$, we have that $K = \Im(q)\mathcal{T}_{X/\mathcal{G}}q(Z)$. Therefore, $\mathcal{T}_X(2^X) \subset \Im(q)(2^{X/\mathcal{G}})$.

Now, suppose that $\mathcal{T}_X(2^X) = \Im(q)(2^{X/\mathcal{G}})$. Let $\Gamma \in 2^{X/\mathcal{G}}$. Then, by Theorem 2.1.51 (c), $\mathcal{T}_{X/\mathcal{G}}(\Gamma) = q\mathcal{T}_X q^{-1}(\Gamma) = 2^q \circ \mathcal{T}_X \circ \Im(q)(\Gamma)$. Since $\Im(q)(\Gamma) \in \mathcal{T}_X(2^X)$, there exists $Z \in 2^X$ such that $\mathcal{T}_X(Z) = \Im(q)(\Gamma)$. Hence, $\mathcal{T}_X \circ \Im(q)(\Gamma) = \mathcal{T}_X\mathcal{T}_X(Z) = \mathcal{T}_X(Z) = \Im(q)(\Gamma)$ (the second equality is true by Theorems 1.6.22 and 2.3.18). Thus, $\mathcal{T}_{X/\mathcal{G}}(\Gamma) = 2^q \circ \Im(q)(\Gamma) = \Gamma$. This implies that $X/\mathcal{G}$ is locally connected, Theorem 2.1.37. Hence, by Theorem 3.3.12, $\mathcal{T}_X$ is continuous.

If $\mathcal{T}_X$ is continuous for X, by the proof of Theorem 3.2.9, $\mathcal{T}_X(2^X) = \Im(q)(2^{X/\mathcal{G}})$. □

3.4 Prajs' Mutual Aposyndetic Decomposition

We begin with the definition of the relation that provides the base for Prajs' theorem.

3.4.1 Definition. Let X be a continuum, and let x_1 and x_2 be two points of X. We say that x_1 and x_2 are related, $x_1 \sim x_2$, provided that X is not mutually aposyndetic at them; that is, if W_{x_1} and W_{x_2} are two subcontinua of X such that $x_1 \in \mathrm{Int}_X(W_{x_1})$ and $x_2 \in \mathrm{Int}_X(W_{x_2})$, then we have that $W_{x_1} \cap W_{x_2} \neq \emptyset$.

3.4.2 Theorem. *Let X be a continuum with the property of Kelley. Then the relation "$\sim$" is an equivalence relation.*

Proof. It is clear that the relation is reflexive and symmetric. We show that it is transitive. Suppose that there exist three points x_1, x_2 and x_3 in X such that $x_1 \sim x_2$, $x_2 \sim x_3$, but $x_1 \not\sim x_3$. Since $x_1 \not\sim x_3$, there exist two disjoint subcontinua W_1 and W_3 such that $x_1 \in \mathrm{Int}_X(W_1)$ and $x_3 \in \mathrm{Int}_X(W_3)$. By Theorem 2.1.67, $\mathcal{T}(W_1) \cap \mathcal{T}(W_3) = \emptyset$. Since $x_1 \sim x_2$, for each subcontinuum W of X such that $x_2 \in \mathrm{Int}_X(W)$, we obtain that $W \cap W_1 \neq \emptyset$. This implies that $x_2 \in \mathcal{T}(W_1)$. Similarly, $x_2 \in \mathcal{T}(W_3)$. Thus, $\mathcal{T}(W_1) \cap \mathcal{T}(W_3) \neq \emptyset$, a contradiction. Therefore, $x_1 \sim x_3$, and "$\sim$" is an equivalence relation. □

3.4.3 Notation. We denote the equivalence class of x with respect to "$\sim$" by Q_x.

3.4.4 Lemma. *Let X be a continuum with the property of Kelley. If x is a point of X, then Q_x is a closed subset of X.*

Proof. Let x be a point of X, and let $z \in X \setminus Q_x$. Then there exist two disjoint subcontinua W_x and W_z such that $x \in \mathrm{Int}(W_x)$ and $z \in \mathrm{Int}(W_z)$. Hence, $\mathrm{Int}(W_z) \subset X \setminus Q_x$. Therefore, Q_x is a closed subset of X. □

3.4.5 Lemma. *Let X be a continuum with the property of Kelley, and let x be a point of X. If $h\colon X \twoheadrightarrow X$ is a homeomorphism, then $h(Q_x) = Q_{h(x)}$.*

Proof. Let $z \in X \setminus h(Q_x)$. Since h is a homeomorphism, $h^{-1}(z) \in X \setminus Q_x$. Hence, there exist two disjoint subcontinua W_x and W_z of X such that $h^{-1}(z) \in \mathrm{Int}_X(W_z)$ and $x \in \mathrm{Int}_X(W_x)$. Thus, $h(W_z)$ and $h(W_x)$ are disjoint subcontinua of X such that $z \in \mathrm{Int}_X(h(W_z))$ and $h(x) \in \mathrm{Int}_X(h(W_x))$. This implies that $z \in X \setminus Q_{h(x)}$.

Now, let $z \in X \setminus Q_{h(x)}$. Then there exist two disjoint subcontinua W_z and W_x such that $z \in \mathrm{Int}_X(W_z)$ and $h(x) \in \mathrm{Int}_X(W_x)$. Hence, $h^{-1}(W_z)$ and $h^{-1}(W_x)$ are two disjoint subcontinua of X such that $h^{-1}(z) \in \mathrm{Int}_X(h^{-1}(W_z))$ and $x \in \mathrm{Int}_X(h^{-1}(W_x))$. Thus, $h^{-1}(z) \in X \setminus Q_x$. This implies that $z \in X \setminus h(Q_x)$. Therefore, $h(Q_x) = Q_{h(x)}$. □

3.4.6 Corollary. *If X is a continuum with the property of Kelley and $\mathcal{Q} = \{Q_x \mid x \in X\}$, then the group of homeomorphisms of X, $\mathcal{H}(X)$, respects $\mathcal{Q}$.*

3.4.7 Lemma. *If X is a continuum with the property of Kelley and x is a point of X, then*

$$Q_x = \bigcap\{\mathcal{T}(W) \mid W \textit{ is a subcontinuum of } X \textit{ and } x \in \mathrm{Int}(W)\}.$$

Proof. Let $z \in X \setminus Q_x$. Then there exist two disjoint subcontinua W_x and W_z of X such that $x \in \mathrm{Int}(W_x)$ and $z \in \mathrm{Int}(W_z)$. Hence, $z \in X \setminus \mathcal{T}(W_x)$. Thus,

$$z \in X \setminus \bigcap\{\mathcal{T}(W) \mid W \text{ is a subcontinuum of } X \text{ and } x \in \mathrm{Int}(W)\}.$$

Let $z \in X \setminus \bigcap\{\mathcal{T}(W) \mid W$ is a subcontinuum of X and $x \in \mathrm{Int}(W)\}$. Then there exists a subcontinuum W_x of X such that $x \in \mathrm{Int}(W_x)$ and $z \in X \setminus \mathcal{T}(W_x)$. Thus, there exists a subcontinuum W_z of X such that $z \in \mathrm{Int}(W_z) \subset W_z \subset X \setminus W_x$. This implies that $z \in X \setminus Q_x$. □

3.4.8 Lemma. *Let X be a continuum with the property of Kelley. If W is a subcontinuum of X with nonempty interior, then $\mathcal{T}(W) = q^{-1}q\mathcal{T}(W)$, where $\mathcal{Q} = \{Q_x \mid x \in X\}$, and $q\colon X \twoheadrightarrow X/\mathcal{Q}$ is the quotient map.*

Proof. Let W be a subcontinuum of X with nonempty interior. Clearly, $\mathcal{T}(W) \subset q^{-1}q\mathcal{T}(W)$. Let $x \in X \setminus \mathcal{T}(W)$. Since $\mathcal{T}$ is idempotent on closed sets (Theorem 2.3.18), $x \in X \setminus \mathcal{T}^2(W)$. Thus, there exists a subcontinuum W_x of X such that $x \in \mathrm{Int}(W_x) \subset W_x \subset X \setminus \mathcal{T}(W)$. By Corollary 1.6.21, there exists a subcontinuum K_W of X such that $\mathcal{T}(W) \subset \mathrm{Int}(K)$ and $W_x \cap K_W = \emptyset$. This implies that $Q_x \cap \mathcal{T}(W) = \emptyset$. Hence, $q(x) \cap q\mathcal{T}(W) = \emptyset$. Thus, $q^{-1}q(x) \cap q^{-1}q\mathcal{T}(W) = \emptyset$, and $x \in X \setminus q^{-1}q\mathcal{T}(W)$. Therefore, $\mathcal{T}(W) = q^{-1}q\mathcal{T}(W)$. □

We are ready to state and prove Janusz R. Prajs' Mutual Aposyndesis Decomposition Theorem.

3.4.9 Theorem. *Let X be a decomposable continuum with the uniform property of Effros. If $\mathcal{Q} = \{Q_x \mid x \in X\}$, then the following hold:*

(1) *$\mathcal{Q}$ is a continuous decomposition of X.*
(2) *The elements of $\mathcal{Q}$ are homogeneous and mutually homeomorphic closed subsets of X.*
(3) *The quotient map $q\colon X \twoheadrightarrow X/\mathcal{Q}$ is uniformly completely regular.*
(4) *The quotient space, $X/\mathcal{Q}$, is a mutually aposyndetic homogeneous continuum.*

Proof. By Theorem 1.4.59, X is a homogeneous continuum. By Theorem 1.6.22, X has the property of Kelley. By Theorem 3.4.2, $\mathcal{Q}$ is a decomposition of X. By Corollary 3.4.6, the group of homeomorphisms of X respects $\mathcal{Q}$. Hence, by Theorem 3.3.6, $\mathcal{Q}$ is a continuous decomposition of X, the elements of $\mathcal{Q}$ are homogeneous and mutually homeomorphic closed subsets of X, the quotient map is uniformly completely regular and the quotient space is a homogeneous continuum.

We need to prove that $X/\mathcal{Q}$ is mutually aposyndetic. To this end, let χ_1 and χ_2 be two distinct elements of $X/\mathcal{Q}$. Let x_1 and x_2 be points of X such that $q(x_1) = \chi_1$ and $q(x_2) = \chi_2$. Then, X is mutually aposyndetic at x_1 and x_2. Hence, there exist two disjoint subcontinua W_{x_1} and W_{x_2} of X such that $x_1 \in \text{Int}_X(W_{x_1})$ and $x_2 \in \text{Int}_X(W_{x_2})$. By Theorem 2.1.67, we have that $\mathcal{T}_X(W_{x_1}) \cap \mathcal{T}_X(W_{x_2}) = \emptyset$. Since $\mathcal{Q}$ is a continuous decomposition, q is open. Thus, by Lemma 3.4.8, $q\mathcal{T}_X(W_{x_1}) \cap q\mathcal{T}_X(W_{x_2}) = \emptyset$, $\chi_1 \in \text{Int}_{X/\mathcal{Q}}(q\mathcal{T}_X(W_{x_1}))$ and $\chi_2 \in \text{Int}_{X/\mathcal{Q}}(q\mathcal{T}_X(W_{x_2}))$. Therefore, $X/\mathcal{Q}$ is mutually aposyndetic. □

3.4.10 Remark. Since every Effros homogeneous continuum has the uniform property of Effros, Theorem 1.4.60, Theorem 3.4.9 is true for Effros homogeneous continua.

The next theorem gives us a sufficient condition to have that Jones' decomposition coincides with Prajs' decomposition.

3.4.11 Theorem. *Let X be a continuum with the uniform property of Effros. If $\mathcal{G} = \{\mathcal{T}_X(\{x\}) \mid x \in X\}$ is Jones' decomposition, $\mathcal{Q} = \{Q_x \mid x \in X\}$ is Prajs' decomposition and $X/\mathcal{G}$ is mutually aposyndetic, then $\mathcal{T}_X(\{x\}) = Q_x$ for each $x \in X$. In particular, $\mathcal{G} = \mathcal{Q}$.*

Proof. By Theorem 1.4.59, X is a homogeneous continuum. By Theorem 3.3.8, $\mathcal{G}$ is a continuous decomposition of X. Let x_1 be a point of X. Let $x_2 \in X \setminus \mathcal{T}_X(\{x_1\})$. Then $\mathcal{T}_X(\{x_1\}) \cap \mathcal{T}_X(\{x_2\}) = \emptyset$. Let $q\colon X \twoheadrightarrow X/\mathcal{G}$ be the quotient map. Note that $q(x_1) \neq q(x_2)$. Since $X/\mathcal{G}$ is mutually aposyndetic, there exist two disjoint subcontinua $\mathcal{W}_1$ and $\mathcal{W}_2$ of $X/\mathcal{G}$ such that $q(x_1) \in \text{Int}_{X/\mathcal{G}}(\mathcal{W}_1)$ and $q(x_2) \in \text{Int}_{X/\mathcal{G}}(\mathcal{W}_2)$. Then, since q is monotone, $q^{-1}(\mathcal{W}_1)$ and $q^{-1}(\mathcal{W}_2)$ are two disjoint subcontinua of X (Lemma 1.4.46) such that $x_1 \in \text{Int}_X(q^{-1}(\mathcal{W}_1))$ and $x_2 \in \text{Int}_X(q^{-1}(\mathcal{W}_2))$. Hence, X is mutually aposyndetic at x_1 and x_2. Thus, $x_1 \in X \setminus Q_{x_2}$ and $x_2 \in X \setminus Q_{x_1}$. This implies that $Q_{x_1} \subset \mathcal{T}_X(\{x_1\})$. Clearly, $\mathcal{T}_X(\{x_1\}) \subset Q_{x_1}$. Therefore, $\mathcal{T}_X(\{x_1\}) = Q_{x_1}$. □

As a consequence of Theorems 5.1.29 and 3.4.11, we have the following.

3.4.12 Corollary. *Let X be a nonaposyndetic continuum with the uniform property of Effros, let $\mathcal{G} = \{\mathcal{T}_X(\{x\}) \mid x \in X\}$ be Jones' decomposition and let $\mathcal{Q} = \{Q_x \mid x \in X\}$ be Prajs' decomposition. If $\mathcal{T}_X$ is continuous for X, then $\mathcal{T}_X(\{x\}) = Q_x$ for each $x \in X$. In particular, $\mathcal{G} = \mathcal{Q}$.*

3.4.13 Theorem. *Let X be a decomposable continuum with the uniform property of Effros, and let $\mathcal{Q} = \{Q_x \mid x \in X\}$ be Prajs' decomposition. If the elements of $\mathcal{Q}$ are connected, then X is not semi-indecomposable.*

Proof. By Theorem 1.4.59, X is a homogeneous continuum. By Theorem 3.4.9, $\mathcal{Q}$ is a continuous decomposition of X. Suppose that the elements of $\mathcal{Q}$ are connected. Since $X/\mathcal{Q}$ is mutually aposyndetic, $X/\mathcal{Q}$ is not semi-indecomposable. Note that the quotient map q is monotone. Now the rest of the proof is equal to the first part of the proof of Theorem 3.3.11. □

For the next results, we adopt the following notation.

3.4.14 Notation. Given a continuum X with the uniform property of Effros, let $\mathcal{G} = \{\mathcal{T}_X(\{x\}) \mid x \in X\}$ be Jones' decomposition, let $X_J = X/\mathcal{G}$ and let $q_J \colon X \twoheadrightarrow X_J$ be the quotient map. Let $\mathcal{Q} = \{Q_x \mid x \in X\}$ be Prajs' decomposition, let $X_P = X/\mathcal{Q}$ and let $q_P \colon X \twoheadrightarrow X_P$ be the quotient map. For X_J, let $\mathfrak{Q}_J = \{\mathcal{Q}_\zeta \mid \zeta \in X_J\}$ be Prajs' decomposition, let $X_{JP} = X_J/\mathfrak{Q}_J$ and let $q_{JP} \colon X_J \twoheadrightarrow X_{JP}$ be the quotient map.

3.4.15 Theorem. *Let X be a continuum with the uniform property of Effros. With the notation in Notation 3.4.14, $q_J(Q_x) = \mathcal{Q}_{q_J(x)}$ for all $x \in X$.*

Proof. By Theorem 1.4.59, X is a homogeneous continuum. Let x be a point of X, and let $\zeta \in X_J \setminus \mathcal{Q}_{q_J(x)}$. Then there exist two disjoint subcontinua $\mathcal{W}_\zeta$ and $\mathcal{W}_{q_J(x)}$ such that $\zeta \in \mathrm{Int}_{X_J}(\mathcal{W}_\zeta)$ and $q_J(x) \in \mathrm{Int}_{X_J}(\mathcal{W}_{q_J(x)})$. Since q_J is monotone (Theorem 3.3.8), we have that $q_J^{-1}(\mathcal{W}_\zeta)$ and $q_J^{-1}(\mathcal{W}_{q_J(x)})$ are disjoint subcontinua of X (Lemma 1.4.46) such that $q_J^{-1}(\zeta) \subset \mathrm{Int}_X(q_J^{-1}(\mathcal{W}_\zeta))$ and $x \in \mathrm{Int}_X(q_J^{-1}(\mathcal{W}_{q_J(x)}))$. Thus, $q_J^{-1}(\zeta) \cap Q_x = \emptyset$. This implies that $\zeta \in X_J \setminus q_J(Q_x)$.

Now, let $\zeta \in X_J \setminus q_J(Q_x)$. Then $q_J^{-1}(\zeta) \cap q_J^{-1}q_J(Q_x) = \emptyset$. In particular, $q_J^{-1}(\zeta) \cap Q_x = \emptyset$. Let $z \in X$ be such that $q_J(z) = \zeta$. Hence, there exist two disjoint subcontinua W_z and W_x of X such that $z \in \mathrm{Int}_X(W_z)$ and $x \in \mathrm{Int}_X(W_x)$. Since q_J is atomic and open (Theorem 3.3.8), $q_J(W_z)$ and $q_J(W_x)$ are two disjoint subcontinua of X_J such that $\zeta = q_J(z) \in \mathrm{Int}_{X_J}(q_J(W_z))$ and $q_J(x) \in \mathrm{Int}_{X_J}(q_J(W_x))$. Thus, $\zeta \in X_J \setminus \mathcal{Q}_{q_J(x)}$. Therefore, $q_J(Q_x) = \mathcal{Q}_{q_J(x)}$. □

3.4.16 Corollary. *Let X be a continuum with the uniform property of Effros. With the notation in Notation 3.4.14, X_{JP} is homeomorphic to X_P.*

Proof. By Theorem 1.4.59, X is a homogeneous continuum. Note that for each $x \in X$, $q_J^{-1}q_J(x) = \mathcal{T}_X(\{x\}) \subset Q_x = q_P^{-1}q_P(x)$. Hence, by the Transgression Theorem [36, 3.2, p. 123], there exists a map $g \colon X_J \twoheadrightarrow X_P$ such that $g \circ q_J = q_P$. Let $\chi \in X_P$, and let $x \in X$ be such that $q_P(x) = \chi$. Then $g^{-1}(\chi) = q_J(Q_x) =$

$\mathcal{Q}_{q_J(x)}$ (the last equality is true by Theorem 3.4.15). Now, let $\alpha \in X_{JP}$, and let $\zeta \in X_J$ be such that $q_{JP}(\zeta) = \alpha$. Then $q_{JP}^{-1}(\alpha) = \mathcal{Q}_\zeta$. Since q_J is surjective, there exists $z \in X$ such that $q_J(z) = \zeta$. Hence, by Theorem 3.4.15, $\mathcal{Q}_\zeta = \mathcal{Q}_{q_J(z)}$. Thus, for each $\chi \in X_P$, there exists $\alpha \in X_{JP}$ such that $g^{-1}(\chi) = q_{JP}^{-1}(\alpha)$, and vice versa. Hence, by the Transgression Theorem, there exist two maps $h\colon X_P \twoheadrightarrow X_{JP}$ and $k\colon X_{JP} \twoheadrightarrow X_P$ such that $h \circ g = q_{JP}$ and $k \circ q_{JP} = g$. Then h and k are inverse of each other. Therefore, X_{JP} is homeomorphic to X_P. □

3.5 Rogers' Terminal Decomposition

We state most of the results needed to prove James T. Rogers', Jr., Terminal Decomposition Theorem for metric homogeneous continua. The proofs of these results may be found in [92, Section 5.3].

We begin with a discussion of Poincaré model of the hyperbolic plane $\mathbb{H}$ ([2, 41] and [117]).

Let $\mathbb{H}$ be the interior of the closed unit disk D in $\mathbb{R}^2$, and let $\mathcal{S}^1$ be its boundary.

3.5.1 Definition. A *geodesic* in $\mathbb{H}$ is the intersection of $\mathbb{H}$ and a circle C in $\mathbb{R}^2$ that intersects $\mathcal{S}^1$ orthogonally (straight lines through the origin are considered circles centred at ∞).

3.5.2 Definition. A *reflection* in a geodesic $C \cap \mathbb{H}$ is a Euclidean inversion in the circle. A *hyperbolic isometry* of $\mathbb{H}$ is a composition of reflections in geodesics.

3.5.3 Definition. The set $\mathbb{H}$ with the family of isometries is the *Poincaré model of the hyperbolic plane*. The boundary $\mathcal{S}^1$ of $\mathbb{H}$, which is not in $\mathbb{H}$, is called the *circle at* ∞.

Let $\mathbb{F}$ be a double torus. The universal covering space of $\mathbb{F}$ may be chosen to be $\mathbb{H}$ ([2] and [117]) and the group of covering homeomorphisms to be a subgroup of the orientation-preserving isometries of $\mathbb{H}$. Each covering homeomorphism (except the identity map $1_{\mathbb{H}}$) is a hyperbolic isometry. The pertinent property for us is that a hyperbolic isometry, when extended to the circle at ∞, has exactly two fixed points in $\mathcal{S}^1$ and none in $\mathbb{H}$ ([2, Theorem 9-3, p. 132] and [117, p. 410]). The universal covering map $\sigma\colon \mathbb{H} \twoheadrightarrow \mathbb{F}$ may be chosen to be a local isometry.

3.5.4 Definition. A *geodesic in* $\mathbb{F}$ is the image under σ of a geodesic in $\mathbb{H}$. A geodesic is *simple* if it has no transverse intersection.

3.5.5 Definition. A *simple closed geodesic* is a geodesic in $\mathbb{F}$ that is a simple closed curve. A simple closed curve in $\mathbb{F}$ is *essential* if it does not bound a disk.

Assume that the universal covering space $(\mathbb{H}, \sigma)$ of $\mathbb{F}$ is constructed with the following properties. The geodesic $\mathbb{H} \cap x$-axis maps to a simple closed geodesic C_1 under σ. The geodesic $\mathbb{H} \cap y$-axis maps to a simple closed geodesic C_2 under σ. Furthermore, $C_1 \cup C_2$ is a figure-eight W, and $C_1 \cap C_2 = \{v\}$.

3.5.6 Theorem. *Let $\sigma\colon \mathbb{H} \twoheadrightarrow \mathbb{F}$ be the universal covering map of $\mathbb{F}$. Let M be a subspace of $\mathbb{F}$, and let $\widetilde{M} = \sigma^{-1}(M)$. Suppose $\mathbb{L}$ is a component of $\widetilde{M}$. If $\mathbb{L}$ is compact, then $\sigma|_{\mathbb{L}}$ is one-to-one.*

3.5.7 Theorem. *Let $\sigma\colon \mathbb{H} \twoheadrightarrow \mathbb{F}$ be the universal covering map of $\mathbb{F}$. Let M be a subspace of $\mathbb{F}$, and let $\widetilde{M} = \sigma^{-1}(M)$. Suppose $\mathbb{L}$ is a component of $\widetilde{M}$. If M is arcwise connected, then $\sigma(\mathbb{L}) = M$.*

3.5.8 Definition. Let $\mathcal{Q}$ be the Hilbert cube, and let $\sigma \times 1_{\mathcal{Q}}\colon \mathbb{H} \times \mathcal{Q} \twoheadrightarrow \mathbb{F} \times \mathcal{Q}$ be the universal covering map of $\mathbb{F} \times \mathcal{Q}$. Let X be a metric continuum essentially embedded in $W \times \mathcal{Q}$ (this means that the embedding is not homotopic to a constant map; note that this eliminates some continua from consideration). Let $f\colon X \twoheadrightarrow W$ be the projection map. Let $\widetilde{X} = (\sigma \times 1_{\mathcal{Q}})^{-1}(X)$, and let $\tilde{f}\colon \widetilde{X} \twoheadrightarrow \widetilde{W}$ be the projection map. If $\mathbb{K}$ is a component of $\widetilde{X}$, then the set $\mathbb{E}(\mathbb{K}) = \{z \in \mathcal{S}^1 \mid z$ is a (Euclidean) limit point of $\tilde{f}(\mathbb{K})\}$ is called the *set of ends of* $\mathbb{K}$.

3.5.9 Notation. Let $\widetilde{W} = \sigma^{-1}(W)$. Then $\widetilde{W}$ is the universal covering space of W, the "infinite snowflake" in Figure 3.1. The set $\mathrm{Cl}(\widetilde{W}) \cap \mathcal{S}^1$ is a Cantor set; call it $\mathfrak{Z}$ [113, p. 341].

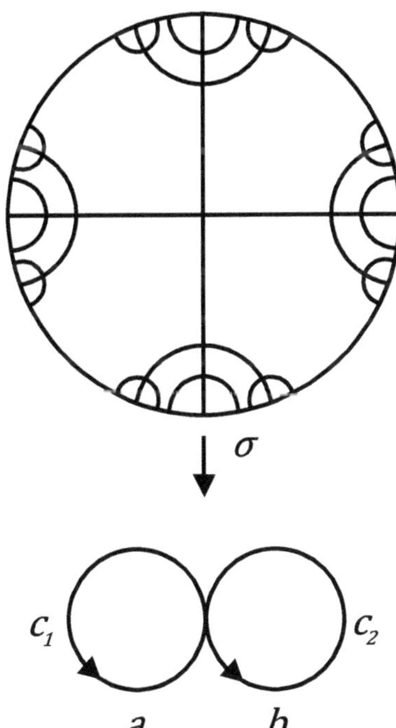

Fig. 3.1 Infinite snowflake

3.5.10 Lemma. *Let X be a metric continuum. If $\ell\colon X \to W$ is a map that is not homotopic to a constant map, then X can be essentially embedded in $W \times \mathcal{Q}$.*

3.5.11 Theorem. *Let $\sigma \times 1_{\mathcal{Q}}\colon \mathbb{H} \times \mathcal{Q} \twoheadrightarrow \mathbb{F} \times \mathcal{Q}$ be the universal covering map of $\mathbb{F} \times \mathcal{Q}$. Let X be a metric continuum essentially embedded in $W \times \mathcal{Q}$. If $\widetilde{X} = (\sigma \times 1_{\mathcal{Q}})^{-1}(X)$, then no component of $\widetilde{X}$ is compact.*

3.5.12 Theorem. *Let $\sigma \times 1_{\mathcal{Q}}\colon \mathbb{H} \times \mathcal{Q} \twoheadrightarrow \mathbb{F} \times \mathcal{Q}$ be the universal covering map of $\mathbb{F} \times \mathcal{Q}$. Let X be a homogeneous metric continuum essentially embedded in $W \times \mathcal{Q}$, and let $\widetilde{X} = (\sigma \times 1_{\mathcal{Q}})^{-1}(X)$. Then there exists $\delta > 0$ such that if $\tilde{x} \in \mathbb{K}$, $\tilde{x}' \in \mathbb{K}'$ and $\tilde{d}(\tilde{x}, \tilde{x}') < \delta$ (where $\mathbb{K}$ and $\mathbb{K}'$ are components of $\widetilde{X}$), then $\mathbb{E}(\mathbb{K}) = \mathbb{E}(\mathbb{K}')$.*

Compactify $\mathbb{H} \times \mathcal{Q}$ with $\mathcal{S}^1 \times \mathcal{Q}$. Shrink each set of the form $\{z\} \times \mathcal{Q}$, where $z \in \mathcal{S}^1$, to a point to obtain another compactification of $\mathbb{H} \times \mathcal{Q}$. This time the remainder is $\mathcal{S}^1$. Let $\pi\colon (\mathbb{H} \times \mathcal{Q}) \cup \mathcal{S}^1 \twoheadrightarrow \mathbb{H} \cup \mathcal{S}^1$ be the map of this latter compactification onto the disk D obtained by naturally extending the projection map. Note that the restriction of π to $\widetilde{X}$ is just $\tilde{f}$. (If $\tilde{x} \in \widetilde{X}$, then $\sigma \circ \tilde{f}(\tilde{x}) = f \circ (\sigma \times 1_{\mathcal{Q}})(\tilde{x}) \in W$. Hence, $\tilde{f}(\tilde{x}) \in \widetilde{W}$.)

3.5.13 Theorem. *Let $\sigma \times 1_{\mathcal{Q}}\colon \mathbb{H} \times \mathcal{Q} \twoheadrightarrow \mathbb{F} \times \mathcal{Q}$ be the universal covering map of $\mathbb{F} \times \mathcal{Q}$. Let X be a homogeneous metric continuum essentially embedded in $W \times \mathcal{Q}$. If $\widetilde{X} = (\sigma \times 1_{\mathcal{Q}})^{-1}(X)$ and $\mathbb{K}$ is a component of $\widetilde{X}$, then $\mathbb{E}(\mathbb{K})$ is either a two-point set or a Cantor set. Furthermore, $\mathbb{E}(\mathbb{K})$ contains a dense subset, each point of which is a fixed point of a hyperbolic isometry φ such that $\varphi \times 1_{\mathcal{Q}}$ is a covering homeomorphism.*

3.5.14 Theorem. *Let $\sigma \times 1_{\mathcal{Q}}\colon \mathbb{H} \times \mathcal{Q} \twoheadrightarrow \mathbb{F} \times \mathcal{Q}$ be the universal covering map of $\mathbb{F} \times \mathcal{Q}$. Let X be a nonaposyndetic homogeneous metric continuum essentially embedded in $W \times \mathcal{Q}$. Let $\widetilde{X} = (\sigma \times 1_{\mathcal{Q}})^{-1}(X)$, and let $\mathbb{K}$ be a component of $\widetilde{X}$. Let $\mathbb{Y} = \mathrm{Cl}_{(\mathbb{H} \times \mathcal{Q}) \cup \mathcal{S}^1}(\mathbb{K})$. Then $\mathbb{Y} \setminus \mathbb{K} = \mathbb{E}(\mathbb{K})$, and $\mathbb{Y}$ is connected im kleinen at each point of $\mathbb{E}(\mathbb{K})$.*

3.5.15 Corollary. *Let $\sigma \times 1_{\mathcal{Q}}\colon \mathbb{H} \times \mathcal{Q} \twoheadrightarrow \mathbb{F} \times \mathcal{Q}$ be the universal covering map of $\mathbb{F} \times \mathcal{Q}$. Let X be a nonaposyndetic homogeneous metric continuum essentially embedded in $W \times \mathcal{Q}$. Let $\widetilde{X} = (\sigma \times 1_{\mathcal{Q}})^{-1}(X)$, and let $\mathbb{K}$ be a component of $\widetilde{X}$. Let $\mathbb{Y} = \mathrm{Cl}_{(\mathbb{H} \times \mathcal{Q}) \cup \mathcal{S}^1}(\mathbb{K})$. Then $\mathbb{Y}$ is aposyndetic at each point of $\mathbb{E}(\mathbb{K})$.*

3.5.16 Theorem. *Let $\sigma \times 1_{\mathcal{Q}}\colon \mathbb{H} \times \mathcal{Q} \twoheadrightarrow \mathbb{F} \times \mathcal{Q}$ be the universal covering map of $\mathbb{F} \times \mathcal{Q}$. Let X be a nonaposyndetic homogeneous metric continuum essentially embedded in $W \times \mathcal{Q}$. Let $\widetilde{X} = (\sigma \times 1_{\mathcal{Q}})^{-1}(X)$, and let $\mathbb{K}$ be a component of $\widetilde{X}$. Let $\mathbb{Y} = \mathrm{Cl}_{(\mathbb{H} \times \mathcal{Q}) \cup \mathcal{S}^1}(\mathbb{K})$. Then the collection $\mathcal{G} = \{\mathcal{T}_{\mathbb{Y}}(\{\tilde{x}\}) \mid \tilde{x} \in \mathbb{Y}\}$ is a continuous, terminal decomposition of $\mathbb{Y}$ with the following properties:*

(1) *The quotient space $\mathbb{Y}/\mathcal{G}$ is aposyndetic.*
(2) *$\mathcal{T}_{\mathbb{Y}}(\{\tilde{x}\})$ is degenerate if $\tilde{x} \in \mathbb{Y} \setminus \mathbb{K}$.*
(3) *The nondegenerate elements of $\mathcal{G}$ are mutually homeomorphic, cell-like, indecomposable, homogeneous metric continua of the same dimension as $\mathbb{Y}$.*
(4) *$\mathbb{K}/\mathcal{G}'$ is homogeneous, where $\mathcal{G}' = \{\mathcal{T}_{\mathbb{Y}}(\{\tilde{x}\}) \mid \tilde{x} \in \mathbb{K}\}$.*

3.5.17 Theorem. *Let $\sigma \times 1_{\mathcal{Q}} \colon \mathbb{H} \times \mathcal{Q} \twoheadrightarrow \mathbb{F} \times \mathcal{Q}$ be the universal covering map of $\mathbb{F} \times \mathcal{Q}$. Let X be a nonaposyndetic homogeneous metric continuum essentially embedded in $W \times \mathcal{Q}$. Let $\widetilde{X} = (\sigma \times 1_{\mathcal{Q}})^{-1}(X)$, and let $\mathbb{K}$ be a component of $\widetilde{X}$. Let $\mathbb{Y} = \mathrm{Cl}_{(\mathbb{H}\times\mathcal{Q})\cup\mathcal{S}^1}(\mathbb{K})$. If $\tilde{x} \in \mathbb{K}$, then $(\sigma \times 1_{\mathcal{Q}})|_{\mathcal{T}_{\mathbb{Y}}(\{\tilde{x}\})}$ is one-to-one.*

3.5.18 Theorem. *Let $\sigma \times 1_{\mathcal{Q}} \colon \mathbb{H} \times \mathcal{Q} \twoheadrightarrow \mathbb{F} \times \mathcal{Q}$ be the universal covering map of $\mathbb{F} \times \mathcal{Q}$. Let X be a nonaposyndetic homogeneous metric continuum essentially embedded in $W \times \mathcal{Q}$. Let $\widetilde{X} = (\sigma \times 1_{\mathcal{Q}})^{-1}(X)$, and let $\mathbb{K}_1$ and $\mathbb{K}_2$ be components of $\widetilde{X}$. Let $\mathbb{Y}_j = \mathrm{Cl}_{(\mathbb{H}\times\mathcal{Q})\cup\mathcal{S}^1}(\mathbb{K}_j)$, $j \in \{1,2\}$. If $\tilde{x}_1 \in \mathbb{K}_1$ and $\tilde{x}_2 \in \mathbb{K}_2$ are such that $(\sigma\times 1_{\mathcal{Q}})(\tilde{x}_1) = (\sigma\times 1_{\mathcal{Q}})(\tilde{x}_2)$, then $\mathcal{T}_{\mathbb{Y}_1}(\{\tilde{x}_1\})$ is homeomorphic to $\mathcal{T}_{\mathbb{Y}_2}(\{\tilde{x}_2\})$, and $(\sigma \times 1_{\mathcal{Q}})(\mathcal{T}_{\mathbb{Y}_1}(\{\tilde{x}_1\})) = (\sigma \times 1_{\mathcal{Q}})(\mathcal{T}_{\mathbb{Y}_2}(\{\tilde{x}_2\}))$.*

3.5.19 Lemma. *Let $\sigma \times 1_{\mathcal{Q}} \colon \mathbb{H} \times \mathcal{Q} \twoheadrightarrow \mathbb{F} \times \mathcal{Q}$ be the universal covering map of $\mathbb{F} \times \mathcal{Q}$. Let X be a nonaposyndetic homogeneous metric continuum essentially embedded in $W \times \mathcal{Q}$. Let $\widetilde{X} = (\sigma \times 1_{\mathcal{Q}})^{-1}(X)$. If $\widetilde{Z}$ is a subcontinuum of $\widetilde{X}$ such that $(\sigma \times 1_{\mathcal{Q}})|_{\widetilde{Z}}$ is one-to-one, then there exists an open subset $\widetilde{U}$ of $\mathbb{H} \times \mathcal{Q}$ such that $\widetilde{Z} \subset \widetilde{U}$ and $(\sigma \times 1_{\mathcal{Q}})|_{\widetilde{U}}$ is one-to-one.*

3.5.20 Theorem. *Let $\sigma \times 1_{\mathcal{Q}} \colon \mathbb{H} \times \mathcal{Q} \twoheadrightarrow \mathbb{F} \times \mathcal{Q}$ be the universal covering map of $\mathbb{F} \times \mathcal{Q}$. Let X be a nonaposyndetic homogeneous metric continuum essentially embedded in $W \times \mathcal{Q}$. Let $\widetilde{X} = (\sigma \times 1_{\mathcal{Q}})^{-1}(X)$, and let $\mathbb{K}$ be a component of $\widetilde{X}$. Let $\mathbb{Y} = \mathrm{Cl}_{(\mathbb{H}\times\mathcal{Q})\cup\mathcal{S}^1}(\mathbb{K})$. If $\tilde{x} \in \mathbb{K}$, then $(\sigma\times 1_{\mathcal{Q}})(\mathcal{T}_{\mathbb{Y}}(\{\tilde{x}\}))$ is a maximal terminal proper, cell-like subcontinuum of X.*

We are ready to state and prove James T. Rogers', Jr., Terminal Decomposition Theorem.

3.5.21 Theorem. *Let $\sigma \times 1_{\mathcal{Q}} \colon \mathbb{H} \times \mathcal{Q} \twoheadrightarrow \mathbb{F} \times \mathcal{Q}$ be the universal covering map of $\mathbb{F} \times \mathcal{Q}$. Let X be a homogeneous metric continuum that admits a map into the figure-eight W that is not homotopic to a constant map. Hence, we may consider X essentially embedded in $\mathbb{F} \times \mathcal{Q}$. Let $\widetilde{X} = (\sigma \times 1_{\mathcal{Q}})^{-1}(X)$. If $\mathcal{G} = \{(\sigma \times 1_{\mathcal{Q}})(\mathcal{T}_{\mathbb{Y}}(\{\tilde{x}\})) \mid \tilde{x} \in \mathbb{K}$, where $\mathbb{K}$ is a component of $\widetilde{X}$, and $\mathbb{Y} = \mathrm{Cl}_{(\mathbb{H}\times\mathcal{Q})\cup\mathcal{S}^1}(\mathbb{K})\}$, then $\mathcal{G}$ is a continuous decomposition such that the following hold:*

(1) *$\mathcal{G}$ is a monotone and terminal decomposition of X.*
(2) *The elements of $\mathcal{G}$ are mutually homeomorphic, indecomposable, cell-like terminal, homogeneous metric continua.*
(3) *The quotient space, $X/\mathcal{G}$, is a homogeneous metric continuum.*
(4) *$X/\mathcal{G}$ does not contain any proper, nondegenerate terminal subcontinuum.*
(5) *If X is decomposable, then $X/\mathcal{G}$ is an aposyndetic metric continuum; in fact, the decomposition $\mathcal{G}$ is F. Burton Jones' decomposition (Theorem 3.3.8).*
(6) *If the elements of $\mathcal{G}$ are nondegenerate, then they have the same dimension as X and $X/\mathcal{G}$ is one-dimensional.*

Proof. Observe that, by [92, Theorem 4.2.31], X has the property of Effros. Also, note that, by Theorem 3.5.20, $\mathcal{G}$ is a collection of maximal proper terminal cell-like subcontinua of X.

We show that $\mathcal{G}$ is a decomposition of X. Let $\tilde{x}_1$ and $\tilde{x}_2$ be two elements of $\widetilde{X}$ such that $(\sigma \times 1_{\mathcal{Q}})(\mathcal{T}_{\mathbb{Y}_1}(\{\tilde{x}_1\})) \cap (\sigma \times 1_{\mathcal{Q}})(\mathcal{T}_{\mathbb{Y}_2}(\{\tilde{x}_2\})) \neq \emptyset$ (here, $\mathbb{K}_j$ is the component of $\widetilde{X}$ such that $\tilde{x}_j \in \mathbb{K}_j$ and $\mathbb{Y}_j = \mathrm{Cl}_{(\mathbb{H}\times\mathcal{Q})\cup\mathcal{S}^1}(\mathbb{K}_j)$, $j \in \{1,2\}$). Let $z \in (\sigma\times 1_{\mathcal{Q}})(\mathcal{T}_{\mathbb{Y}_1}(\{\tilde{x}_1\}))\cap(\sigma\times 1_{\mathcal{Q}})(\mathcal{T}_{\mathbb{Y}_2}(\{\tilde{x}_2\}))$. Then there exists $\tilde{z}_j \in \mathcal{T}_{\mathbb{Y}_j}(\{\tilde{x}_j\})$ such that $(\sigma \times 1_{\mathcal{Q}})(\tilde{z}_j) = z$, $j \in \{1,2\}$. Hence, by Theorem 3.5.18, $\mathcal{T}_{\mathbb{Y}_1}(\{\tilde{x}_1\})$ is homeomorphic to $\mathcal{T}_{\mathbb{Y}_2}(\{\tilde{x}_2\})$ and $(\sigma \times 1_{\mathcal{Q}})(\mathcal{T}_{\mathbb{Y}_1}(\{\tilde{x}_1\})) = (\sigma \times 1_{\mathcal{Q}})(\mathcal{T}_{\mathbb{Y}_2}(\{\tilde{x}_2\}))$. Since $\sigma \times 1_{\mathcal{Q}}$ is a covering map,

$$X = \bigcup\{(\sigma \times 1_{\mathcal{Q}})(\mathcal{T}_{\mathbb{Y}}(\{\tilde{x}\})) \mid \tilde{x} \in \mathbb{K}, \text{ where } \mathbb{K} \text{ is a component of } \widetilde{X}, \text{ and } \mathbb{Y} = \mathrm{Cl}_{(\mathbb{H}\times\mathcal{Q})\cup\mathcal{S}^1}(\mathbb{K})\}.$$

Therefore, $\mathcal{G}$ is a decomposition of X.

Now, we prove that the homeomorphism group of X, $\mathcal{H}(X)$, respects $\mathcal{G}$ and then apply Theorem 3.3.6 to conclude that $\mathcal{G}$ is continuous.

Let $h \in \mathcal{H}(X)$, and let $\tilde{x}_1$ and $\tilde{x}_2$ be two points of $\widetilde{X}$ such that

$$h((\sigma \times 1_{\mathcal{Q}})(\mathcal{T}_{\mathbb{Y}_1}(\{\tilde{x}_1\}))) \cap (\sigma \times 1_{\mathcal{Q}})(\mathcal{T}_{\mathbb{Y}_2}(\{\tilde{x}_2\})) \neq \emptyset.$$

Since $(\sigma \times 1_{\mathcal{Q}})(\mathcal{T}_{\mathbb{Y}_1}(\{\tilde{x}_1\}))$ is terminal, $h((\sigma \times 1_{\mathcal{Q}})(\mathcal{T}_{\mathbb{Y}_1}(\{\tilde{x}_1\})))$ is terminal (Corollary 1.4.54). Thus, since $(\sigma \times 1_{\mathcal{Q}})(\mathcal{T}_{\mathbb{Y}_2}(\{\tilde{x}_2\}))$ is a maximal terminal subcontinuum,

$$h((\sigma \times 1_{\mathcal{Q}})(\mathcal{T}_{\mathbb{Y}_1}(\{\tilde{x}_1\}))) \subset (\sigma \times 1_{\mathcal{Q}})(\mathcal{T}_{\mathbb{Y}_2}(\{\tilde{x}_2\})).$$

This implies that $(\sigma \times 1_{\mathcal{Q}})(\mathcal{T}_{\mathbb{Y}_1}(\{\tilde{x}_1\})) \subset h^{-1}((\sigma \times 1_{\mathcal{Q}})(\mathcal{T}_{\mathbb{Y}_2}(\{\tilde{x}_2\})))$. Hence, $(\sigma \times 1_{\mathcal{Q}})(\mathcal{T}_{\mathbb{Y}_1}(\{\tilde{x}_1\})) = h^{-1}((\sigma \times 1_{\mathcal{Q}})(\mathcal{T}_{\mathbb{Y}_2}(\{\tilde{x}_2\})))$. Consequently, $h((\sigma \times 1_{\mathcal{Q}})(\mathcal{T}_{\mathbb{Y}_1}(\{\tilde{x}_1\}))) = (\sigma \times 1_{\mathcal{Q}})(\mathcal{T}_{\mathbb{Y}_2}(\{\tilde{x}_2\}))$. Therefore, $\mathcal{H}(X)$ respects $\mathcal{G}$.

Since $\mathcal{H}(X)$ respects $\mathcal{G}$, by Theorem 3.3.6, the elements of $\mathcal{G}$ are mutually homeomorphic homogeneous continua, and the quotient space $X/\mathcal{G}$ is homogeneous. The fact that the nondegenerate elements of $\mathcal{G}$ have the same dimension as X follows from [112, Theorem 8]. The proof of the indecomposability of the elements of $\mathcal{G}$ is similar to the one given in Theorem 3.2.7. The fact that $X/\mathcal{G}$ is one-dimensional may be found in [115, Theorem 6].

Next, we show that $X/\mathcal{G}$ does not contain nondegenerate proper terminal subcontinua. To this end, let Γ be a nondegenerate terminal proper subcontinuum of $X/\mathcal{G}$. Let $q\colon X \twoheadrightarrow X/\mathcal{G}$ be the quotient map. Then q is a monotone map. Thus, $q^{-1}(\Gamma)$ is a subcontinuum of X (Lemma 1.4.46). Note that

$$q^{-1}(\Gamma) = \bigcup\{(\sigma \times 1_{\mathcal{Q}})(\mathcal{T}_{\mathbb{Y}}(\{\tilde{x}\})), \mid \tilde{x} \in \mathbb{K}, \text{ where } \mathbb{K} \text{ is a component of } \widetilde{X},\ \mathbb{Y} = \mathrm{Cl}_{(\mathbb{H}\times\mathcal{Q})\cup\mathcal{S}^1}(\mathbb{K}) \text{ and } q((\sigma \times 1_{\mathcal{Q}})(\mathcal{T}_{\mathbb{Y}}(\{\tilde{x}\}))) \in \Gamma\}.$$

Since Γ is a nondegenerate proper subcontinuum of $X/\mathcal{G}$, $q^{-1}(\Gamma)$ is a proper subcontinuum of X, and if $q((\sigma \times 1_{\mathcal{Q}})(\mathcal{T}_{\mathbb{Y}}(\{\tilde{x}\}))) \in \Gamma$, $(\sigma \times 1_{\mathcal{Q}})(\mathcal{T}_{\mathbb{Y}}(\{\tilde{x}\})) \subsetneq q^{-1}(\Gamma)$. Hence, $q^{-1}(\Gamma)$ is not a terminal subcontinuum of X (each element $(\sigma \times 1_{\mathcal{Q}})(\mathcal{T}_{\mathbb{Y}}(\{\tilde{x}\}))$ of $\mathcal{G}$ is a maximal terminal proper subcontinuum of X). Thus, there exists a subcontinuum Z of X such that $Z \cap q^{-1}(\Gamma) \neq \emptyset$, $Z \setminus q^{-1}(\Gamma) \neq \emptyset$ and $q^{-1}(\Gamma) \setminus Z \neq \emptyset$. Since $q(Z \cap q^{-1}(\Gamma)) = q(Z) \cap \Gamma$, $q(Z)$ is a subcontinuum of $X/\mathcal{G}$ such that $q(Z) \cap \Gamma \neq \emptyset$. Since Γ is a terminal subcontinuum of $X/\mathcal{G}$, either $q(Z) \subset \Gamma$ or $\Gamma \subset q(Z)$.

If $q(Z) \subset \Gamma$, then $q^{-1}(q(Z)) \subset q^{-1}(\Gamma)$. Hence, $Z \subset q^{-1}(\Gamma)$, a contradiction. Suppose, then, that $\Gamma \subset q(Z)$. This implies that $q^{-1}(\Gamma) \subset q^{-1}q(Z)$. Thus, for each $x \in q^{-1}(\Gamma)$, there exists $z \in Z$ such that $q(z) = q(x)$. Hence, $q^{-1}q(x) \cap Z \neq \emptyset$. Since $q^{-1}q(x)$ is a terminal subcontinuum of X, either $Z \subset q^{-1}q(x)$ or $q^{-1}q(x) \subset Z$. If $Z \subset q^{-1}q(x)$, then $Z \subset q^{-1}(\Gamma)$, a contradiction. Thus, $q^{-1}q(x) \subset Z$. Since x is an arbitrary point of $q^{-1}(\Gamma)$, $q^{-1}(\Gamma) \subset Z$, a contradiction. Therefore, $X/\mathcal{G}$ does not contain nondegenerate terminal proper subcontinua.

Next, suppose X is decomposable. Let $\mathcal{G}' = \{\mathcal{T}_X(\{x\}) \mid x \in X\}$ be Jones' decomposition (Theorem 3.3.8). We show that $\mathcal{G} = \mathcal{G}'$. Thus, by Theorem 3.3.8, $X/\mathcal{G}$ is aposyndetic.

Let $(\sigma \times 1_{\mathcal{Q}})(\mathcal{T}_{\mathbb{Y}}(\{\tilde{x}\})) \in \mathcal{G}$. Let $x \in (\sigma \times 1_{\mathcal{Q}})(\mathcal{T}_{\mathbb{Y}}(\{\tilde{x}\}))$.Then $(\sigma \times 1_{\mathcal{Q}})(\mathcal{T}_{\mathbb{Y}}(\{\tilde{x}\})) \cap \mathcal{T}_X(\{x\}) \neq \emptyset$. Since both of these continua are maximal terminal subcontinua of X (Corollary 3.3.3), $(\sigma \times 1_{\mathcal{Q}})(\mathcal{T}_{\mathbb{Y}}(\{\tilde{x}\})) = \mathcal{T}_X(\{x\})$. Thus, $(\sigma \times 1_{\mathcal{Q}})(\mathcal{T}_{\mathbb{Y}}(\{\tilde{x}\})) \in \mathcal{G}'$. Hence, $\mathcal{G} \subset \mathcal{G}'$.

Let $\mathcal{T}_X(\{x\}) \in \mathcal{G}'$. Since $\sigma \times 1_{\mathcal{Q}}$ is a covering map, there exists $\tilde{x} \in \widetilde{X}$ such that $(\sigma \times 1_{\mathcal{Q}})(\tilde{x}) = x$. Let $\mathbb{K}$ be the component of $\widetilde{X}$ such that $\tilde{x} \in \mathbb{K}$. Then $(\sigma \times 1_{\mathcal{Q}})(\mathcal{T}_{\mathbb{Y}}(\{\tilde{x}\})) \cap \mathcal{T}_X(\{x\}) \neq \emptyset$. Since both of these continua are maximal terminal subcontinua of X, $(\sigma \times 1_{\mathcal{Q}})(\mathcal{T}_{\mathbb{Y}}(\{\tilde{x}\})) = \mathcal{T}_X(\{x\})$. Thus, $\mathcal{T}_X(\{x\}) \in \mathcal{G}$. Hence, $\mathcal{G}' \subset \mathcal{G}$.

Therefore, $\mathcal{G} = \mathcal{G}'$. □

3.6 Other Decomposition Theorems

We start with the next technical lemma.

3.6.1 Lemma. *Let X be a continuum of type λ. Let $\mathcal{G}$ be the finest monotone upper semicontinuous decomposition of X such that each element of $\mathcal{G}$ is nowhere dense and $X/\mathcal{G}$ is an arc, and let $q\colon X \twoheadrightarrow [0,1]$ be the quotient map. If $x \in X$ and $A \subset q^{-1}(q(x))$, then $\mathcal{T}(A) \subset q^{-1}(q(x))$.*

Proof. Let $y \in X \setminus q^{-1}q(x)$. Then $q(y) \neq q(x)$. Hence, there exists a closed subinterval $[r,t]$ of $[0,1]$ such that $q(y) \in \mathrm{Int}_{[0,1]}([r,t])$ and $q(x) \in [0,1] \setminus [r,t]$. This implies that $q^{-1}q(x) \cap q^{-1}([r,t]) = \emptyset$. Since q is monotone, Lemma 1.4.46, $q^{-1}([r,t])$ is a subcontinuum of X. By construction, $y \in \mathrm{Int}_X(q^{-1}([r,t])$. Therefore, $y \in X \setminus \mathcal{T}(A)$, and $\mathcal{T}(A) \subset q^{-1}q(x)$. □

Our first decomposition theorem is as follows.

3.6.2 Theorem. *If X is a continuum of type λ, then $\{\mathcal{T}^2(\{x\}) \mid x \in X\}$ is the finest monotone upper semicontinuous decomposition $\mathcal{G}$ of X such that each element of $\mathcal{G}$ is nowhere dense and $X/\mathcal{G}$ is an arc.*

Proof. Let $\mathcal{G}$ be the finest monotone upper semicontinuous decomposition of X such that each element of $\mathcal{G}$ is nowhere dense and $X/\mathcal{G}$ is an arc, and let $q\colon X \twoheadrightarrow [0,1]$ be the quotient map. Observe that $\mathcal{G} = \{q^{-1}q(x) \mid x \in X\}$.

Let $x \in X$. Note that $\mathcal{T}(\{x\}) \subset q^{-1}q(x)$, Lemma 3.6.1. By [119, Theorem 18, p. 26], there exists $z \in q^{-1}q(x)$ such that $\mathcal{T}(\{z\}) = q^{-1}q(x)$. Since X is $\mathcal{T}$-symmetric, Corollary 2.2.3, and $x \in \mathcal{T}(\{z\})$, we have that $z \in \mathcal{T}(\{x\})$. Hence, by Lemma 3.6.1, $q^{-1}q(x) = \mathcal{T}(\{z\}) \subset \mathcal{T}^2(\{x\}) \subset q^{-1}q(x)$. Thus, $\mathcal{T}^2(\{x\}) = q^{-1}q(x)$. Since x is an arbitrary point of X, $\mathcal{G} = \{\mathcal{T}^2(\{x\}) \mid x \in X\}$. □

3.6.3 Notation. Let Z be a set, and let $\mathcal{G}$ be a decomposition. For each point z of Z, let $G(z)$ be the unique element of $\mathcal{G}$ that contains z. If $\mathcal{H}$ is another decomposition of Z, we say that $\mathcal{G}$ refines $\mathcal{H}$, written $\mathcal{G} < \mathcal{H}$, if and only if $G(z) \subset H(z)$ for all $z \in Z$. If $\{\mathcal{G}_\gamma\}_{\gamma\in\Gamma}$ is a collection of decompositions of Z, let the common intersection element be $G(z) = \bigcap_{\gamma\in\Gamma}\{G_\gamma(z) \mid G_\gamma(z) \in \mathcal{G}_\gamma\}$. Note that $\mathcal{G}_\Gamma = \{G(z) \mid z \in Z\}$ is a decomposition of Z. Also, $\mathcal{G}_\Gamma$ is precisely the greatest lower bound of $\{\mathcal{G}_\gamma\}_{\gamma\in\Gamma}$ under the partial order $<$ on the family of all decompositions of Z.

3.6.4 Definition. Let Z be a set, and let $\mathfrak{G} = \{\mathcal{G}_\gamma\}_{\gamma\in\Gamma}$ be the family of all decompositions of Z that have a certain property of P. We say that $\mathcal{G}$ is the *core decomposition of Z with respect to P* if and only if $\mathcal{G} = \mathcal{G}_\Gamma$ and $\mathcal{G} \in \mathfrak{G}$.

Our second decomposition theorem is a special case of [46, Theorem 2.5].

3.6.5 Theorem. *If X is a continuum, then there exists a core decomposition $\mathcal{G}$ of X with respect to the property: "$\mathcal{G}$ is upper semicontinuous with $\mathcal{T}$-closed elements" (Definition 4.1.1).*

3.6.6 Theorem. *If X is a continuum, then there exists a unique decomposition $\mathcal{G}$ of X such that*

(1) *$\mathcal{G}$ is the core decomposition of X with respect to being upper semicontinuous with $\mathcal{T}_X$-closed elements (Definition 4.1.1);*
(2) *$\mathcal{G}$ is monotone and $X/\mathcal{G}$ is aposyndetic;*
(3) *$\mathcal{G}$ is the core decomposition of X with respect to being monotone with $X/\mathcal{G}$ aposyndetic.*

Proof. Let $\mathcal{G}$ be the core decomposition of X with respect to being upper semicontinuous with $\mathcal{T}_X$-closed elements, given by Theorem 3.6.5. Hence, $\mathcal{G}$ satisfies (1). Let $q\colon X \twoheadrightarrow X/\mathcal{G}$ be the quotient map, and let $\chi \in X/\mathcal{G}$. Since q is monotone, by Theorem 2.1.51 (c), $\mathcal{T}_{X/\mathcal{G}}(\{\chi\}) = q\mathcal{T}_X q^{-1}(\{\chi\}) = qq^{-1}(\{\chi\}) = \{\chi\}$ (the elements of $\mathcal{G}$ are $\mathcal{T}_X$-closed sets). Thus, by Theorem 2.1.34, $X/\mathcal{G}$ is aposyndetic.

Let $\mathcal{D} = \{D \mid D$ is a component of an element of $\mathcal{G}\}$. Then, by Theorem 1.1.35, $\mathcal{D}$ is an upper semicontinuous decomposition of X, and, by Theorem 4.1.15, the elements of $\mathcal{D}$ are $\mathcal{T}_X$-closed sets. By the definitions of $\mathcal{D}$ and $\mathcal{G}$, both inequalities

$\mathcal{D} < \mathcal{G}$ and $\mathcal{G} < \mathcal{D}$ hold. Hence, $\mathcal{G} = \mathcal{D}$, and $\mathcal{G}$ is monotone. Thus, $\mathcal{G}$ satisfies (2). If $\mathcal{H}$ is any monotone decomposition of X such that $X/\mathcal{H}$ is aposyndetic, then, by Theorem 3.6.5, the elements of $\mathcal{H}$ are $\mathcal{T}_X$-closed. Hence, $\mathcal{G} < \mathcal{H}$. Thus, $\mathcal{G}$ satisfies (3). □

A special case for metric θ-continua of Theorem 3.6.6 is the following [45, Theorem 6.1]:

3.6.7 Theorem. *If X is a metric θ-continuum, then there exists a unique decomposition $\mathcal{G}$ such that $\mathcal{G}$ has the following three characterizations:*

(1) *$\mathcal{G}$ is the core decomposition of X with respect to being upper semicontinuous with $\mathcal{T}$-closed elements (Definition 4.1.1);*
(2) *$\mathcal{G}$ is the core decomposition of X with respect to being monotone, and $X/\mathcal{G}$ is aposyndetic;*
(3) *$\mathcal{G}$ is the core decomposition of X with respect to being monotone, and $X/\mathcal{G}$ is a graph.*

Proof. By Theorem 3.6.6, there exists a unique core decomposition $\mathcal{G}$ of X such that $\mathcal{G}$ satisfies (1) and (2). We show (3). By [92, Theorem 1.7.3] and [45, Theorem 3.1], $X/\mathcal{G}$ is a metric θ-continuum, and, by [45, Theorem 3.11], $X/\mathcal{G}$ is locally connected. In fact, by [45, Theorem 4.7], $X/\mathcal{G}$ is a graph. Thus, $\mathcal{G}$ is monotone, and $X/\mathcal{G}$ is a graph. Let $\mathcal{H}$ be any monotone decomposition of X such that $X/\mathcal{H}$ is a graph. Then, in particular, $X/\mathcal{H}$ is aposyndetic. Hence, by (2), $\mathcal{G} < \mathcal{H}$. Therefore, $\mathcal{G}$ satisfies (3). □

The next result is [46, Lemma 3.6].

3.6.8 Theorem. *Let X be a continuum, and let $n \in \mathbb{N}$. If $\mathcal{K} = \{\mathcal{T}^n(\{x\}) \mid x \in X\}$ is a decomposition of X and $\mathcal{T}^\infty(\{x\}) = \mathcal{T}^n(\{x\})$ (Definition 4.4.1) for all $x \in X$, then $\mathcal{K}$ is the core decomposition $\mathcal{G}$ of Theorem 3.6.6.*

3.6.9 Theorem. *Let X be a metric θ-continuum. If $n \in \mathbb{N}$ is such that $\mathcal{T}^{n+1}(\{x\}) = \mathcal{T}^n(\{x\})$ for every point x in X, then $\mathcal{G} = \{\mathcal{T}^n(\{x\}) \mid x \in X\}$ is the core decomposition of Theorem 3.6.6.*

Proof. By Theorem 2.4.8, $\mathcal{G} = \{\mathcal{T}^n(\{x\}) \mid x \in X\}$ is a decomposition of X. Hence, by Theorem 3.6.8, $\mathcal{G}$ is the core decomposition of Theorem 3.6.6. □

A proof of the following theorem may be found in [54, Theorem 2]:

3.6.10 Theorem. *Let X be a metric θ_n-continuum. Then X admits a monotone, upper semicontinuous decomposition $\mathcal{G}$ such that the elements of $\mathcal{G}$ have void interior and the quotient space $X/\mathcal{G}$ is a graph if and only if* $\mathrm{Int}(\mathcal{T}(H)) = \emptyset$ *for every subcontinuum H of X with void interior. Furthermore, $\mathcal{G} = \{\mathcal{T}^{2n}(\{x\}) \mid x \in X\}$.*

References for Chapter 3

Section 3.1: [11, 84, 86, 90, 98].
Section 3.2: [43, 53, 73, 83, 84, 86, 90, 91, 97].
Section 3.3: [35, 90–92, 110].
Section 3.4: [90, 91, 110].
Section 3.5: [2, 41, 92, 112–115, 117].
Section 3.6: [43, 45, 46, 54].

Chapter 4
$\mathcal{T}$-Closed Sets

The family of $\mathcal{T}$-closed sets have been considered by several authors, for example [46] and [122]. We introduce the family of $\mathcal{T}$-closed sets of a continuum X and present its main properties. We also give a characterization of $\mathcal{T}$-closed sets. We consider minimal $\mathcal{T}$-closed sets and the set function $\mathcal{T}^\infty$. We introduce the $\mathcal{T}$-growth bound of a continuum.

4.1 Main Properties of $\mathcal{T}$-Closed Sets

We start with a couple of definitions and give some results that provide examples of families of $\mathcal{T}$-closed sets.

4.1.1 Definition. Given a continuum X, a subset A of X is a *$\mathcal{T}$-closed set* provided that $\mathcal{T}(A) = A$. We denote the family of $\mathcal{T}$-closed sets of a continuum X by $\mathfrak{T}(X)$. Note that $X \in \mathfrak{T}(X)$.

4.1.2 Definition. A continuum X *is m-aposyndetic* provided that for each subset K of X with m points, $K \in \mathfrak{T}(X)$. We say X *is countable closed aposyndetic* if for each countable closed subset K of X, $K \in \mathfrak{T}(X)$.

4.1.3 Theorem. *If X is a continuum and A is a closed subset of X such that $\mathcal{T}(A)$ is totally disconnected, then $A \in \mathfrak{T}(X)$.*

Proof. Let A be a closed subset of X such that $\mathcal{T}(A)$ is a totally disconnected subset of X. By Remark 2.1.5, $A \subset \mathcal{T}(A)$. By Corollary 2.1.20, each component of $\mathcal{T}(A)$ intersects $\mathrm{Cl}(A) = A$. Since all the components of $\mathcal{T}(A)$ are singletons, $\mathcal{T}(A) \subset A$. □

4.1.4 Corollary. *Let X be a continuum such that $\mathcal{T}(2^X) = 2^X$. If K is a nonempty closed and totally disconnected subset of X, then $K \in \mathfrak{T}(X)$.*

S. Macías, *Set Function $\mathcal{T}$*, Developments in Mathematics 67,
https://doi.org/10.1007/978-3-030-65081-0_4

Proof. Let K be a nonempty closed and totally disconnected subset of X. Since $\mathcal{T}(2^X) = 2^X$, there exists $L \in 2^X$ such that $\mathcal{T}(L) = K$. Since K is totally disconnected, by Theorem 4.1.3, $\mathcal{T}(L) = L$. Thus, $K = \mathcal{T}(L) = L$, and $K \in \mathfrak{T}(X)$. □

The next theorem tells us that $\mathcal{T}$ of a product of continua behaves like the identity map at the product of two proper closed subsets. This result was already proved in Corollary 2.3.26, but we include another proof here.

4.1.5 Theorem. *Let X and Y be continua. If A and B are proper closed subsets of X and Y, respectively, then $A \times B \in \mathfrak{T}(X \times Y)$.*

Proof. Let A and B be two proper closed subsets of X and Y, respectively. By Remark 2.1.5, $A \times B \subset \mathcal{T}(A \times B)$.

Let $(x, y) \in (X \times Y) \setminus (A \times B)$; without loss of generality, we assume that $x \in X \setminus A$. Then there exists an open subset U of X such that $x \in U \subset \mathrm{Cl}(U) \subset X \setminus A$. Hence, $(\mathrm{Cl}(U) \times Y) \cap (A \times B) = \emptyset$.

Let $z \in Y \setminus B$. Then $(X \times \{z\}) \cap (A \times B) = \emptyset$. Thus, $(x, y) \in (\mathrm{Cl}(U) \times Y) \cup (X \times \{z\}) \subset (X \times Y) \setminus (A \times B)$. Since $(\mathrm{Cl}(U) \times Y) \cup (X \times \{z\})$ is a continuum containing (x, y) in its interior, $(x, y) \in (X \times Y) \setminus \mathcal{T}(A \times B)$. Therefore, $A \times B \in \mathfrak{T}(X \times Y)$. □

As a consequence of Theorem 2.1.34 and Theorem 4.1.5, we have the following corollary:

4.1.6 Corollary. *If X and Y are continua, then $X \times Y$ is an aposyndetic continuum.*

4.1.7 Theorem. *Let X and Y be continua. If A and B are two closed totally disconnected subsets of X and Y, respectively, then for each closed subset K of $A \times B$, $K \in \mathfrak{T}(X)$.*

Proof. Let K be a closed subset of $A \times B$. By Theorem 4.1.5, $\mathcal{T}(A \times B) = A \times B$. Since K is a closed subset of $A \times B$, by Proposition 2.1.7, $\mathcal{T}(K) \subset \mathcal{T}(A \times B) = A \times B$. Hence, $\mathcal{T}(K)$ is totally disconnected. By Theorem 4.1.3, $K \in \mathfrak{T}(X)$. □

4.1.8 Theorem. *Let X and Y be continua. If K is a countable closed subset of $X \times Y$, then $K \in \mathfrak{T}(X \times Y)$. In particular, $X \times Y$ is m-aposyndetic for each $m \in \mathbb{N}$.*

Proof. Let K be a countable closed subset of $X \times Y$. Let $\pi_X : X \times Y \twoheadrightarrow X$ and $\pi_Y : X \times Y \twoheadrightarrow Y$ be the projection maps. Since K is countable and closed, $\pi_X(K)$ and $\pi_Y(K)$ are countable closed subsets of X and Y, respectively. Then $K \subset \pi_X(K) \times \pi_Y(K)$. Thus, by Theorem 4.1.7, $K \in \mathfrak{T}(X \times Y)$. □

4.1.9 Theorem. *If X is a metric continuum and K is a countable closed subset of $K(X)$, then $K \in \mathfrak{T}(K(X))$.*

Proof. Let K be a countable closed subset of $K(X)$. We assume first that $\nu_X \notin K$. Then $q^{-1}(K)$ is a countable closed subset of $X \times [0, 1]$. Hence, by Theorem 4.1.8, $\mathcal{T}_{X \times [0,1]} q^{-1}(K) = q^{-1}(K)$. Since q is monotone, by Theorem 2.1.51 (c), we have that $\mathcal{T}_{K(X)}(K) = q\mathcal{T}_{X \times [0,1]} q^{-1}(K)$. Therefore, $\mathcal{T}_{K(X)}(K) = qq^{-1}(K) = K$.

If $\nu_X \in K$ and ν_X is an isolated point of K, then $K \setminus \{\nu_X\}$ is a closed subset of $K(X)$. Then, by the previous paragraph, $\mathcal{T}_{K(X)}(K \setminus \{\nu_X\}) = K \setminus \{\nu_X\}$. Hence, we have that $K \in \mathfrak{T}(K(X))$.

Next, we assume that $\nu_X \in K$. Then note that $q^{-1}(K) = (X \times \{1\}) \cup \{(x_n, t_n)\}_{n=1}^{\infty}$, also suppose that $\mathrm{Cl}_{X\times[0,1]}(\{(x_n, t_n)\}_{n=1}^{\infty}) \setminus \{(x_n, t_n)\}_{n=1}^{\infty} \subset X \times \{1\}$. Let $(x, t) \in (X \times [0, 1]) \setminus q^{-1}(K)$. We assume that $t \in (0, 1)$, the case $t = 0$ is similar. Since $q^{-1}(K)$ is closed and $X \times [0, 1]$ is metric, there exist an open set U of X such that $x \in U$ and an $\varepsilon > 0$ such that $0 < t - \varepsilon < t + \varepsilon < 1$ and $\mathrm{Cl}_X(U) \times [t - \varepsilon, t + \varepsilon] \subset (X \times [0, 1]) \setminus q^{-1}(K)$. Let $s \in [t - \varepsilon, t + \varepsilon] \setminus \{t_n\}_{n=1}^{\infty}$, and let $W = (X \times \{s\}) \cup (\mathrm{Cl}_X(U) \times [t - \varepsilon, t + \varepsilon])$. Then W is a subcontinuum of $X \times [0, 1]$ such that $(x, t) \in \mathrm{Int}_{X\times[0,1]}(W) \subset W \subset (X \times [0, 1]) \setminus q^{-1}(K)$. Hence, $(x, t) \in (X \times [0, 1]) \setminus \mathcal{T}_{X\times[0,1]} q^{-1}(K)$. Thus, $\mathcal{T}_{X\times[0,1]} q^{-1}(K) = q^{-1}(K)$. Since q is monotone, by Theorem 2.1.51 part (c), we have that $\mathcal{T}_{K(X)}(K) = q\mathcal{T}_{X\times[0,1]} q^{-1}(K)$. Therefore, $\mathcal{T}_{K(X)}(K) = qq^{-1}(K) = K$. □

4.1.10 Theorem. *If X is a metric continuum and K is a countable closed subset of $\Sigma(X)$, then $K \in \mathfrak{T}(\Sigma(X))$.*

Proof. Let K be a countable closed subset of $\Sigma(X)$. Let ν^+ and ν^- be the vertexes of $\Sigma(X)$. We assume first that $\{\nu^+, \nu^-\} \cap K = \emptyset$. Then $q^{-1}(K)$ is a countable closed subset of $X \times [0, 1]$. Hence, by Theorem 4.1.8, $\mathcal{T}_{X\times[0,1]} q^{-1}(K) = q^{-1}(K)$. Since q is monotone, by Theorem 2.1.51 (c), $\mathcal{T}_{\Sigma(X)}(K) = q\mathcal{T}_{X\times[0,1]} q^{-1}(K)$. Therefore, $\mathcal{T}_{\Sigma(X)}(K) = qq^{-1}(K) = K$.

Next, we assume that $\nu^+ \in K$ and $\nu^- \in \Sigma(X) \setminus K$. If ν^+ is an isolated point of K, a similar argument to the one given in Theorem 4.1.9 shows that $\mathcal{T}_{\Sigma(X)}(K) = K$. Suppose that $q^{-1}(K) = (X \times \{1\}) \cup \{(x_n, t_n)\}_{n=1}^{\infty}$, also assume that $\mathrm{Cl}_{X\times[0,1]}(\{(x_n, t_n)\}_{n=1}^{\infty}) \setminus \{(x_n, t_n)\}_{n=1}^{\infty} \subset X \times \{1\}$. Let $(x, t) \in (X \times [0, 1]) \setminus q^{-1}(K)$. Without loss of generality, we assume that $t \neq 0$. Since $q^{-1}(K)$ is closed and $X \times [0, 1]$ is metric, there exist an open set U of X such that $x \in U$ and an $\varepsilon > 0$ such that $0 < t - \varepsilon < t + \varepsilon < 1$ and $\mathrm{Cl}_X(U) \times [t - \varepsilon, t + \varepsilon] \subset (X \times [0, 1]) \setminus q^{-1}(K)$. Let $s \in [t - \varepsilon, t + \varepsilon] \setminus \{t_n\}_{n=1}^{\infty}$, and let $W = (X \times \{s\}) \cup (\mathrm{Cl}_X(U) \times [t - \varepsilon, t + \varepsilon])$. Then W is a subcontinuum of $X \times [0, 1]$ such that $(x, t) \in \mathrm{Int}_{X\times[0,1]}(W) \subset W \subset (X \times [0, 1]) \setminus q^{-1}(K)$. Hence, $(x, t) \in (X \times [0, 1]) \setminus \mathcal{T}_{X\times[0,1]} q^{-1}(K)$. Thus, $\mathcal{T}_{X\times[0,1]} q^{-1}(K) = q^{-1}(K)$. Since q is monotone, by Theorem 2.1.51 (c), we have that $\mathcal{T}_{\Sigma(X)}(K) = q\mathcal{T}_{X\times[0,1]} q^{-1}(K)$. Therefore, $\mathcal{T}_{\Sigma(X)}(K) = qq^{-1}(K) = K$.

Now, suppose that $\{\nu^+, \nu^-\} \subset K$. If either ν^+ or ν^- or both are isolated points of K, a similar argument to the one given in Theorem 4.1.9 shows that $\mathcal{T}_{\Sigma(X)}(K) = K$. Assume that $q^{-1}(K) = (X \times \{1\}) \cup (X \times \{0\}) \cup \{(x_n, t_n)\}_{n=1}^{\infty} \cup \{(x'_n, t'_n)\}_{n=1}^{\infty}$. Also suppose that $\mathrm{Cl}_{X\times[0,1]}(\{(x_n, t_n)\}_{n=1}^{\infty}) \setminus \{(x_n, t_n)\}_{n=1}^{\infty} \subset X \times \{1\}$ and $\mathrm{Cl}_{X\times[0,1]}(\{(x'_n, t'_n)\}_{n=1}^{\infty}) \setminus \{(x'_n, t'_n)\}_{n=1}^{\infty} \subset X \times \{0\}$. Let $(x, t) \in (X \times (0, 1)) \setminus q^{-1}(K)$. Since $X \times [0, 1]$ is metric, there exist an open set U of X and an $\varepsilon > 0$ such that $0 < t - \varepsilon < t + \varepsilon < 1$ and $\mathrm{Cl}_X(U) \times [t - \varepsilon, t + \varepsilon] \subset (X \times [0, 1]) \setminus q^{-1}(K)$. Let $s \in [t - \varepsilon, t + \varepsilon] \setminus (\{t_n\}_{n=1}^{\infty} \cup \{t'_n\}_{n=1}^{\infty})$ and let

$W = (X \times \{s\}) \cup (\mathrm{Cl}_X(U) \times [t - \varepsilon, t + \varepsilon])$. Then W is a subcontinuum of $X \times [0,1]$ such that $(x,t) \in \mathrm{Int}_{X\times[0,1]}(W) \subset W \subset (X \times [0,1]) \setminus q^{-1}(K)$. Hence, $(x,t) \in (X \times [0,1]) \setminus \mathcal{T}_{X\times[0,1]} q^{-1}(K)$. Thus, $\mathcal{T}_{X\times[0,1]} q^{-1}(K) = q^{-1}(K)$. Since q is monotone, by Theorem 2.1.51 (c), we have that $\mathcal{T}_{\Sigma(X)}(K) = q\mathcal{T}_{X\times[0,1]} q^{-1}(K)$. Therefore, $\mathcal{T}_{\Sigma(X)}(K) = qq^{-1}(K) = K$. □

4.1.11 Theorem. *If X is a metric continuum, then $\mathfrak{T}(X)$ is a G_δ subset of 2^X.*

Proof. Given $\varepsilon > 0$, define

$$\Theta_\varepsilon = \{A \in 2^X \mid \mathcal{T}(A) \subset \mathcal{V}_\varepsilon(A)\}.$$

Note that if $A \in \Theta_\varepsilon$, then $\mathcal{H}(A, \mathcal{T}(A)) < \varepsilon$. If $A \in \Theta_\varepsilon$, then define

$$\mathcal{M}_\varepsilon(A) = \bigcup_{0<s<1} \left(\mathcal{V}^{\mathcal{H}}_{s\varepsilon}(A) \cap \{B \in 2^X \mid \mathcal{T}(B) \subset \mathcal{V}_{(1-s)\varepsilon}(A)\}\right).$$

Note that $A \in \mathcal{M}_\varepsilon(A)$ and $\mathcal{M}_\varepsilon(A)$ is an open subset of 2^X (the second set is open in 2^X since $\mathcal{T}$ is upper semicontinuous, Theorem 5.1.1).

We show that $\mathcal{M}_\varepsilon(A) \subset \Theta_\varepsilon$. Let $B \in \mathcal{M}_\varepsilon(A)$. Then there exists $s \in (0,1)$ such that $\mathcal{H}(B, A) < s\varepsilon$ and $\mathcal{T}(B) \subset \mathcal{V}_{(1-s)\varepsilon}(A)$. We need to prove that $\mathcal{T}(B) \subset \mathcal{V}_\varepsilon(B)$. Let $x \in \mathcal{T}(B)$. Then, since $\mathcal{T}(B) \subset \mathcal{V}_{(1-s)\varepsilon}(A)$, there exists $a \in A$ such that $d(x,a) < (1-s)\varepsilon$. Hence, since $\mathcal{H}(B, A) < s\varepsilon$, there exists $b \in B$ such that $d(b,a) < s\varepsilon$. By the triangle inequality, we obtain that $d(x,b) \leq d(x,a) + d(a,b) < (1-s)\varepsilon + s\varepsilon = \varepsilon$. Thus, $x \in \mathcal{V}_\varepsilon(B)$, and $\mathcal{T}(B) \subset \mathcal{V}_\varepsilon(B)$. Hence, A is an interior point of Θ_ε. Therefore, Θ_ε is open.

Observe that

$$\mathfrak{T}(X) = \bigcap_{n=1}^{\infty} \Theta_{\frac{1}{n}}.$$

Therefore, $\mathfrak{T}(X)$ is a G_δ subset of 2^X. □

The next example shows that there exist continua X such that $\mathfrak{T}(X)$ is not connected.

4.1.12 Example. Let X be an irreducible metric continuum such that X is the union of two sequences of indecomposable continua, $\{Y_n\}_{n=1}^\infty$ and $\{Z_n\}_{n=1}^\infty$, both converging to the singleton $\{p\}$. Then $|\mathfrak{T}(X)| = 4$, namely

$$\mathfrak{T}(X) = \left\{X, \{p\}, \bigcup_{n=1}^{\infty} Y_n \cup \{p\}, \bigcup_{n=1}^{\infty} Z_n \cup \{p\}\right\}.$$

Therefore, $\mathfrak{T}(X)$ is not connected.

The following example shows that there exists a metric continuum X such that $\mathcal{T}$ is idempotent and $\mathfrak{T}(X)$ is not closed in 2^X.

4.1.13 Example. Let X be the topologist sine curve (Example 1.4.9). For each positive integer n, let

$$A_n = \left\{ \left(\frac{2}{\pi(2n+1)}, \sin\left(\frac{\pi(2n+1)}{2} \right) \right) \right\}$$

and let $A_0 = \{(0,0)\}$. Then $\{A_n\}_{n=1}^{\infty}$ converges to A_0, $\mathcal{T}(A_n) = A_n$ for all n and $\mathcal{T}(A_0) = \{0\} \times [-1,1]$. Therefore, $\mathfrak{T}(X)$ is not closed in 2^X. Note that $\mathcal{T}$ is idempotent on closed sets.

4.1.14 Theorem. *Let X be a continuum. If $A \in \mathfrak{T}(X)$ and K is a component of A, then $K \in \mathfrak{T}(X)$.*

Proof. Let $A \in \mathfrak{T}(X)$ and let K be a component of A. Then $\mathcal{T}(K) \subset \mathcal{T}(A) = A$, and $\mathcal{T}(K)$ is connected, Theorem 2.1.27. Since K is a component of A, $\mathcal{T}(K) \subset K$. Therefore, $K \in \mathfrak{T}(X)$. □

Theorem 4.1.14 may be generalized as follows:

4.1.15 Theorem. *Let X be a continuum and let $A \in \mathfrak{T}(X)$. If B is a closed subset of X that is the union of some components of A, then $B \in \mathfrak{T}(X)$.*

Proof. Let B be a closed subset of X that is the union of some components of A. Suppose $B = \bigcup_{\lambda\in\Lambda} K_\lambda$. Then $B = \bigcup_{\lambda\in\Lambda} K_\lambda \subset \mathcal{T}\left(\bigcup_{\lambda\in\Lambda} K_\lambda\right) \subset \mathcal{T}(A) = A$. Since $\bigcup_{\lambda\in\Lambda} K_\lambda \subset \mathcal{T}\left(\bigcup_{\lambda\in\Lambda} K_\lambda\right)$, each component of $\mathcal{T}\left(\bigcup_{\lambda\in\Lambda} K_\lambda\right)$ must intersect $\bigcup_{\lambda\in\Lambda} K_\lambda$, Corollary 2.1.20. Let L be a component of $\mathcal{T}\left(\bigcup_{\lambda\in\Lambda} K_\lambda\right)$. Hence, there exists $\lambda_0 \in \Lambda$ such that $K_{\lambda_0} \cap L \neq \emptyset$. This implies that $K_{\lambda_0} \cup L$ is a connected subset of A. Since K_{λ_0} is a component of A, we have that $L \subset K_{\lambda_0}$. Thus, $\mathcal{T}\left(\bigcup_{\lambda\in\Lambda} K_\lambda\right) \subset \bigcup_{\lambda\in\Lambda} K_\lambda$. Therefore, $B \in \mathfrak{T}(X)$. □

4.1.16 Theorem. *Let X be a continuum. If $A \in \mathfrak{T}(X)$, then for each separation $P \cup Q$ of $X \setminus A$, we have that $A \cup P \in \mathfrak{T}(X)$.*

Proof. Let $A \in \mathfrak{T}(X)$ and let $P\cup Q$ be a separation of $X \setminus A$. Let $x \in X\setminus(A\cup P)$. Then $x \in Q$. Since $x \in X \setminus A$ and $A \in \mathfrak{T}(X)$, there exists a subcontinuum W of X such that $x \in \mathrm{Int}(W) \subset W \subset X \setminus A$. Since $W \subset P \cup Q$, $Q \cap W \neq \emptyset$ and P and Q are separated, we obtain that $W \subset Q$. Hence, $W \cap (A \cup P) = \emptyset$. Thus, $x \in X \setminus \mathcal{T}(A \cup P)$. Therefore, $A \cup P \in \mathfrak{T}(X)$. □

A similar argument to the one given for Theorem 4.1.16 proves the following:

4.1.17 Theorem. *Let X be a continuum. If $A \in \mathfrak{T}(X)$ and K is a component of $\mathrm{Cl}(X \setminus A)$, then $A \cup K \in \mathfrak{T}(X)$.*

The next simple theorem is very important for the family of minimal $\mathcal{T}$-closed sets.

4.1.18 Theorem. *Let X be a continuum. If $\{A_\lambda\}_{\lambda\in\Lambda}$ is a family of elements of $\mathfrak{T}(X)$, then $\bigcap_{\lambda\in\Lambda} A_\lambda \in \mathfrak{T}(X)$.*

Proof. Since

$$\bigcap_{\lambda\in\Lambda} A_\lambda \subset \mathcal{T}\left(\bigcap_{\lambda\in\Lambda} A_\lambda\right) \subset \bigcap_{\lambda\in\Lambda} \mathcal{T}(A_\lambda) = \bigcap_{\lambda\in\Lambda} A_\lambda,$$

we have that $\bigcap_{\lambda\in\Lambda} A_\lambda \in \mathfrak{T}(X)$. □

The following example shows that the union of two $\mathcal{T}$-closed sets is not necessarily a $\mathcal{T}$-closed set.

4.1.19 Example. Let X be the suspension over the closure of the harmonic sequence (Example 1.4.23). Then $\mathcal{T}(\{a\}) = \{a\}$, $\mathcal{T}(\{b\}) = \{b\}$ and $\mathcal{T}(\{a, b\})$ is the limit segment joining a and b.

Note that the characterization of aposyndetic and locally connected continua in terms of the set function $\mathcal{T}$ given in Theorems 2.1.34, 2.1.37 and 2.1.38, can be written in terms of the family of $\mathcal{T}$-closed sets as follows:

4.1.20 Theorem. *A continuum X is aposyndetic if and only if $\mathcal{F}_1(X) \subset \mathfrak{T}(X)$.*

4.1.21 Theorem. *A continuum X is locally connected if and only if $\mathfrak{T}(X) = 2^X$.*

4.1.22 Theorem. *A continuum X is locally connected if and only if $\mathcal{C}_1(X) \subset \mathfrak{T}(X)$.*

From Theorem 2.1.44, we obtain:

4.1.23 Theorem. *If X is an indecomposable, then $\mathfrak{T}(X) = \{X\}$.*

The next example shows that the converse of Theorem 4.1.23 is not true:

4.1.24 Example. Let X be two Knaster continua glued by its end point, Example 2.3.35. Then X is a decomposable continuum and $\mathfrak{T}(X) = \{X\}$.

4.1.25 Theorem. *Let X and Y be continua and let $f : X \twoheadrightarrow Y$ be a monotone map. If $B \in \mathfrak{T}(Y)$, then $f^{-1}(B) \in \mathfrak{T}(X)$.*

Proof. Let $x \in X \setminus f^{-1}(B)$. Then $f(x) \in Y \setminus B$. Since $B \in \mathfrak{T}(Y)$, there exists a subcontinuum W of Y such that $f(x) \in \mathrm{Int}_Y(W) \subset W \subset Y \setminus B$. Hence, since f is monotone, $f^{-1}(W)$ is a subcontinuum of X (Lemma 1.4.46) and $x \in f^{-1}(x) \subset \mathrm{Int}_X(f^{-1}(W)) \subset f^{-1}(W) \subset f^{-1}(Y \setminus B) \subset X \setminus f^{-1}(B)$. Thus, $x \in X \setminus \mathcal{T}_X f^{-1}(B)$. Therefore, $f^{-1}(B) \in \mathfrak{T}(X)$. □

As a consequence of Theorem 4.1.25, we have:

4.1.26 Corollary. *Let X and Y be continua and let $f : X \twoheadrightarrow Y$ be a monotone map. Then $|\mathfrak{T}(Y)| \leq |\mathfrak{T}(X)|$.*

The next example shows that the monotone image of a $\mathcal{T}$-closed set is not necessarily $\mathcal{T}$-closed.

4.1.27 Example. Let

$$Z = (\{0\} \times [-1, 2]) \cup \left\{ \left(x, \sin\left(\frac{1}{x}\right)\right) \mid x \in \left(0, \frac{\pi}{2}\right] \right\}$$

and let X be as in Example 4.1.13. Note that $X \subset Z$. Let $f\colon Z \twoheadrightarrow X$ be given by

$$f((x, y)) = \begin{cases} (x, y), & \text{if } (x, y) \in X; \\ (0, 1), & \text{if } (x, y) \in Z \setminus X. \end{cases}$$

Then f is a monotone retraction. Note that $A = \{0\} \times \left[\frac{3}{2}, 2\right] \in \mathfrak{T}(Z)$, $f(A) = \{(0, 1)\}$ and $\mathcal{T}_X(\{(0, 1)\}) = \{0\} \times [-1, 1]$. Hence, $f(A) \notin \mathfrak{T}(X)$.

4.1.28 Theorem. *Let X and Y be continua and let $f\colon X \twoheadrightarrow Y$ be a quasi-monotone map. If $B \in \mathfrak{T}(Y)$, then $f^{-1}(B) \in \mathfrak{T}(X)$.*

Proof. Let $B \in \mathfrak{T}(Y)$. By Remark 2.1.5, $f^{-1}(B) \subset \mathcal{T}_X f^{-1}(B)$. By Theorem 2.1.59 part (1), $\mathcal{T}_X f^{-1}(B) \subset f^{-1}\mathcal{T}_Y(B)$. Since $B \in \mathfrak{T}(Y)$, $\mathcal{T}_Y(B) = B$. Hence, $\mathcal{T}_X f^{-1}(B) \subset f^{-1}(B)$. Therefore, $f^{-1}(B) \in \mathfrak{T}(X)$. □

4.1.29 Corollary. *Let X and Y be continua and let $f\colon X \twoheadrightarrow Y$ be a quasi-monotone map. Then $|\mathfrak{T}(Y)| \leq |\mathfrak{T}(X)|$.*

4.1.30 Lemma. *Let X and Y be continua and let $f\colon X \twoheadrightarrow Y$ be an atomic map. If $A \in \mathfrak{T}(X)$, then $A = f^{-1}(f(A))$.*

Proof. Let $A \in \mathfrak{T}(X)$. We know that $A \subset f^{-1}f(A)$. Let $x \in f^{-1}f(A)$. There exists $a \in A$ such that $f(a) = f(x)$. Thus, by Lemma 2.1.55, $x \in f^{-1}f(a) \subset \mathcal{T}_X(\{a\}) \subset \mathcal{T}(A) = A$. Therefore, $A = f^{-1}f(A)$. □

4.1.31 Theorem. *Let X and Y be continua and let $f\colon X \twoheadrightarrow Y$ be an atomic map. If $A \in \mathfrak{T}(X)$, then $f(A) \in \mathfrak{T}(Y)$.*

Proof. Let $A \in \mathfrak{T}(X)$. By Theorem 2.1.51 (a), we have that $\mathcal{T}_Y f(A) \subset f\mathcal{T}_X f^{-1}f(A)$. By Lemma 4.1.30, $A = f^{-1}f(A)$. Thus, $\mathcal{T}_Y f(A) \subset f\mathcal{T}_X(A) = f(A)$. Hence, $\mathcal{T}_Y f(A) = f(A)$. Therefore, $f(A) \in \mathfrak{T}(Y)$. □

4.1.32 Theorem. *Let X and Y be continua and let $f\colon X \twoheadrightarrow Y$ be an atomic map. Then $|\mathfrak{T}(X)| = |\mathfrak{T}(Y)|$.*

Proof. Note that atomic maps are monotone [92, Theorem 8.1.24]. Also observe that by Theorem 4.1.31, we have that $|\mathfrak{T}(X)| \leq |\mathfrak{T}(Y)|$. By Corollary 4.1.26, we know that $|\mathfrak{T}(X)| \geq |\mathfrak{T}(Y)|$. Therefore, $|\mathfrak{T}(X)| = |\mathfrak{T}(Y)|$. □

The following example shows that there exist monotone maps that are not atomic, between decomposable continua with the same cardinality of $\mathcal{T}$-closed sets.

4.1.33 Example. Let Z be the Knaster indecomposable continuum [92, Example 2.4.7], let X be the union of three copies of Z glued by their end points and let

Y be two copies of Z also glued by their end points. Let v be the common point of the two copies of Z in Y. Let $f: X \twoheadrightarrow Y$ be the map that sends two of the copies of Z in X homeomorphically to the two copies of Z in Y and sends the third copy of Z in X to $\{v\}$. Then f is a monotone map, f is not an atomic map, and $|\mathfrak{T}(X)| = |\mathfrak{T}(Y)| = 1$.

4.1.34 Definition. Let X be a compactum, let A be subset of X and let λ be an ordinal number. If λ is not a limit ordinal, then $\mathcal{T}^{\lambda+1}(A) = \mathcal{T}\mathcal{T}^{\lambda}(A)$. If λ is a limit ordinal, then $\mathcal{T}^{\lambda}(A) = \mathrm{Cl}\left(\bigcup_{\gamma<\lambda} \mathcal{T}^{\gamma}(A)\right)$.

4.1.35 Theorem. *Let X and Y be continua and let $f: X \twoheadrightarrow Y$ be a monotone map. If A is a subset of X, then $f\mathcal{T}_X^{\alpha}(A) \subset \mathcal{T}_Y^{\alpha} f(A)$ for all ordinals α.*

Proof. The proof is done by transfinite induction. The case $\alpha = 1$ follows from part (b) of Theorem 2.1.51. Now, suppose that $f\mathcal{T}_X^{\alpha}(A) \subset \mathcal{T}_Y^{\alpha} f(A)$ for some ordinal α. Then $f\mathcal{T}_X^{\alpha+1}(A) \subset \mathcal{T}_Y^{\alpha} f\mathcal{T}_X(A) \subset \mathcal{T}_Y^{\alpha+1} f(A)$.

Let γ be a limit ordinal and suppose that for each $\alpha < \gamma$, $f\mathcal{T}_X^{\alpha}(A) \subset \mathcal{T}_Y^{\alpha} f(A)$. Then $\bigcup_{\alpha<\gamma} f\mathcal{T}_X^{\alpha}(A) \subset \bigcup_{\alpha<\gamma} \mathcal{T}_Y^{\alpha} f(A)$. Hence, since f is a closed map,

$$f\mathcal{T}_X^{\gamma}(A) = f\left(\mathrm{Cl}\left(\bigcup_{\alpha<\gamma} \mathcal{T}_X^{\alpha}(A)\right)\right) = \mathrm{Cl}\left(f\left(\bigcup_{\alpha<\gamma} \mathcal{T}_X^{\alpha}(A)\right)\right) =$$

$$\mathrm{Cl}\left(\bigcup_{\alpha<\gamma} f\mathcal{T}_X^{\alpha}(A)\right) \subset \mathrm{Cl}\left(\bigcup_{\alpha<\gamma} \mathcal{T}_Y^{\alpha} f(A)\right) = \mathcal{T}_Y^{\gamma} f(A).$$

Therefore, $f\mathcal{T}_X^{\alpha}(A) \subset \mathcal{T}_Y^{\alpha} f(A)$ for all ordinals α. □

With essentially the same proof as in Theorem 4.1.35, using Theorem 2.1.59 part (1) for case $n = 1$, we have

4.1.36 Theorem. *Let X and Y be continua and let $f: X \twoheadrightarrow Y$ be a quasi-monotone map. If A is a subset of X, then $f\mathcal{T}_X^{\alpha}(A) \subset \mathcal{T}_Y^{\alpha} f(A)$ for all ordinals α.*

4.2 Characterizations

Now, we give a characterization of the elements of the family of $\mathcal{T}$-closed sets. We present two necessary conditions for a set to be $\mathcal{T}$-closed that, in general, are not sufficient, but we prove that these conditions are sufficient for the class of metric continua with the property of Kelley. We also give a complete characterization for metric continua.

4.2.1 Theorem. *Let X be a continuum and let A be a closed subset of X. Then $A \in \mathfrak{T}(X)$ if and only if for every open subset U of X containing A, there exists*

an open subset V of X such that $A \subset V \subset U$ and $X \setminus V$ only has finitely many components.

Proof. Suppose $A \in \mathfrak{T}(X)$ and let U be an open subset of X such that $A \subset U$. Since $A \in \mathfrak{T}(X)$, for each $x \in X \setminus U$, there exists a subcontinuum W_x of X such that $x \in \operatorname{Int}(W_x) \subset W_x \subset X \setminus A$. Since $X \setminus U$ is compact, there exist $x_1, \dots, x_n$ in $X \setminus U$ such that $X \setminus U \subset \bigcup_{j=1}^n \operatorname{Int}(W_{x_j})$. Let $V = \bigcup_{j=1}^n (X \setminus W_{x_j})$. Then $A \subset V \subset U$ and $X \setminus V = \bigcup_{j=1}^n W_{x_j}$. Hence, $X \setminus V$ only has finitely many components.

Now, assume the condition stated is satisfied. Let $x \in X \setminus A$. Since A is closed, there exist an open subset U of X such that $A \subset U$ and $x \in X \setminus U$. By our assumption, there exists an open subset V of X such that $A \subset V \subset U$ and $X \setminus V$ only has finitely many components. Let L be the component of $X \setminus V$ containing x. By Lemma 1.4.1, $x \in \operatorname{Int}(L)$. Thus, $x \in X \setminus \mathcal{T}(A)$. Therefore, $A \in \mathfrak{T}(X)$. □

As a consequence of this, we obtain the following:

4.2.2 Corollary. *Let X be a continuum and let $A \in \mathfrak{T}(X)$. If $Z = X \times \{0,1\}/R$, where R is the equivalence relation that identifies $(a,0)$ with $(a,1)$ for all $a \in A$, then $q(A \times \{0,1\}) \in \mathfrak{T}(Z)$, where $q\colon X \times \{0,1\} \twoheadrightarrow Z$ is the quotient map.*

Proof. Let U be an open subset of Z containing $q(A \times \{0,1\})$. Then $q^{-1}(U)$ is an open subset of $X \times \{0,1\}$ containing $A \times \{0,1\}$. Since $A \in \mathfrak{T}(X)$, by Theorem 4.2.1, there exists an open subset V of X such that $A \times \{0,1\} \subset V \times \{0,1\} \subset q^{-1}(U)$ and $(X \times \{0,1\}) \setminus (V \times \{0,1\})$ only has finitely many components. Then $q(V \times \{0,1\})$ is an open subset of Z, $q(A \times \{0,1\}) \subset q(V \times \{0,1\}) \subset U$ and $Z \setminus q(V \times \{0,1\})$ only has finitely many components. Therefore, by Theorem 4.2.1, $q(A \times \{0,1\}) \in \mathfrak{T}(Z)$. □

Note the following property of $\mathcal{T}$-closed sets, which is a consequence of Theorem 2.3.8:

4.2.3 Theorem. *Let X be a continuum. If $A \in \mathfrak{T}(X)$, then every component of $X \setminus A$ is open and continuumwise connected.*

The next example shows that the converse of Theorem 4.2.3 is not true. First, we introduce the following:

4.2.4 Notation. If χ_1 and χ_2 are two elements of $\mathbb{R}^n$, $\overline{\chi_1\chi_2}$ denotes the convex arc in $\mathbb{R}^n$ joining χ_1 and χ_2.

4.2.5 Example. In $\mathbb{R}^2$, for each $n \in \mathbb{N}$, let $A_n = \overline{(0,0)(1,\frac{1}{n})}$, let $A_0 = \overline{(0,0)(1,0)}$ and let $Z = \bigcup_{n=0}^\infty A_n$ (Z is known as the *harmonic fan*). For every $n \in \mathbb{N}$, note that the point $(\frac{1}{2^n}, \frac{1}{2^n n}) \in A_n$. Given $n \in \mathbb{N}$, let B_n be a semi-circle in $\mathbb{R}^3$ joining the points $(\frac{1}{2^n}, \frac{1}{2^n n})$ and $(\frac{1}{2^n}, 0)$ and such that $B_n \cap Z = \{(\frac{1}{2^n}, \frac{1}{2^n n}), (\frac{1}{2^n}, 0)\}$. Let $X = Z \cup (\bigcup_{n=1}^\infty B_n)$. Note that $\{(0,0)\}$ is a closed subset of X such that $X \setminus \{(0,0)\}$ is open and arcwise connected, but $\mathcal{T}(\{(0,0)\}) = A_0$. Hence, $\{(0,0)\}$ is not a $\mathcal{T}$-closed set of X. Note that $\mathcal{T}$ is idempotent on closed sets for X, but $\mathcal{T}$ is not idempotent for X.

The next example shows that the converse of Theorem 4.2.3 is not true even for metric continua for which $\mathcal{T}$ is idempotent.

4.2.6 Example. Let X be the metric continuum shown in the left part of Figure 4.1. The continua X_n, shown in the right part of that figure, are disjoint copies of the topologist sine curve, Example 1.4.9, converging to the arc $\overline{ap}$. The point a_n is an end point of X_n and the sequence $\{a_n\}_{n=1}^{\infty}$ converges to the point a. The arcs $\overline{a_n a_{n+1}}$ are free arcs of X and the sequence $\{\overline{a_n a_{n+1}}\}_{n=1}^{\infty}$ converges to $\{a\}$. Let $A = \{p\}$. Then $X \setminus A$ is an open subset of X and it is continuumwise connected. Since $X_n \setminus \{a_n\}$ is an open component of X and X is locally connected at each point of $\overline{a_n a_{n+1}}$, it is enough to show that Theorem 2.3.4 holds for each point of $\overline{ap}$. Let q be a point of $\overline{ap}$ and let K be a subcontinuum of X such that $q \in \text{Int}(K)$. Since $q \in \overline{ap}$, the sequence $\{X_n\}_{n=1}^{\infty}$ converges to $\overline{ap}$ and $q \in \text{Int}(K)$, there exists $m \in \mathbb{N}$ such that $X_m \subset K$ for all $n \geq m$. Since K is a subcontinuum of X, $\overline{ap} \subset K$ and $\overline{a_n a_{n+1}} \subset K$ for every $n \geq m$. Let $W = \overline{ap} \cup \left(\bigcup_{n=m+1}^{\infty} \overline{a_n a_{n+1}}\right) \cup \left(\bigcup_{n=m+1}^{\infty} X_n\right)$. Then W is a subcontinuum of X, $p \in \text{Int}(W) \subset W \subset \text{Int}(K)$. Therefore, $\mathcal{T}$ is idempotent on X. Since each subcontinuum of X containing a in its interior must contain p, $\mathcal{T}(A) \neq A$. Hence, $A \notin \mathfrak{T}(X)$.

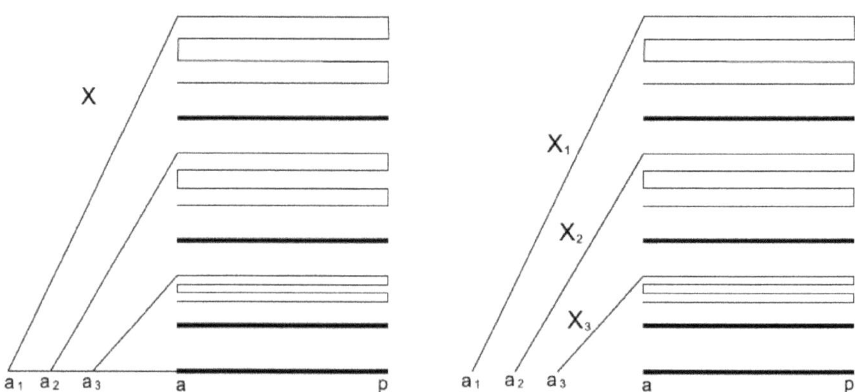

Fig. 4.1 Family of topologist sine curves

The converse of Theorem 4.2.3 is true for metric continua with the property of Kelley.

4.2.7 Theorem. *Let X be a metric continuum with the property of Kelley. Then $A \in \mathfrak{T}(X)$ if and only if each component of $X \setminus A$ is open and continuumwise connected.*

Proof. If $A \in \mathfrak{T}(X)$, then, by Theorem 4.2.3, each component of $X \setminus A$ is open and continuumwise connected.

Suppose A is a closed subset of X such that each component of $X \setminus A$ is open and continuumwise connected. Let $x_0 \in X \setminus A$ and let L be the component of $X \setminus A$ containing x_0. For each $n \in \mathbb{N}$, let W_n be the component of $X \setminus \mathcal{V}_{\frac{1}{n}}(A)$ containing

x_0. Then $W_n \subset L$. Let $x_1 \in L$. Since L is continuumwise connected, there exists a subcontinuum K of X such that $\{x_0, x_1\} \subset K \subset L$. Hence, there exists $n \in \mathbb{N}$ such that $K \subset W_n$. Thus, $L = \bigcup_{n=1}^{\infty} W_n$. Note that $W_n \subset W_{n+1}$ for every n. Since L is open, by the Baire Category Theorem [129, Theorem 25.3], there exists m such that $\text{Int}(W_m) \neq \emptyset$. Let $w \in \text{Int}(W_m)$ and let $\varepsilon > 0$ be such that $\mathcal{V}_\varepsilon(w) \subset W$ and $d(W_m, A) > \varepsilon$. By the proof of Theorem 1.6.20, there exists a subcontinuum K_0 of X such that $\mathcal{V}_\delta(x_0) \subset K_0$. By construction, we see that $K_0 \cap A = \emptyset$. Thus, $x_0 \in X \setminus \mathcal{T}(A)$. Therefore, $A \in \mathfrak{T}(X)$. □

Since homogeneous metric continua have the property of Kelley [92, Theorem 4.2.36], from Theorem 4.2.7, we have:

4.2.8 Corollary. *Let X be a homogeneous metric continuum. Then $A \in \mathfrak{T}(X)$ if and only if each component of $X \setminus A$ is open and continuumwise connected.*

The next result characterizes $\mathcal{T}$-closed sets for metric continua.

4.2.9 Theorem. *Let X be a metric continuum and let A be a closed subset of X. Then $A \in \mathfrak{T}(X)$ if and only if each component of $X \setminus A$ is open and it is an exhausted metric σ-continuum.*

Proof. Let A be a closed subset of X, suppose that $A \in \mathfrak{T}(X)$ and let L be a component of $X \setminus A$. By Theorem 4.2.3, L is an open subset of X. To see that L is an exhausted σ-continuum, let x be a point of L. Since $L \cap A = L \cap \mathcal{T}(A) = \emptyset$, there exists a subcontinuum W of X such that $x \in \text{Int}(W) \subset W \subset X \setminus A$. Hence, $W \subset L$. Thus, part (4) of Theorem 1.4.75 is satisfied. Therefore, L is an exhausted metric σ-continuum.

Now, assume that each component of $X \setminus A$ is open and it is an exhausted metric σ-continuum. Let $x \in X \setminus A$ and let L be the component of $X \setminus A$ that contains x. Then, by part (4) of Theorem 1.4.75, there exists a subcontinuum W of X such that $x \in \text{Int}(W) \subset W \subset L$. Hence, $W \cap A = \emptyset$ and $x \in X \setminus \mathcal{T}(A)$. Therefore, $A \in \mathfrak{T}(X)$. □

4.2.10 Theorem. *Let X be a type λ continuum and let $q\colon X \twoheadrightarrow [0,1]$ be the quotient map of the finest upper semicontinuous decomposition of X. Then $A \in \mathfrak{T}(X)$ if and only if there exists a nonempty closed subset B of $[0,1]$ such that $A = q^{-1}(B)$.*

Proof. Let B be a closed subset of $[0,1]$ and let $x \in X \setminus q^{-1}(B)$. Then $q(x) \in [0,1] \setminus B$. Thus, there exists an interval $[a,b]$ such that $q(x) \in (a,b)$ and $[a,b] \cap B = \emptyset$. This implies, since q is monotone, that $q^{-1}([a,b])$ is a subcontinuum of X (Lemma 1.4.46) such that $x \in \text{Int}(q^{-1}([a,b]))$ and $q^{-1}([a,b]) \cap q^{-1}(B) = \emptyset$. Therefore, $\mathcal{T}q^{-1}(B) = q^{-1}(B)$, and $q^{-1}(B) \in \mathfrak{T}(X)$.

Let $A \in \mathfrak{T}(X)$. Since X is a type λ continuum, by Theorem 3.6.2, $\mathcal{G} = \{\mathcal{T}^2(\{x\}) \mid x \in X\}$ is the finest monotone upper semicontinuous decomposition of X. Thus, since $A \in \mathfrak{T}(X)$, $q^{-1}q(A) = \bigcup\{\mathcal{T}^2(\{a\}) \mid a \in A\} \subset \mathcal{T}^2(A) = A$. Therefore, $A = q^{-1}q(A)$. □

The next theorem extends Theorem 4.2.10 to atomic maps between continua.

4.2.11 Theorem. *Let X and Y be continua, let $f\colon X \twoheadrightarrow Y$ be an atomic map, and let A be a nonempty closed subset of X. Then $A \in \mathfrak{T}(X)$ if and only if there exists a nonempty closed subset B of Y such that $B \in \mathfrak{T}(Y)$ and $A = f^{-1}(B)$.*

Proof. Assume that $A \in \mathfrak{T}(X)$. By Lemma 4.1.30, $A = f^{-1}f(A)$, and by Theorem 4.1.31, $f(A) \in \mathfrak{T}(Y)$. Hence, if $B = f(A)$, we are done.

Suppose that there exists $B \in \mathfrak{T}(Y)$ and $A = f^{-1}(B)$. Since atomic maps are monotone [92, Theorem 8.1.24], by Theorem 4.1.25, $A = f^{-1}(B) \in \mathfrak{T}(X)$. □

4.3 Minimal $\mathcal{T}$-closed Sets

4.3.1 Definition. Let X be a continuum and let A be a subset of X. Then A is a *minimal $\mathcal{T}$-closed set of X* provided that $A \in \mathfrak{T}(X)$ and $A' \notin \mathfrak{T}(X)$ for each proper closed subset A' of A. The family of minimal $\mathcal{T}$-closed sets of X is denoted by $\mathfrak{MT}(X)$.

As a consequence of Theorems 4.1.14 and 4.1.18, we have:

4.3.2 Theorem. *Let X be a continuum. If $A \in \mathfrak{MT}(X)$, then A is a subcontinuum of X.*

Let us note the following consequence of Theorem 4.1.20:

4.3.3 Theorem. *A continuum X is aposyndetic if and only if $\mathcal{F}_1(X) = \mathfrak{MT}(X)$.*

In the next example we present a metric continuum X such that all the elements of $\mathfrak{MT}(X)$ are nondegenerate.

4.3.4 Example. Let Z be the cone over the ternary Cantor set, with vertex ν_Z (Example 2.1.15). Remove each point of the base of Z and compactify each ray with an arc as a remainder, obtaining a space homeomorphic to the topologist $\sin\left(\frac{1}{x}\right)$-curve (Example 1.4.9) in such a way that we obtain a continuum X. Then $\mathfrak{MT}(X)$ consists of the family of limit bars of all the topologist $\sin\left(\frac{1}{x}\right)$-curves.

The next example shows that there exists a metric continuum X such that $\mathfrak{MT}(X)$ is not closed in $\mathcal{C}_1(X)$.

4.3.5 Example. Let X be the continuum of Example 4.3.4. Shrink the limit bar of the topologist $\sin\left(\frac{1}{x}\right)$-curve of the leg corresponding to 0 and identify this point with the vertex ν_Z, obtaining a continuum Y. Then $\{\nu_Z\} \in \mathrm{Cl}_{\mathcal{C}_1(Y)}(\mathfrak{MT}(Y))$ but $\mathcal{T}_Y(\{\nu_Z\}) = Y$.

4.4 The Set Function $\mathcal{T}^{\infty}$

4.4.1 Definition. Given a continuum X, we define the set function $\mathcal{T}^{\infty}$ as follows: If A is a subset of X, then

$$\mathcal{T}^{\infty}(A) = \bigcap\{B \in \mathfrak{T}(X) \mid A \subset B\}.$$

Note that, by Theorem 4.1.18, $\mathcal{T}^{\infty}(A) \in \mathfrak{T}(X)$.

4.4.2 Remark. Observe that $\mathcal{T} \circ \mathcal{T}^{\infty} = \mathcal{T}^{\infty} \circ \mathcal{T} = \mathcal{T}^{\infty}$. Also note that, $\mathcal{T}^{\infty} = \mathcal{T}$ if and only if $\mathcal{T}$ is idempotent. It follows from its definition that $\mathcal{T}^{\infty}$ is idempotent.

4.4.3 Remark. The set function $\mathcal{T}^{\infty}$ is not necessarily upper semicontinuous as can be seen from Example 4.1.12. It is not lower semicontinuous either since it would imply that $\mathcal{T}$ would be continuous when it is idempotent.

4.4.4 Theorem. *If X is a continuum, then there exists an ordinal number α such that $\mathcal{T}^{\infty} = \mathcal{T}^{\alpha}$.*

Proof. Given the continuum X, let γ be the first ordinal greater than the cardinality of X. Then given any subset A of X, for the transfinite sequence $\{\mathcal{T}^{\beta}(A)\}_{\beta<\gamma}$, there must be an ordinal $\beta_A < \gamma$ such that $\mathcal{T}^{\beta_A+1}(A) = \mathcal{T}^{\beta_A}(A)$, since $\{\mathcal{T}^{\beta}(A)\}_{\beta<\gamma}$ cannot increase γ times when X has fewer points than that. Then $\alpha = \sup\{\beta_A \mid A \subset X\}$ is such that $\mathcal{T}^{\infty} = \mathcal{T}^{\alpha}$. □

As consequence of Theorems 4.1.20, 4.1.21, 4.1.22 and 4.1.23, we obtain:

4.4.5 Theorem. *A continuum X is aposyndetic if and only if $\mathcal{T}^{\infty}(\{x\}) = \{x\}$ for all $x \in X$.*

4.4.6 Theorem. *A continuum X is locally connected if and only if $\mathcal{T}^{\infty}(A) = A$ for each nonempty closed subset A of X.*

4.4.7 Theorem. *A continuum X is locally connected if and only if $\mathcal{T}^{\infty}(A) = A$ for every subcontinuum A of X.*

4.4.8 Theorem. *If X is an indecomposable continuum, then $\mathcal{T}^{\infty}(A) = X$ for all nonempty closed subsets A of X.*

4.4.9 Remark. Note that the converse of Theorem 4.4.8 is not true by Example 4.1.24.

As a consequence of Theorems 4.1.35 and 4.4.4, we have:

4.4.10 Theorem. *Let X and Y be continua and let $f: X \twoheadrightarrow Y$ be a monotone map. If A is a subset of X, then $f\mathcal{T}_X^{\infty}(A) \subset \mathcal{T}_Y^{\infty} f(A)$.*

As a consequence of Theorems 4.1.36 and 4.4.4, we have:

4.4.11 Theorem. *Let X and Y be continua and let $f: X \twoheadrightarrow Y$ be a quasi-monotone map. If A is a subset of X, then $f\mathcal{T}_X^{\infty}(A) \subset \mathcal{T}_Y^{\infty} f(A)$.*

4.4.12 Corollary. *Let X and Y be continua and let $f\colon X \twoheadrightarrow Y$ be a monotone map. If Y is aposyndetic, then $f|_{\mathcal{T}_X^\infty(\{x\})}$ is a constant map for each $x \in X$.*

Proof. Let x be a point of X. By Theorem 4.4.10, $f\mathcal{T}_X^\infty(\{x\}) \subset \mathcal{T}_Y^\infty(\{f(x)\}) = \{f(x)\}$ (the last equality holds by Theorem 4.4.5). □

4.4.13 Corollary. *Let X and Y be continua and let $f\colon X \twoheadrightarrow Y$ be a quasi-monotone map. If Y is aposyndetic, then $f|_{\mathcal{T}_X^\infty(\{x\})}$ is a constant map for each $x \in X$.*

In the next example we show that inclusion cannot be changed to equality in Theorem 4.4.10.

4.4.14 Example. Let Y and Z be indecomposable continua and let $X = Y \times Z$. Note that the projection maps $\pi_Y\colon X \twoheadrightarrow Y$ and $\pi_Z\colon X \twoheadrightarrow Z$ are monotone and open. Let A be a nonempty proper closed subset of Y and let z_0 be a point of Z. On one hand, by Theorems 4.1.5 and 4.4.4, we have that $\mathcal{T}_X^\infty(A \times \{z_0\}) = A \times \{z_0\}$. Hence, $\pi_Z\mathcal{T}_X^\infty(A \times \{z_0\}) = \{z_0\}$. On the other hand, by Theorem 4.4.8, $\mathcal{T}_Z^\infty\pi_Z(A \times \{z_0\}) = Z$. Therefore, $\pi_Z\mathcal{T}_X^\infty(A \times \{z_0\}) \subsetneq \mathcal{T}_Z^\infty\pi_Z(A \times \{z_0\})$.

The next example proves that Theorem 4.4.10 is not true for open maps.

4.4.15 Example. Let X be as in Example 4.1.13, let $Y = [-1, 1]$ and let $f\colon X \twoheadrightarrow Y$ be the map given by $f((x, y)) = y$. Then f is an open map, $f\mathcal{T}_X^\infty(\{(0,0)\}) = Y$ and $\mathcal{T}_Y^\infty f(\{(0,0)\}) = \{0\}$.

In the next example we give a monotone map $f\colon X \twoheadrightarrow Y$ between metric continua such that $\mathcal{T}_Y^\infty(B) \not\subset f\mathcal{T}_X^\infty f^{-1}(B)$ for some subset B of Y.

4.4.16 Example. Let Z_1 be the Cantor fan (Example 2.1.15), with vertex $\nu_{Z_1} = (\frac{1}{2}, 1)$. Let Z_2 be a copy of Z_1, with vertex $\nu_{Z_2} = (1,0)$ in the base of Z_1 such that $Z_1 \cap Z_2 = \{(1,0)\}$. Let $Y = Z_1 \cup Z_2$. Let $X = Z_1 \cup A \cup Z_3$, where Z_3 is a copy of Z_1, with vertex $\nu_{Z_3} = (1, -1)$ and $A = \{1\} \times [-1, 0]$ is the convex arc joining the point $(1,0)$ of Z_1 with ν_{Z_3}. Let $f\colon X \twoheadrightarrow Y$ be the map that shrinks the arc A to the point $(1,0)$. Then f is a monotone map, $\mathcal{T}_Y^\infty(\{\nu_{Z_1}\}) = Y$ and $f\mathcal{T}_X^\infty f^{-1}(\{\nu_{Z_1}\}) = Z_1$.

4.4.17 Theorem. *Let X and Y be continua and let $f\colon X \twoheadrightarrow Y$ be an open monotone map. If B is a subset of Y, then $\mathcal{T}_X^\alpha f^{-1}(B) = f^{-1}\mathcal{T}_Y^\alpha(B)$ for each ordinal number α.*

Proof. We do the proof by transfinite induction. Since f is open and monotone, by part (e) of Theorem 2.1.51, we obtain that $\mathcal{T}_X f^{-1}(B) = f^{-1}\mathcal{T}_Y(B)$.

Suppose $\mathcal{T}_X^\alpha f^{-1}(B) = f^{-1}\mathcal{T}_Y^\alpha(B)$ for some ordinal α. Then $\mathcal{T}_X^{\alpha+1} f^{-1}(B) = \mathcal{T}_X f^{-1}\mathcal{T}_Y^\alpha(B) = f^{-1}\mathcal{T}_Y^\alpha(B)$.

Let γ be a limit ordinal and suppose that for every $\alpha < \gamma$, we have $\mathcal{T}_X^\alpha f^{-1}(B) = f^{-1}\mathcal{T}_Y^\alpha(B)$. Then we have that $\bigcup_{\alpha<\gamma} \mathcal{T}_X^\alpha f^{-1}(B) = \bigcup_{\alpha<\gamma} f^{-1}\mathcal{T}_Y^\alpha(B)$. This implies that,

$$\mathcal{T}_X^\gamma f^{-1}(B) = \text{Cl}\left(\bigcup_{\alpha<\gamma} \mathcal{T}_X^\alpha f^{-1}(B)\right) =$$

$$\text{Cl}\left(\bigcup_{\alpha<\gamma} f^{-1}\mathcal{T}_Y^\alpha(B)\right) = f^{-1}\left(\text{Cl}\left(\bigcup_{\alpha<\gamma} \mathcal{T}_Y^\alpha(B)\right)\right) = f^{-1}\mathcal{T}_Y^\gamma(B).$$

Therefore, $\mathcal{T}_X^\alpha(f^{-1}(B)) = f^{-1}(\mathcal{T}_Y^\alpha(B))$ for each ordinal number α. □

With the same proof given for Theorem 4.4.17 and using Theorem 2.1.59 part (3) for case $n = 1$, we have:

4.4.18 Theorem. *Let X and Y be continua and let $f\colon X \twoheadrightarrow Y$ be an open quasi-monotone map. If B is a subset of Y, then $\mathcal{T}_X^\alpha f^{-1}(B) = f^{-1}\mathcal{T}_Y^\alpha(B)$ for each ordinal number α.*

From Theorems 4.4.4 and 4.4.17, we have:

4.4.19 Theorem. *Let X and Y be continua and let $f\colon X \twoheadrightarrow Y$ be an open monotone map. If B is a subset of Y, then $\mathcal{T}_X^\infty f^{-1}(B) = f^{-1}\mathcal{T}_Y^\infty(B)$.*

As a consequence of Theorems 4.4.4 and 4.4.18, we have

4.4.20 Theorem. *Let X and Y be continua and let $f\colon X \twoheadrightarrow Y$ be an open quasi-monotone map. If B is a subset of Y, then $\mathcal{T}_X^\infty f^{-1}(B) = f^{-1}\mathcal{T}_Y^\infty(B)$.*

4.4.21 Definition. A continuum X is $\mathcal{T}^\infty$*-additive* provided that for each nonempty closed subset $\mathcal{A}$ of 2^X whose union is closed in X, $\mathcal{T}^\infty\left(\bigcup\{A \mid A \in \mathcal{A}\}\right) = \bigcup\{\mathcal{T}^\infty(A) \mid A \in \mathcal{A}\}$.

4.4.22 Definition. A continuum X is $\mathcal{T}^\infty$*-symmetric* (*point* $\mathcal{T}^\infty$*-symmetric*) if for each two nonempty closed subsets A and B of X (for any two points p and q of X), we have that $A \cap \mathcal{T}^\infty(B) = \emptyset$ if and only if $\mathcal{T}^\infty(A) \cap B = \emptyset$ ($p \in \mathcal{T}^\infty(\{q\})$ if and only if $q \in \mathcal{T}^\infty(\{p\})$).

4.4.23 Theorem. *If X is a $\mathcal{T}^\infty$-symmetric continuum, then X is $\mathcal{T}^\infty$-additive.*

Proof. Let X be a $\mathcal{T}^\infty$-symmetric continuum and let $\mathcal{A}$ be a closed subset of 2^X whose union is closed in X. Note that $\bigcup\{\mathcal{T}^\infty(A) \mid A \in \mathcal{A}\} \subset \mathcal{T}^\infty\left(\bigcup\{A \mid A \in \mathcal{A}\}\right)$. Let $x \in X \setminus \bigcup\{\mathcal{T}^\infty(A) \mid A \in \mathcal{A}\}$. Then, by symmetry, $\mathcal{T}^\infty(\{x\}) \cap A = \emptyset$ for all $A \in \mathcal{A}$. Hence, $\mathcal{T}^\infty(\{x\}) \cap \left(\bigcup\{A \mid A \in \mathcal{A}\}\right) = \emptyset$. Thus, by symmetry, $\{x\} \cap \mathcal{T}^\infty(\bigcup\{A \mid A \in \mathcal{A}\}) = \emptyset$. As a consequence of this, $x \in X \setminus \mathcal{T}^\infty\left(\bigcup\{A \mid A \in \mathcal{A}\}\right)$ and $\bigcup\{\mathcal{T}^\infty(A) \mid A \in \mathcal{A}\} = \mathcal{T}^\infty\left(\bigcup\{A \mid A \in \mathcal{A}\}\right)$. Therefore, X is $\mathcal{T}^\infty$-additive. □

4.4.24 Theorem. *A continuum X is $\mathcal{T}^\infty$-symmetric if and only if X is point $\mathcal{T}^\infty$-symmetric and $\mathcal{T}^\infty$-additive.*

Proof. If X is a $\mathcal{T}^\infty$-symmetric continuum, then clearly X is point $\mathcal{T}^\infty$-symmetric and, by Theorem 4.4.23, X is $\mathcal{T}^\infty$-additive.

Suppose X is point $\mathcal{T}^\infty$-symmetric and $\mathcal{T}^\infty$-additive. Let A and A' be two closed subsets of X. Assume that $A \cap \mathcal{T}^\infty(A') = \emptyset$ and $\mathcal{T}^\infty(A) \cap A' \neq \emptyset$. Let $a' \in \mathcal{T}^\infty(A) \cap A'$. Then $a' \in A'$ and $a' \in \mathcal{T}^\infty(A)$. Note that $a' \in X \setminus A$. Since X is $\mathcal{T}^\infty$-additive, $\mathcal{T}^\infty(A) = \bigcup\{\mathcal{T}^\infty(\{a\}) \mid a \in A\}$. Since $a' \in \mathcal{T}^\infty(A)$, there exists $a \in A$ such that $a' \in \mathcal{T}^\infty(\{a\})$. This implies, by point $\mathcal{T}^\infty$-symmetry, that $a \in \mathcal{T}^\infty(\{a'\})$. Hence, since $a' \in A'$, $a \in \mathcal{T}^\infty(\{a'\}) \subset \mathcal{T}^\infty(A')$. Thus, $a \in A \cap \mathcal{T}^\infty(A')$, a contradiction to our assumption. Therefore, $\mathcal{T}^\infty(A) \cap A' = \emptyset$ and X is $\mathcal{T}^\infty$-symmetric. □

4.4.25 Definition. The *pseudo-arc* is the only hereditarily indecomposable chainable metric continuum [92, Definition 2.4.3].

4.4.26 Remark. R H Bing proved that any two hereditarily indecomposable chainable metric continua are homeomorphic [13, Theorem 21].

4.4.27 Definition. By a *"harmonic fan" of pseudo-arcs* we mean the following: Consider the harmonic fan $Z = \bigcup_{n=0}^{\infty} A_n$ of Example 4.2.5 defined in $\mathbb{R}^3$. For each $n \in \mathbb{N}$, replace A_n with a pseudo-arc, X_n, in such a way that the sequence $\{X_n\}_{n=1}^{\infty}$ converges to the arc A_0. Then a "harmonic fan" of pseudo-arcs is a space homeomorphic to $X = A_0 \cup (\bigcup_{n=1}^{\infty} X_n)$.

In the next example, we present a $\mathcal{T}^\infty$-additive metric continuum X such that it is not $\mathcal{T}$-additive.

4.4.28 Example. In $\mathbb{R}^3$, let $p = (0,0,0)$ and let $q = (1,0,0)$. With vertex p, let $Z_1 = A_0 \cup (\bigcup_{n=1}^{\infty} X_n)$ be a "harmonic fan" of pseudo-arcs converging to the convex arc $A_0 = \overline{p\left(\frac{1}{3},0,0\right)}$. With vertex q, let $Z_2 = A_0' \cup (\bigcup_{n=1}^{\infty} X_n')$ be a "harmonic fan" of pseudo-arcs converging to the convex arc $A_0' = \overline{\left(\frac{2}{3},0,0\right) q}$. For each $n \in \mathbb{N}$, let X_n'' be a pseudo-arc such that $|X_n \cap X_n''| = |X_n' \cap X_n''| = 1$ and the sequence $\{X_n''\}_{n=1}^{\infty}$ converges to the convex arc $\overline{\left(\frac{1}{3},0,0\right)\left(\frac{2}{3},0,0\right)}$. Let $X = Z_1 \cup Z_2 \cup (\bigcup_{n=1}^{\infty} X_n'') \cup \overline{\left(\frac{1}{3},0,0\right)\left(\frac{2}{3},0,0\right)}$. Then X is a continuum such that $\mathcal{T}^\infty(\{x\}) = X$ for all $x \in X$. Hence, X is $\mathcal{T}^\infty$-additive. However, $\mathcal{T}(\{p,q\}) \neq \mathcal{T}(\{p\}) \cup \mathcal{T}(\{q\})$. Thus, X is not $\mathcal{T}$-additive.

In the next example, we give a $\mathcal{T}^\infty$-symmetric metric continuum X such that it is not $\mathcal{T}$-symmetric.

4.4.29 Example. In $\mathbb{R}^3$, let $p = (1,0,0)$ and let $q = \left(\frac{1}{2},0,0\right)$. With vertex p, let $Z_1' = A_0' \cup (\bigcup_{n=1}^{\infty} X_n)$ be a "harmonic fan" of pseudo-arcs converging to the convex arc $A_0' = \overline{p\left(\frac{3}{4},0,0\right)}$. For each $n \in \mathbb{N}$, let X_n' be a pseudo-arc such that $|X_n \cap X_n'| = 1$ and the sequence $\{X_n'\}_{n=1}^{\infty}$ converges to the convex arc $\overline{\left(\frac{3}{4},0,0\right) q}$. Let $Z_1 = Z_1' \cup (\bigcup_{n=1}^{\infty} X_n') \cup \overline{\left(\frac{3}{4},0,0\right) q}$. For each integer $n \geq 2$, with vertex $\left(\frac{1}{n},0,0\right)$, let Z_n be a "harmonic fan" of pseudo-arcs converging to the convex arc $\overline{\left(\frac{1}{n},0,0\right)\left(\frac{1}{n+1},0,0\right)}$. Note that the sequence of continua $\{Z_n\}_{n=1}^{\infty}$ converges to $\{(0,0,0)\}$. Let $X' = \{(0,0,0)\} \cup (\bigcup_{n=1}^{\infty} Z_n)$. Now, identify the point $(0,0,0)$ with a point $z \in X_1 \setminus \{p\}$ and let X be this quotient space. Then X is a continuum such that $\mathcal{T}^\infty(\{x\}) = X$ for all $x \in X$. Hence, X is $\mathcal{T}^\infty$-symmetric. However, $q \in \mathcal{T}(\{p\})$ and $p \in X \setminus \mathcal{T}(\{q\})$. Thus, X is not $\mathcal{T}$-symmetric.

4.4.30 Remark. Observe that the metric continuum X of Example 4.4.29 is a $\mathcal{T}^\infty$-symmetric continuum for which $\mathcal{T}^\infty = \mathcal{T}^\omega$.

In the next example, we present a $\mathcal{T}$-symmetric metric continuum X such that it is not $\mathcal{T}^\infty$-symmetric.

4.4.31 Example. Let X be the metric continuum defined in Example 4.1.12 and consider the subcontinuum Z of X defined by $Z = \{p\} \cup (\bigcup_{n=1}^\infty Z_n)$. Then Z is a $\mathcal{T}$-symmetric continuum. Note that for each $z \in Z \setminus \{p\}$, $\mathcal{T}^\infty(\{z\}) = Z$ and $\mathcal{T}^\infty(\{p\}) = \{p\}$. Hence, Z is not $\mathcal{T}^\infty$-symmetric.

4.4.32 Theorem. *If X is a $\mathcal{T}$-additive continuum, then X is $\mathcal{T}^\infty$-additive.*

Proof. Let $\mathcal{A}$ be a nonempty closed subset of 2^X whose union is closed in X. First note that $\bigcup\{\mathcal{T}^\infty(A) \mid A \in \mathcal{A}\} \in \mathfrak{T}(X)$. To see this, observe that since X is $\mathcal{T}$-additive,

$$\mathcal{T}\left(\bigcup\{\mathcal{T}^\infty(A) \mid A \in \mathcal{A}\}\right) =$$

$$\bigcup\{\mathcal{T}(\mathcal{T}^\infty(A)) \mid A \in \mathcal{A}\} = \bigcup\{\mathcal{T}^\infty(A) \mid A \in \mathcal{A}\},$$

the last equality follows from Remark 4.4.2.

If $\mathcal{T}^\infty\left(\bigcup\{A \mid A \in \mathcal{A}\}\right) \neq \bigcup\{\mathcal{T}^\infty(A) \mid A \in \mathcal{A}\}$, we would have that $\bigcup\{\mathcal{T}^\infty(A) \mid A \in \mathcal{A}\} \notin \mathfrak{T}(X)$, a contradiction to the previous paragraph. Therefore, X is $\mathcal{T}^\infty$-additive. □

4.4.33 Corollary. *If X is either a weakly irreducible, an irreducible or a hereditarily unicoherent continuum, then X is $\mathcal{T}^\infty$-additive.*

Proof. If X is a weakly irreducible continuum, then, by Theorems 2.2.2 and 2.2.11, X is $\mathcal{T}$-additive. If X is an irreducible continuum, then, by Corollary 2.2.3 and Theorem 2.2.11, X is $\mathcal{T}$-additive. If X is a hereditarily unicoherent continuum, then, by Theorem 2.2.12, X is $\mathcal{T}$-additive. The corollary now follows from Theorem 4.4.32. □

4.5 $\mathcal{T}$-growth Bound

We introduce the $\mathcal{T}$-growth bound of a continuum X. We show that the $\mathcal{T}$-growth bound of a type λ continuum is at most three, we also prove that if X is a continuum with finite $\mathcal{T}$-growth bound and f is a monotone open map from X onto a continuum Y, then the $\mathcal{T}$-growth bound of Y is at most the $\mathcal{T}$-growth bound of X.

4.5.1 Definition. Given a continuum X, the $\mathcal{T}$*-growth bound of* X is the smallest ordinal α such that for all subsets A of X, $\mathcal{T}^\infty(A) = \mathcal{T}^\alpha(A)$. The $\mathcal{T}$-growth bound of X is denoted by $\mathcal{T}GB(X)$.

4.5.2 Theorem. *If X is a locally connected continuum, then $\mathcal{T}GB(X) = 1$.*

4.5.3 Theorem. *If X is a continuum for which $\mathcal{T}$ is idempotent, then $\mathcal{T}GB(X)=1$.*

4.5.4 Theorem. *If X is a homogeneous continuum with the property of Kelley, then $\mathcal{T}GB(X) \leq 2$.*

Proof. Since X is a homogeneous continuum with the property of Kelley, $\mathcal{T}$ is idempotent on closed sets, Theorem 2.3.18. Therefore, $\mathcal{T}GB(X) \leq 2$. □

4.5.5 Remark. Let Z be the product of two homogeneous indecomposable metric continua. Then Z has the property of Kelley [92, Theorem 4.2.36]. Hence, $\mathcal{T}_Z$ is idempotent on closed sets, Theorem 2.3.18. Also, $\mathcal{T}_Z$ is not idempotent, Corollary 2.3.23. Thus, $\mathcal{T}GB(Z) = 2$. Note that for a locally connected homogeneous continuum W, we have that $\mathcal{T}GB(W) = 1$ (Theorem 4.5.2).

4.5.6 Lemma. *If X is a type λ continuum, then $\mathcal{T}^4(A) = \mathcal{T}^3(A)$ for all subsets A of X.*

Proof. Let L be a nonempty closed subset of X. Then, by Theorem 4.2.10, we have that $q^{-1}q(L) \in \mathfrak{T}(X)$. Hence, $\mathcal{T}^2(L) \subset q^{-1}q(L)$. By Theorem 3.6.2, $\mathcal{G} = \{\mathcal{T}^2(\{x\}) \mid x \in X\}$ is the finest monotone upper semicontinuous decomposition of X. Thus, $q^{-1}q(L) = \bigcup\{\mathcal{T}^2(\{\ell\}) \mid \ell \in L\} \subset \mathcal{T}^2(L)$. Therefore, $\mathcal{T}^2(L) = q^{-1}q(L)$, and $\mathcal{T}^3(L) = \mathcal{T}q^{-1}q(L) = \mathcal{T}^2(L)$.

Now let A be a nonempty subset of X. Then $\mathcal{T}(A)$ is a nonempty closed subset of X and $\mathcal{T}^3(A) = \mathcal{T}^2\mathcal{T}(A) = q^{-1}q\mathcal{T}(A)$. Therefore, $\mathcal{T}^4(A) = \mathcal{T}q^{-1}q\mathcal{T}(A) = \mathcal{T}^3(A)$. □

4.5.7 Corollary. *If X is a type λ continuum, then $\mathcal{T}GB(X) \leq 3$.*

4.5.8 Remark. Note that if X is a continuously irreducible continuum (Definition 1.4.65), then $\mathcal{T}$ is continuous for X, Theorem 5.2.9. Thus, $\mathcal{T}$ is idempotent on X, Theorem 5.1.6. Therefore, $\mathcal{T}GB(X) = 1$ (Theorem 4.5.3).

4.5.9 Theorem. *Let X and Y be continua and let $f\colon X \twoheadrightarrow Y$ be an open monotone map. If $\mathcal{T}GB(X) < \infty$, then $\mathcal{T}GB(Y) \leq \mathcal{T}GB(X)$.*

Proof. Suppose $\mathcal{T}GB(X) = n$ and $\mathcal{T}GB(Y) \geq n+1$. Then there exists a subset B of Y such that $B \subsetneq \mathcal{T}_Y(B) \subsetneq \mathcal{T}_Y^2(B) \subsetneq \cdots \subsetneq \mathcal{T}_Y^{n+1}(B)$. Then $f^{-1}(B) \subsetneq f^{-1}\mathcal{T}_Y(B) \subsetneq f^{-1}\mathcal{T}_Y^2(B) \subsetneq \cdots \subsetneq f^{-1}\mathcal{T}_Y^{n+1}(B)$. Since f is open and monotone, by Theorem 4.4.17, we have that $f^{-1}\mathcal{T}_Y^n(B) = \mathcal{T}_X^n f^{-1}(B)$. Hence, $f^{-1}(B) \subsetneq \mathcal{T}_X f^{-1}(B) \subsetneq \mathcal{T}_X^2 f^{-1}(B) \subsetneq \cdots \subsetneq \mathcal{T}_X^{n+1} f^{-1}(B)$. Thus, $\mathcal{T}GB(X) \geq n+1$, a contradiction. Therefore, $\mathcal{T}GB(Y) \leq \mathcal{T}GB(X)$. □

The next examples show that the requirement for f to be open and monotone is necessary. Before presenting Example 4.5.12, we need the following:

4.5.10 Lemma. *Let Z_1 be the Cantor fan, Example 2.1.15. Then $\mathcal{T}GB(Z_1) = 2$.*

Proof. Let z be a point of $Z_1 \setminus \{\nu_{Z_1}\}$. Let b_z be the basis point of the coning arc of Z_1 containing z, and let $L_z = \{(1-t)z + tb_z \mid t \in [0,1]\}$.

Observe that $\mathcal{T}(\{\nu_{Z_1}\}) = Z_1$, Example 2.1.15. Also note that if $A \subset Z_1 \setminus \{\nu_{Z_1}\}$, then $\mathcal{T}(A) = \bigcup\{L_a \mid a \in \text{Cl}(A)\}$. In addition, if $\mathcal{T}(A) \subset Z_1 \setminus \{\nu_{Z_1}\}$, then $\mathcal{T}^2(A) = \mathcal{T}(A)$. On the other hand, if $\nu_{Z_1} \in \mathcal{T}(A)$, then $\mathcal{T}^2(A) = Z_1$ and $\mathcal{T}^3(A) = \mathcal{T}^2(A)$. Therefore, $\mathcal{T}GB(Z_1) = 2$. □

4.5.11 Corollary. *If X and Y are the spaces described in Example 4.4.16, then $\mathcal{T}GB(X) = 2$ and $\mathcal{T}GB(Y) = 3$.*

Proof. Recall that $X = Z_1 \cup A \cup Z_3$. Let B be a subset of X. Observe the following:

(1) If $B \subset Z_1$, then $\mathcal{T}_X(B) = \mathcal{T}_{Z_1}(B)$.
(2) If $B \subset A \setminus \{\nu_{Z_3}\}$, then $\mathcal{T}_X(B) = \mathcal{T}_A(B)$.
(3) If $\nu_{Z_3} \in B \subset A$, then $\mathcal{T}_X(B) = \mathcal{T}_A(B) \cup Z_3$.
(4) If $B \subset Z_3$, then $\mathcal{T}_X(B) = \mathcal{T}_{Z_3}(B)$.

We show that $\mathcal{T}_X(B) = \mathcal{T}_X(B \cap Z_1) \cup \mathcal{T}_X(B \cap A) \cup \mathcal{T}_X(B \cap Z_3)$. By Proposition 2.1.7, $\mathcal{T}_X(B \cap Z_1) \cup \mathcal{T}_X(B \cap A) \cup \mathcal{T}_X(B \cap Z_3) \subset \mathcal{T}_X(B)$. Let $x \in X \setminus [\mathcal{T}_X(B \cap Z_1) \cup \mathcal{T}_X(B \cap A) \cup \mathcal{T}_X(B \cap Z_3)]$. Then there exist three subcontinua W_1, W_2 and W_3 of X such that $x \in \text{Int}_X(W_1) \subset W_1 \subset X \setminus (Z_1 \cap B)$, $x \in \text{Int}_X(W_2) \subset W_2 \subset X \setminus (A \cap B)$, and $x \in \text{Int}_X(W_3) \subset W_3 \subset X \setminus (Z_3 \cap B)$. Let $W = W_1 \cap W_2 \cap W_3$. Since X is hereditarily unicoherent, W is a subcontinuum of X, $x \in \text{Int}_X(W)$ and $W \cap [(Z_1 \cap B) \cup (A \cap B) \cup (Z_3 \cap B)] = \emptyset$. Since $B = (Z_1 \cap B) \cup (A \cap B) \cup (Z_3 \cap B)$, we obtain that $W \cap B = \emptyset$. Thus, $x \in X \setminus \mathcal{T}_X(B)$. Therefore, $\mathcal{T}_X(B) = \mathcal{T}_X(B \cap Z_1) \cup \mathcal{T}_X(B \cap A) \cup \mathcal{T}_X(B \cap Z_3)$.

Since X is hereditarily unicoherent, X is $\mathcal{T}_X$-additive, Theorem 2.2.12. Hence, it follows from Lemma 4.5.10, (1), (2), (3) and (4) and the fact that A is an arc that: if B is a subset of X, then

$$\mathcal{T}_X^3(B) = \mathcal{T}_X^3(B \cap Z_1) \cup \mathcal{T}_X^3(B \cap A) \cup \mathcal{T}_X^3(B \cap Z_3) =$$

$$\mathcal{T}_X^2(B \cap Z_1) \cup \mathcal{T}_X^2(B \cap A) \cup \mathcal{T}_X^2(B \cap Z_3) = \mathcal{T}_X^2(B).$$

Thus, $\mathcal{T}GB(X) \leq 2$. Let $C = \left\{(1-t)(1,0) + t\left(\frac{1}{2}, 1\right) \mid t \in [0,1)\right\}$. Then $\mathcal{T}_X(C) = \left\{(1-t)(1,0) + t\left(\frac{1}{2}, 1\right) \mid t \in [0,1]\right\}$, $\mathcal{T}_X^2(C) = Z_1$, and $\mathcal{T}_X^3(C) = \mathcal{T}_X^2(C)$. Therefore, $\mathcal{T}GB(X) = 2$.

Recall $Y = Z_1 \cup Z_2$. Let $C = \left\{t\left(\frac{1}{2}, 1\right) \mid t \in [0,1)\right\}$. Then $\mathcal{T}_Y(C) = \left\{t\left(\frac{1}{2}, 1\right) \mid t \in [0,1]\right\}$, $\mathcal{T}_Y^2(C) = Z_1$, $\mathcal{T}_Y^3(C) = Y$ and $\mathcal{T}_Y^4(C) = \mathcal{T}_Y^3(C)$. Thus, $\mathcal{T}GB(Y) \geq 3$.

Let B be a subset of Y. Note the following:

(5) If $B \subset Z_1 \setminus \{\nu_{Z_2}\}$, then $\mathcal{T}_Y(B) = \mathcal{T}_{Z_1}(B)$.
(6) If $\nu_{Z_2} \in B \subset Z_1$, then $\mathcal{T}_Y(B) = \mathcal{T}_{Z_1}(B) \cup Z_2$.
(7) If $B \subset Z_2$, then $\mathcal{T}_Y(B) = \mathcal{T}_{Z_2}(B)$.

We prove that $\mathcal{T}_Y(B) = \mathcal{T}_Y(B \cap Z_1) \cup \mathcal{T}_Y(B \cap Z_2)$. By Proposition 2.1.7, $\mathcal{T}_Y(B \cap Z_1) \cup \mathcal{T}_Y(B \cap Z_2) \subset \mathcal{T}_Y(B)$. Let $y \in X \setminus [\mathcal{T}_Y(B \cap Z_1) \cup \mathcal{T}_Y(B \cap Z_2)]$. Then there exist two subcontinua K_1 and K_2 of Y such that $y \in \mathrm{Int}_Y(K_1) \subset K_1 \subset Y \setminus (B \cap Z_1)$ and $y \in \mathrm{Int}_Y(K_2) \subset K_2 \subset Y \setminus (B \cap Z_2)$. Let $K = K_1 \cap K_2$. Then K is a subcontinuum of Y such that $y \in \mathrm{Int}_Y(K)$ and $K \cap [(B \cap Z_1) \cup (B \cap Z_2)] = \emptyset$. Since $B = (B \cap Z_1) \cup (B \cap Z_2)$, we have that $K \cap B = \emptyset$. Thus, $y \in Y \setminus \mathcal{T}_Y(B)$. Therefore, $\mathcal{T}_Y(B) = \mathcal{T}_Y(B \cap Z_1) \cup \mathcal{T}_Y(B \cap Z_2)$.

Since Y is hereditarily unicoherent, Y is $\mathcal{T}_Y$-additive, Theorem 2.2.12. Hence, it follows from Lemma 4.5.10, (5), (6) and (7) that: if B is a subset of Y, then

$$\mathcal{T}_Y^4(B) = \mathcal{T}_Y^4(B \cap Z_1) \cup \mathcal{T}_Y^4(B \cap Z_2) = \mathcal{T}_Y^3(B \cap Z_1) \cup \mathcal{T}_Y^3(B \cap Z_2) = \mathcal{T}_Y^3(B).$$

Therefore, $\mathcal{T}GB(Y) = 3$. □

4.5.12 Example. Let $f \colon X \twoheadrightarrow Y$ be the monotone map given in Example 4.4.16. Note that, by Corollary 4.5.11, $\mathcal{T}GB(Y) = 3$ and $\mathcal{T}GB(X) = 2$.

4.5.13 Example. Let

$$X_1 = \{0\} \times [-1, 3] \cup \left\{ (x, y) \in \mathbb{R}^2 \;\middle|\; y = \sin\left(\frac{1}{x}\right) \text{ and } 0 < x \leq \frac{2}{\pi} \right\} \cup$$

$$\left\{ (x, y) \in \mathbb{R}^2 \;\middle|\; y = 2 + \sin\left(\frac{1}{x}\right) \text{ and } -\frac{2}{\pi} \leq x < 0 \right\}.$$

Then X_1 is an irreducible continuum between the points $\left(-\frac{2}{\pi}, 0\right)$ and $\left(\frac{2}{\pi}, 0\right)$. Let X_2 be the union of two copies of X_1 glued by the point $\left(\frac{2}{\pi}, 0\right)$; i.e., $X_2 = X_1 \vee X_1$ [24, p. 392]. Let $f_1^2 \colon X_2 \twoheadrightarrow X_1$ be the projection map. Suppose we have defined the spaces $X_1, \ldots, X_n$ and maps $f_k^{k+1} \colon X_{k+1} \twoheadrightarrow X_k$, $k \in \{1, \ldots, n-1\}$. Let $X_{n+1} = X_n \vee X_n$ and let $f_n^{n+1} \colon X_{n+1} \twoheadrightarrow X_n$ be the projection map. Hence, we have an inverse sequence $\{X_n, f_n^{n+1}\}_{n=1}^{\infty}$ of continua with open bonding maps. Let $X_\infty = \varprojlim \{X_n, f_n^{n+1}\}_{n=1}^{\infty}$. Then $\{X_n, f_n^{n+1}\}_{n=1}^{\infty}$ is an indecomposable inverse sequence (see [92, Definition 2.1.18] for the definition). Thus, X_∞ is an indecomposable continuum [92, Theorem 2.1.19]. Since the bonding maps f_n^{n+1} are open, the projection maps $f_n \colon X_\infty \twoheadrightarrow X_n$ are open [92, Corollary 2.1.10]. Since X_∞ is indecomposable, $\mathcal{T}GB(X_\infty) = 1$. Meanwhile, $\mathcal{T}GB(X_1) = 2$.

The next example gives a one-dimensional aposyndetic metric continuum whose $\mathcal{T}$-growth bound is infinite.

4.5.14 Example. Let $\{X_n\}_{n=1}^{\infty}$ be a sequence of suspensions over the Cantor ternary set. For each $n \in \mathbb{N}$, let ν_n^+ and ν_n^- be the vertexes of X_n. Assume that, for each positive integer n, $|X_n \cap X_{n+1}| = 2$, $\{\nu_n^+, \nu_n^-\} \cap (X_n \cap X_{n+1}) = \emptyset$ and $\{\nu_{n+1}^+, \nu_{n+1}^-\} \subset X_n \cap X_{n+1}$. Also assume that $\lim \operatorname{diam}(X_n) = 0$ and the sequence $\{X_n\}_{n=1}^{\infty}$ converges to $\{p\}$. Let $X = \{p\} \cup \left(\bigcup_{n=1}^{\infty} X_n\right)$. Then X is a one-dimensional aposyndetic continuum and $\mathcal{T}^n(\{\nu_1^+, \nu_1^-\}) = \bigcup_{j=1}^{n} X_n$ for every $n \in \mathbb{N}$ (compare with Example 4.1.19). Therefore, $\mathcal{T}GB(X) = \infty$.

References for Chapter 4

Section 4.1: [9, 39, 44, 46, 89, 92, 96].
Section 4.2: [1, 9, 39, 46, 89].
Section 4.3: [9].
Section 4.4: [9, 46, 96].
Section 4.5: [9, 24, 92].

Chapter 5
Continuity of $\mathcal{T}$

Let X be a continuum. By Remark 2.1.5, the image of any subset of X under $\mathcal{T}$ is a closed subset of X. Then we may restrict the domain of $\mathcal{T}$ to the hyperspace, 2^X, of nonempty closed subsets of X. Since 2^X has a topology, we may ask if $\mathcal{T}\colon 2^X \to 2^X$ is continuous. The answer to this question is negative, as can be seen from Example 2.1.15. On the other hand, by Theorems 2.1.37 and 2.1.44, $\mathcal{T}$ is continuous for locally connected continua and for indecomposable continua, respectively. In this chapter we present results related to the continuity of $\mathcal{T}$ and examples of classes of decomposable nonlocally connected metric continua for which $\mathcal{T}$ is continuous. In particular, we show that if a continuum X is almost connected im kleinen at each of its points and $\mathcal{T}$ is continuous, then X is locally connected.

5.1 Main Properties

We present the properties given by David P. Bellamy [3] for continua for which the set function $\mathcal{T}$ is continuous. We give a characterization of the continuity of $\mathcal{T}$ in terms of the idempotency of $\mathcal{T}$ on closed sets and a characterization of continua with the uniform property of Effros for which $\mathcal{T}$ is continuous. We begin with the next theorem that says that $\mathcal{T}$ is always upper semicontinuous.

5.1.1 Theorem. *Let X be a continuum. If U is an open subset of X, then $\mathcal{U} = \{A \in 2^X \mid \mathcal{T}(A) \subset U\}$ is open in 2^X; that is, $\mathcal{T}$ is upper semicontinuous.*

Proof. Let U be an open subset of X, and let $\mathcal{U} = \{A \in 2^X \mid \mathcal{T}(A) \subset U\}$. We show that $\mathcal{U}$ is open in 2^X. Let $B \in \mathrm{Cl}_{2^X}(2^X \setminus \mathcal{U})$. Then there exists a net $\{B_\lambda\}_{\lambda\in\Lambda}$ of elements of $2^X \setminus \mathcal{U}$ converging to B [40, 1.6.3] and [104, Theorem 4]. Note that for each $\lambda \in \Lambda$, $\mathcal{T}(B_\lambda) \cap (X \setminus U) \neq \emptyset$. Let $x_\lambda \in \mathcal{T}(B_\lambda) \cap (X \setminus U)$. Without loss of generality, we assume that $\{x_\lambda\}_{\lambda\in\Lambda}$ converges to a point $x \in X$ [40, 3.1.23 and 1.6.1]. Note that $x \in X \setminus U$. We assert that $x \in \mathcal{T}(B)$. Suppose $x \in X \setminus \mathcal{T}(B)$.

S. Macías, *Set Function $\mathcal{T}$*, Developments in Mathematics 67,
https://doi.org/10.1007/978-3-030-65081-0_5

Then there exists a subcontinuum W of X such that $x \in \operatorname{Int}(W) \subset W \subset X \setminus B$. Since $\{B_\lambda\}_{\lambda\in\Lambda}$ converges to B and $\{x_\lambda\}_{\lambda\in\Lambda}$ converges to x, there exists $\lambda_0 \in \Lambda$ such that for each $\lambda \geq \lambda_0$, $B_\lambda \subset X \setminus W$ and $x_\lambda \in \operatorname{Int}(W)$. Let $\lambda \geq \lambda_0$. Then $x \in \operatorname{Int}(W) \subset W \subset X \setminus B_\lambda$. This implies that $x_\lambda \in X \setminus \mathcal{T}(B_\lambda)$, a contradiction. Thus, $x \in \mathcal{T}(B) \cap (X \setminus U)$. Hence, $B \in 2^X \setminus \mathcal{U}$. Therefore, $2^X \setminus \mathcal{U}$ is closed in 2^X and $\mathcal{U}$ is open. □

The next two theorems give sufficient conditions for $\mathcal{T}$ to be continuous for a continuum.

5.1.2 Theorem. *Let X and Z be continua, where Z is locally connected. If $f\colon X \twoheadrightarrow Z$ is a surjective, monotone and open map such that for each proper subcontinuum W of X, $f(W) \neq Z$, then $\mathcal{T}_X$ is continuous for X.*

Proof. Let A be a closed subset of X. Since Z is locally connected, $\mathcal{T}_Z f(A) = f(A)$ (Theorem 2.1.37). Hence, $f\mathcal{T}_X f^{-1}f(A) = f(A)$, Theorem 2.1.51 part (c). Thus,

$$f^{-1}f\mathcal{T}_X f^{-1}f(A) = f^{-1}f(A)$$

and

$$\mathcal{T}_X f^{-1}f(A) \subset f^{-1}f(A).$$

Then $\mathcal{T}_X f^{-1}f(A) = f^{-1}f(A)$, Remark 2.1.5. Since $A \subset f^{-1}f(A)$, $\mathcal{T}_X(A) \subset \mathcal{T}_X f^{-1}f(A) = f^{-1}f(A)$ (the inclusion holds by Proposition 2.1.7).

Suppose there exists $x \in f^{-1}f(A) \setminus \mathcal{T}_X(A)$. Then there exists a subcontinuum W of X such that $x \in \operatorname{Int}(W) \subset W \subset X \setminus A$. Since f is open, $f(x) \in \operatorname{Int}(f(W))$. For each $z \in Z \setminus f(\operatorname{Int}(W))$, there exists a subcontinuum M_z of Z such that $z \in \operatorname{Int}(M_z) \subset M_z \subset Z \setminus \{f(x)\}$ ($\mathcal{T}_Z(\{z\}) = \{z\}$). Since $Z \setminus f(\operatorname{Int}(W))$ is compact, there exist $z_1, \ldots, z_n$ in $Z \setminus f(\operatorname{Int}(W))$ such that $Z \setminus f(\operatorname{Int}(W)) \subset \bigcup_{j=1}^n \operatorname{Int}(M_{z_j}) \subset \bigcup_{j=1}^n M_{z_j}$. By choosing n as small as possible, we may assume that $M_{z_j} \cap f(W) \neq \emptyset$ for each $j \in \{1, \ldots, n\}$. Then $Z = f(W) \cup \left(\bigcup_{j=1}^n M_{z_j}\right)$, and $f^{-1}f(W) \cup \left(\bigcup_{j=1}^n f^{-1}(M_{z_j})\right) = X$.

For each $j \in \{1, \ldots, n\}$, let $q_j \in f(W) \cap M_{z_j}$. Then there exists $p_j \in W$ such that $f(p_j) = q_j$. Since $p_j \in f^{-1}(M_{z_j})$, $W \cap f^{-1}(M_{z_j}) \neq \emptyset$, $j \in \{1, \ldots, n\}$. Let $Y = W \cup \left(\bigcup_{j=1}^n f^{-1}(M_{z_j})\right)$. Then Y is a subcontinuum of X. We assert that $Y \neq X$. To see this, recall that $x \in f^{-1}f(A)$. Thus, there exists $y \in A$ such that $f(y) = f(x)$. Then $y \in X \setminus W$ and $f(y) \in Z \setminus \left(\bigcup_{j=1}^n M_{z_j}\right)$ (recall the construction of the M_{z_j}'s). Hence, $y \in X \setminus \left(\bigcup_{j=1}^n f^{-1}(M_{z_j})\right)$. However, $f(Y) = Z$, contrary to our hypothesis. Therefore, $\mathcal{T}_X(A) = f^{-1}f(A)$, i.e., $\mathcal{T}_X = \Im(f) \circ 2^f$. Since f is continuous and open, $\Im(f)$ and 2^f are continuous (Theorems 1.6.16 and 1.6.15, respectively). Therefore, $\mathcal{T}_X$ is continuous. □

5.1.3 Definition. Let $f\colon X \to Z$ be a map between continua. We say that f *is* $\mathcal{T}_{XZ}$*-continuous* provided that always $f\mathcal{T}_X(A) \subset \mathcal{T}_Z f(A)$, for every subset A of X or, equivalently, $f^{-1}\mathcal{T}_Z(B) \supset \mathcal{T}_X f^{-1}(B)$ for each subset B of Z.

5.1.4 Theorem. *Let X be a continuum for which $\mathcal{T}_X$ is continuous, and let Z be a continuum. If $f\colon X \twoheadrightarrow Z$ is a $\mathcal{T}_{XZ}$-continuous surjective open map, then $\mathcal{T}_Z$ is continuous for Z.*

Proof. Since f is $\mathcal{T}_{XZ}$-continuous, by Theorem 2.1.51 part (a),

$$f\mathcal{T}_X f^{-1}(B) = \mathcal{T}_Z(B)$$

for each subset B of Z. Thus, $\mathcal{T}_Z = 2^f \circ \mathcal{T}_X \circ \Im(f)$. Since f is continuous and open, $\Im(f)$ and 2^f are continuous (Theorems 1.6.16 and 1.6.15, respectively). Hence, $\mathcal{T}_Z$ is a composition of three maps. Therefore, $\mathcal{T}_Z$ is continuous. □

5.1.5 Notation. Let X be a compactum. If A is any subset of X, then

$$\mathrm{CL}(A) = \{B \in 2^X \mid B \subset A\}.$$

The following theorem tells us that the idempotency of $\mathcal{T}$ is related to its continuity.

5.1.6 Theorem. *If X is a continuum for which $\mathcal{T}$ is continuous, then $\mathcal{T}$ is idempotent on X.*

Proof. Let W be a subcontinuum of X, and let $x \in \mathrm{Int}(W)$. Since $X \setminus \mathrm{Int}(W)$ is a closed subset of X, we have that $\mathrm{CL}(X \setminus \mathrm{Int}(W))$ is closed in 2^X. Then $\mathcal{T}^{-1}(\mathrm{CL}(X \setminus \mathrm{Int}(W)))$ is a closed subset of 2^X, by the continuity of $\mathcal{T}$. Note that

$$\mathrm{CL}(X \setminus W) \subset \mathcal{T}^{-1}(\mathrm{CL}(X \setminus \mathrm{Int}(W)))$$

because $X \setminus W \subset X \setminus \mathrm{Int}(W)$, and Remark 2.1.5.

Since, for $W \neq X$, $X \setminus \mathrm{Int}(W)$ is a limit point of $\mathrm{CL}(X \setminus W)$ (recall that $\mathrm{Cl}(X \setminus W) = X \setminus \mathrm{Int}(W)$), it follows that $\mathcal{T}(X \setminus \mathrm{Int}(W)) \subset X \setminus \mathrm{Int}(W)$. Then there exists a subcontinuum M of X such that $x \in \mathrm{Int}(M) \subset M \subset \mathrm{Int}(W)$. If $W = X$, then it suffices to choose $M = X$. Thus, in either case, the theorem follows from Theorem 2.3.4. □

The next result gives us a characterization of the continuity of the set function $\mathcal{T}$.

5.1.7 Theorem. *Let X be a continuum. Then $\mathcal{T}$ is continuous if and only if $\mathcal{T}$ is idempotent on closed sets and $\mathcal{T}(2^X)$ is closed in 2^X.*

Proof. Suppose $\mathcal{T}$ is continuous. Then, by Theorem 5.1.6, $\mathcal{T}$ is idempotent on all subsets of X and, clearly, $\mathcal{T}(2^X)$ is closed in 2^X.

Suppose $\mathcal{T}$ is idempotent on closed sets and $\mathcal{T}(2^X)$ is closed in 2^X. Let $\{A_\lambda\}_{\lambda\in\Lambda}$ be a net of elements of 2^X converging to $A \in 2^X$. By [40, 3.1.23 and 1.6.1] and [104, Theorem 4], without loss of generality, we assume that $\{\mathcal{T}(A_\lambda)\}_{\lambda\in\Lambda}$ converges to an element $L \in 2^X$. By Lemma 3.1.7, $L \subset \mathcal{T}(A) \subset \mathcal{T}(L)$. Since $\mathcal{T}(2^X)$ is closed in 2^X, there exists $K \in 2^X$ such that $\mathcal{T}(K) = L$. Hence, $\mathcal{T}(L) = \mathcal{T}^2(K) = \mathcal{T}(K) = L$ (the second equality holds because $\mathcal{T}$ is idempotent on closed sets). Therefore, $\mathcal{T}(A) = L$ and $\mathcal{T}$ is continuous [97, Proposition 3.38].

□

As a consequence of Corollary 2.3.16 and Theorem 5.1.6, we have the following.

5.1.8 Corollary. *If X is a continuum for which $\mathcal{T}$ is continuous and $\mathcal{T}\colon 2^X \twoheadrightarrow 2^X$ is surjective, then X is locally connected.*

5.1.9 Theorem. *If X is a continuum for which $\mathcal{T}$ is continuous, then $\mathcal{T}(\{p,q\}) = \mathcal{T}(\{p\}) \cup \mathcal{T}(\{q\})$ for every $p, q \in X$.*

Proof. By Theorem 3.2.9, $\mathcal{G} = \{\mathcal{T}(\{x\}) \mid x \in X\}$ is a decomposition. Hence, X is point $\mathcal{T}$-symmetric.

Let $p \in X$. For each $x \in X$, define

$$A(x) = \{y \in X \mid \mathcal{T}(\{x,y\}) \in \mathcal{C}_1(X)\}$$

and

$$B(x) = \{y \in X \mid \mathcal{T}(\{x,y\}) = \mathcal{T}(\{x\}) \cup \mathcal{T}(\{y\})\}.$$

We assert that $A(x)$ and $B(x)$ are both closed for every $x \in X$. Let $x \in X$, and let $Z(x) \in \{A(x), B(x)\}$. If $y \in \mathrm{Cl}(Z(x))$, then there exists a net $\{y_\lambda\}_{\lambda\in\Lambda}$ of elements of $Z(x)$ converging to y [40, 1.6.3]. Hence, the net $\{\{x, y_\lambda\}\}_{\lambda\in\Lambda}$ converges to $\{x,y\}$ (note that for each $z, w \in X$, $\{z,w\} = \cup(\{\{z\},\{w\}\})$). Since the union function is continuous (Lemma 1.6.13) and $\{\{x,y_\lambda\}\}_{\lambda\in\Lambda}$ converges to $\{x,y\}$, we have that $\{\cup\{\{x\},\{y_\lambda\}\}\}_{\lambda\in\Lambda}$ converges to $\cup\{\{x\},\{y\}\} = \{x,y\}$. Hence, by the continuity of $\mathcal{T}$, the net $\{\mathcal{T}(\{x,y_\lambda\})\}_{\lambda\in\Lambda}$ converges to $\mathcal{T}(\{x,y\})$ [97, Proposition 3.38].

If $Z(x)$ is $A(x)$, then, since $\mathcal{C}_1(X)$ is compact (Theorem 1.6.7), $\mathcal{T}(\{x,y\}) \in \mathcal{C}_1(X)$. Suppose $Z(x)$ is $B(x)$. Then for each $\lambda \in \Lambda$, $\mathcal{T}(\{x,y_\lambda\}) = \mathcal{T}(\{x\}) \cup \mathcal{T}(\{y_\lambda\})$. Hence, $\{\mathcal{T}(\{x,y_\lambda\})\}_{\lambda\in\Lambda} = \{\mathcal{T}(\{x\}) \cup \mathcal{T}(\{y_\lambda\})\}_{\lambda\in\Lambda}$, and this net converges to $\mathcal{T}(\{x\}) \cup \mathcal{T}(\{y\})$ (by the continuity of $\mathcal{T}$ [97, Proposition 3.38] and the union function, Lemma 1.6.13). By the previous paragraph, we have that $\{\mathcal{T}(\{x,y_\lambda\})\}_{\lambda\in\Lambda}$ also converges to $\mathcal{T}(\{x,y\})$. Thus, $\mathcal{T}(\{x,y\}) = \mathcal{T}(\{x\}) \cup \mathcal{T}\{y\})$. Hence, $A(x)$ and $B(x)$ are both closed subsets of X. Note that, by Corollary 2.1.50, $X = A(p) \cup B(p)$.

Let $x \in \mathcal{T}(\{p\})$. Then $\mathcal{T}(\{x\}) \subset \mathcal{T}^2(\{p\}) = \mathcal{T}(\{p\})$ (by Proposition 2.1.7 and the fact that $\mathcal{T}$ is idempotent, Theorem 5.1.6). Since $\mathcal{G} = \{\mathcal{T}(\{x\}) \mid x \in X\}$ is a decomposition, Theorem 3.2.9, and $x \in \mathcal{T}(\{p\})$, $p \in \mathcal{T}(\{x\})$. Hence, $\mathcal{T}(\{p\}) \subset \mathcal{T}(\{x\})$. Therefore, $\mathcal{T}(\{p\}) = \mathcal{T}(\{x\})$. Also, since $\{p,x\} \subset \mathcal{T}(\{p\})$, $\mathcal{T}(\{p,x\}) \subset \mathcal{T}(\{p\})$. Thus, $\mathcal{T}(\{p,x\}) = \mathcal{T}(\{p\})$. Since $\mathcal{T}(\{p\})$ is a continuum (Theorem 2.1.27), $x \in A(p) \cap B(p)$. Now, let $x \in A(p) \cap B(p)$. Then $\mathcal{T}(\{p,x\}) = \mathcal{T}(\{p\}) \cup \mathcal{T}(\{x\})$, and this set is a continuum. Hence, $\mathcal{T}(\{p\}) \cap \mathcal{T}(\{x\}) \neq \emptyset$.

Let $y \in \mathcal{T}(\{p\}) \cap \mathcal{T}(\{x\})$. Then $x \in \mathcal{T}(\{y\}) \subset \mathcal{T}^2(\{p\}) = \mathcal{T}(\{p\})$. Therefore, $A(p) \cap B(p) = \mathcal{T}(\{p\})$.

Now, suppose $B(p) \neq \mathcal{T}(\{p\})$. Let $y \in A(p)$, and let $x \in B(p) \setminus \mathcal{T}(\{p\})$ be arbitrary points. Then $\mathcal{T}(\{p,x\}) = \mathcal{T}(\{p\}) \cup \mathcal{T}(\{x\})$ and $\mathcal{T}(\{p\}) \cap \mathcal{T}(\{x\}) = \emptyset$. Hence, $\mathcal{T}(\{x\}) \cap A(p) = \emptyset$. (Suppose there exists $z \in \mathcal{T}(\{x\}) \cap A(p)$, then, as before, $\mathcal{T}(\{z\}) = \mathcal{T}(\{x\})$. Since $\{p,z\} \subset \mathcal{T}(\{p\}) \cup \mathcal{T}(\{z\}) = \mathcal{T}(\{p\}) \cup \mathcal{T}(\{x\}) \subset \mathcal{T}(\{p,x\})$, $\mathcal{T}(\{p,z\}) \subset \mathcal{T}^2(\{p,x\}) = \mathcal{T}(\{p,x\})$. Similarly, we have that $\mathcal{T}(\{p,x\}) \subset \mathcal{T}(\{p,z\})$. Hence, $\mathcal{T}(\{p,z\}) = \mathcal{T}(\{p,x\})$. Since $z \in A(p)$, $\mathcal{T}(\{p,z\})$ is connected. Thus, $\mathcal{T}(\{p,x\})$ is connected, a contradiction.) Let U be an open set such that $\mathcal{T}(\{x\}) \subset U$ and $\mathrm{Cl}(U) \cap A(p) = \emptyset$. Now, let $q \in \mathrm{Bd}(U)$. Then $q \in (X \setminus A(p)) \cap (X \setminus \mathcal{T}(\{x\}))$. Hence, $q \in X \setminus \mathcal{T}(\{p,x\})$. Since $\mathcal{T}$ is idempotent (Theorem 5.1.6), $q \in X \setminus \mathcal{T}^2(\{p,x\})$. Thus, there exists a subcontinuum W of X such that $q \in \mathrm{Int}(W) \subset W \subset X \setminus \mathcal{T}(\{p,x\})$. Then $W \subset B(p)$, since otherwise $(W \cap A(p)) \cup (W \cap B(p))$ would be a separation of W (recall that $A(p) \cap B(p) = \mathcal{T}(\{p\})$ and $W \cap \mathcal{T}(\{p\}) = \emptyset$). Therefore, $y \in X \setminus W$, and $q \in X \setminus \mathcal{T}(\{x,y\})$ (if $q \in \mathcal{T}(\{x,y\})$, then $\{x,y\} \cap W \neq \emptyset$, a contradiction). Since q is an arbitrary point of $\mathrm{Bd}(U)$,

$$\mathcal{T}(\{x,y\}) = (\mathcal{T}(\{x,y\}) \cap U) \cup (\mathcal{T}(\{x,y\}) \cap (X \setminus \mathrm{Cl}(U))),$$

where $\mathcal{T}(\{x,y\}) \cap U$ and $\mathcal{T}(\{x,y\}) \cap (X \setminus \mathrm{Cl}(U))$ are separated. Thus, $\mathcal{T}(\{x,y\})$ is not connected. Hence, $\mathcal{T}(\{x,y\}) = \mathcal{T}(\{x\}) \cup \mathcal{T}(\{y\})$, Corollary 2.1.50. So, $x \in B(y)$. Thus, $B(p) \setminus \mathcal{T}(\{p\}) \subset B(y)$.

Note that $p \in \mathrm{Cl}(B(p) \setminus \mathcal{T}(\{p\}))$. If not, $p \in \mathrm{Int}(A(p))$, and since $A(p)$ is a continuum and $\mathcal{T}$ is idempotent (Theorem 5.1.6), there exists a continuum M such that $p \in \mathrm{Int}(M) \subset M \subset \mathrm{Int}(A(p))$ (Theorem 2.3.4). Hence, M misses some point $q \in \mathcal{T}(\{p\})$, and $p \in X \setminus \mathcal{T}(\{q\})$, contradicting the point $\mathcal{T}$-symmetry of X. Thus, $p \in \mathrm{Cl}(B(p) \setminus \mathcal{T}(\{p\}))$.

Since $B(y)$ is closed and $p \in \mathrm{Cl}(B(p) \setminus \mathcal{T}(\{p\}))$, it follows that $p \in B(y)$, or that $y \in B(p)$. But $y \in A(p)$, so that $y \in \mathcal{T}(\{p\})$, and $A(p) = \mathcal{T}(\{p\})$. By contrapositive, if $A(p) \neq \mathcal{T}(\{p\})$, then $B(p) = \mathcal{T}(\{p\})$. So, for each $p \in X$, either $A(p) = X$ or $B(p) = X$. Suppose there exists $p \in X$ such that $A(p) = X$. Let $q \in X$ be arbitrary. If $q \in \mathcal{T}(\{p\})$, then $A(q) = A(p) = X$. (As we did before, if $z \in A(p) \cup A(q)$, then, since $\mathcal{T}(\{p\}) = \mathcal{T}(\{q\})$, $\mathcal{T}(\{p,z\}) = \mathcal{T}(\{q,z\})$.) Suppose $q \in X \setminus \mathcal{T}(\{p\})$. Then $q \in A(p)$, so $p \in A(q)$. Since $p \in X \setminus \mathcal{T}(\{q\})$, $A(q) \neq \mathcal{T}(\{q\})$. Hence, $A(q) = X$. Thus, either $A(p) = X$ for every $p \in X$ or $B(p) = X$ for each $p \in X$. If $B(p) = X$ for every $p \in X$, the theorem is proved. So, suppose $A(p) = X$ for all $p \in X$. Then for each pair of points p and q of X, $\mathcal{T}(\{p,q\})$ is a continuum. Hence, by Theorem 2.3.13, X is indecomposable. Thus, by Theorem 2.1.44, $B(p) = X$ for all $p \in X$ in this case also. □

5.1.10 Theorem. *If X is a continuum for which $\mathcal{T}$ is continuous, then X is $\mathcal{T}$-additive.*

Proof. By Theorem 3.2.9, $\mathcal{G} = \{\mathcal{T}(\{x\}) \mid x \in X\}$ is a decomposition. Hence, X is point $\mathcal{T}$-symmetric.

Let $\mathcal{F}(X) = \bigcup_{n=1}^{\infty} \mathcal{F}_n(X)$. Then $\mathcal{F}(X)$ is the family of all finite subsets of X.

We show first that for each $A \in \mathcal{F}(X)$, $\mathcal{T}(A) = \bigcup_{a\in A} \mathcal{T}(\{a\})$. Suppose this is not true. Then there exists an $M \in \mathcal{F}(X)$ such that $\mathcal{T}(M) \neq \bigcup_{p\in M} \mathcal{T}(\{p\})$. Take M with the smallest cardinality. By Theorem 5.1.9, M has at least three elements. We assert that $\mathcal{T}(M)$ is a continuum. If $\mathcal{T}(M)$ is not connected, then there exist two disjoint closed subsets of A and B of X such that $\mathcal{T}(M) = A \cup B$. Then, by Lemma 2.1.46, Corollary 2.1.14 and the minimality of M:

$$\mathcal{T}(M) = A \cup B = \mathcal{T}(M \cap A) \cup \mathcal{T}(M \cap B)$$

$$= \bigcup_{p\in M\cap A} \mathcal{T}(\{p\}) \cup \bigcup_{p\in M\cap B} \mathcal{T}(\{p\}) = \bigcup_{p\in M} \mathcal{T}(\{p\}),$$

contrary to the choice of M. Moreover, if p and q are distinct points of M, $\mathcal{T}(\{p\}) \cap \mathcal{T}(\{q\}) = \emptyset$; otherwise, by point $\mathcal{T}$-symmetry and idempotency, $\mathcal{T}(\{p\}) = \mathcal{T}(\{q\})$ and then, since $M \subset \mathcal{T}(M \setminus \{p\})$:

$$\mathcal{T}(M) \subset \mathcal{T}^2(M \setminus \{p\}) = \mathcal{T}(M \setminus \{p\}) = \bigcup_{r\in M\setminus\{p\}} \mathcal{T}(\{r\}) \subset \mathcal{T}(M).$$

Thus, $\mathcal{T}(M) = \bigcup_{p\in M} \mathcal{T}(\{p\})$. This also contradicts the choice of M.

Now, let $p \in M$ be arbitrary, and let $N = M \setminus \{p\}$. Then N has at least two elements. Since N has cardinality smaller than the cardinality of M, $\mathcal{T}(N) = \bigcup_{r\in N} \mathcal{T}(\{r\})$; that is, $\mathcal{T}(N)$ is a finite union of pairwise disjoint subcontinua. Hence, $\mathcal{T}(N)$ is not a continuum. Now, define

$$L = \left\{ x \in X \;\middle|\; \mathcal{T}(N \cup \{x\}) = \bigcup_{r\in N} \mathcal{T}(\{r\}) \cup \mathcal{T}(\{x\}) \right\}$$

and

$$K = \{x \in X \mid \mathcal{T}(N \cup \{x\}) \in \mathcal{C}_1(X)\}.$$

Note that $N \subset L$ and $p \in K$. Thus, $L \neq \emptyset$ and $K \neq \emptyset$. We claim that L and K are closed subsets of X. To see this, let Z be either L or K. Let $x \in \mathrm{Cl}(Z)$. Then there exists a net $\{x_\lambda\}_{\lambda\in\Lambda}$ of elements of Z converging to x [40, 1.6.3].

If Z is L, since $\mathcal{T}$ and the union map σ (Lemma 1.6.13) are continuous, on one hand, $\{\mathcal{T}(N \cup \{x_\lambda\})\}_{\lambda\in\Lambda}$ converges to $\mathcal{T}(N \cup \{x\})$ [97, Proposition 3.38]. On the other hand, since $\{x_\lambda\}_{\lambda\in\Lambda}$ is contained in L, $\mathcal{T}(N \cup \{x_\lambda\}) = \bigcup_{r\in N} \mathcal{T}(\{r\}) \cup \mathcal{T}(\{x_\lambda\})$, for each $\lambda \in \Lambda$. Hence, $\mathcal{T}(N \cup \{x_\lambda\}_{\lambda\in\Lambda}) = \bigcup_{r\in N} \mathcal{T}(\{r\}) \cup \{\mathcal{T}(\{x_\lambda\})\}_{\lambda\in\Lambda}$, and $\bigcup_{r\in N} \mathcal{T}(\{r\}) \cup \{\mathcal{T}(\{x_\lambda\})\}_{\lambda\in\Lambda}$ converges to $\bigcup_{r\in N} \mathcal{T}(\{r\}) \cup \mathcal{T}(\{x\})$. Therefore, $\mathcal{T}(N \cup \{x\}) = \bigcup_{r\in N} \mathcal{T}(\{r\}) \cup \mathcal{T}(\{x\})$, and L is closed.

If Z is K, since $\mathcal{T}$ and the union map σ are continuous, the sequence $\{\mathcal{T}(N \cup \{x_\lambda\})\}_{\lambda\in\Lambda}$ is a net of continua converging to $\mathcal{T}(N \cup \{x\})$ ([97, Proposition 3.38] and [104, Theorem 4]) and $\mathcal{C}_1(X)$ is closed in 2^X (Theorem 1.6.7), then $\mathcal{T}(N \cup \{x\}) \in \mathcal{C}_1(X)$. Hence, $x \in K$. Therefore, K is closed.

If $y \in K \cap L$, then $\mathcal{T}(N \cup \{y\}) = \bigcup_{r \in N} \mathcal{T}(\{r\}) \cup \mathcal{T}(\{y\})$ and this set is connected. Thus, $\mathcal{T}(\{y\}) \cap \mathcal{T}(\{r\}) \neq \emptyset$ for each $r \in N$. Then, by point $\mathcal{T}$-symmetry and idempotency, $\mathcal{T}(\{y\}) = \mathcal{T}(\{r\})$ for every $r \in N$, a contradiction to the fact that for any two points r_1 and r_2 in N, if $r_1 \neq r_2$, then $\mathcal{T}(\{r_1\}) \neq \mathcal{T}(\{r_2\})$. Thus, $K \cap L = \emptyset$.

Let $x \in X \setminus L$. Then $\mathcal{T}(N \cup \{x\})$ is a continuum; otherwise, there would exist two disjoint closed subsets A and B of X such that $\mathcal{T}(N \cup \{x\}) = A \cup B$. Hence, by Lemma 2.1.46 and Corollary 2.1.14:

$$\mathcal{T}(N \cup \{x\}) = A \cup B = \mathcal{T}((N \cup \{x\}) \cap A) \cup \mathcal{T}((N \cup \{x\}) \cap B)$$

$$= \bigcup_{r \in (N \cup \{x\}) \cap A} \mathcal{T}(\{r\}) \cup \bigcup_{r \subset (N \cup \{x\}) \cap B} \mathcal{T}(\{r\}) = \bigcup_{r \in N} \mathcal{T}(\{r\}) \cup \mathcal{T}(\{x\}),$$

a contradiction to the fact that $x \in X \setminus L$. Therefore, $\mathcal{T}(N \cup \{x\})$ is a continuum. Thus, $x \in K$. Hence, $X = K \cup L$, and K and L are disjoint closed sets, a contradiction to the fact that X is connected. Therefore, $\mathcal{T}(A) = \bigcup_{a \in A} \mathcal{T}(\{a\})$ for each $A \in \mathcal{F}(X)$.

Let $U \in \mathfrak{U}_X$ and let $U' \in \mathfrak{U}_X$ be such that $2U' \subset U$. Since $\mathcal{T}$ is uniformly continuous (Theorem 1.3.15), there exists $V \in \mathfrak{U}_X$ such that if A and B are elements of 2^X and $\rho_{2^X}(A, B) < 2^V$, then $\rho_{2^X}(\mathcal{T}(A), \mathcal{T}(B)) < 2^{U'}$. Let B be a point of 2^X. Since $\mathcal{F}(X)$ is dense in 2^X (Lemma 1.6.9), there exists A in $\mathcal{F}(X)$ such that $\rho_{2^X}(A, B) < 2^V$. This implies that $\rho_{2^{2^X}}(\mathcal{F}_1(A), \mathcal{F}_1(B)) < 2^{2^V}$. Hence, by Theorem 1.6.15, we have that $\rho_{2^{2^X}}(2^{\mathcal{T}}(\mathcal{F}_1(A)), 2^{\mathcal{T}}(\mathcal{F}_1(B)) < 2^{2^{U'}}$. Thus, by Lemma 1.6.13, $\rho_{2^X}\left(\bigcup_{a \in A} \mathcal{T}(\{a\}), \bigcup_{b \in B} \mathcal{T}(\{b\})\right) < 2^{U'}$. Since $\rho_{2^X}(\mathcal{T}(A), \mathcal{T}(B)) < 2^{U'}$, $\rho_{2^X}\left(\bigcup_{a \in A} \mathcal{T}(\{a\}), \bigcup_{b \in B} \mathcal{T}(\{b\})\right) < 2^{U'}$ and $2U' \subset U$, we obtain that $\rho_{2^X}\left(\mathcal{T}(B), \bigcup_{b \in B} \mathcal{T}(\{b\})\right) < 2^U$. Therefore, $\mathcal{T}(B) = \bigcup_{b \in B} \mathcal{T}(\{b\})$. Now, the fact that X is $\mathcal{T}$-additive follows. □

As a consequence of Theorems 2.2.17 and 5.1.10, we have the following.

5.1.11 Corollary. *If X is a continuum for which $\mathcal{T}$ is continuous, then X is $\mathcal{T}$-symmetric.*

Proof. By Theorem 3.2.7, $\mathcal{G} = \{\mathcal{T}(\{x\}) \mid x \in X\}$ is a decomposition. Hence, X is point $\mathcal{T}$-symmetric.

Since $\mathcal{T}$ is continuous for X and X is point $\mathcal{T}$-symmetric, X is $\mathcal{T}$-additive (Theorem 5.1.10). Thus, since X is point $\mathcal{T}$-symmetric and $\mathcal{T}$-additive, X is $\mathcal{T}$-symmetric (Theorem 2.2.17). □

5.1.12 Theorem. *If X is a continuum for which $\mathcal{T}$ is continuous, W is a subcontinuum of X with nonempty interior and U is an open subset of X such that $W \subset U$, then there exists a point $p \in X$ such that $\mathcal{T}(\{p\}) \subset U$.*

Proof. If $X \setminus W$ is connected, let $p \in \text{Int}(W)$. Then $\text{Cl}(X \setminus W)$ is a continuum with every point outside W in its interior such that $\text{Cl}(X \setminus W) \subset X \setminus \{p\}$. Hence, $\mathcal{T}(\{p\}) \subset W \subset U$.

Suppose $X \setminus W$ is not connected. Thus, there exist two nonempty disjoint open subsets M and N of X such that $X \setminus W = M \cup N$. Then if $x \in M$, $\mathcal{T}(\{x\}) \subset \text{Cl}(M)$, because $W \cup N$ is a continuum (Lemma 1.4.32) having every point outside $\text{Cl}(M)$ in its interior and such that $W \cup N \subset X \setminus \{x\}$. Similarly, if $x \in N$, $\mathcal{T}(\{x\}) \subset \text{Cl}(N)$. Now, let

$$A = \{x \in X \mid \mathcal{T}(\{x\}) \cap M \cap (X \setminus U) \neq \emptyset\}$$

and

$$B = \{x \in X \mid \mathcal{T}(\{x\}) \cap M \neq \emptyset\}.$$

Observe that $A \subset B$, $B \neq \emptyset$ (since $M \subset B$), B is open ($\mathcal{G} = \{\mathcal{T}(\{x\}) \mid x \in X\}$ is a continuous decomposition, Theorem 3.2.9) while A is closed. (Note that $X \setminus A = \{x \in X \mid \mathcal{T}(\{x\}) \subset (X \setminus M) \cup U\}$. Since $\text{Cl}(M) \setminus M \subset U$, $(X \setminus M) \cup U$ is open. Thus, $X \setminus A$ is open, since $\mathcal{G}$ is a continuous decomposition.) Since $N \cap B = \emptyset$, $B \neq X$. Thus, $A \neq B$ because X is connected. Let $x \in B \setminus A$. Then $\mathcal{T}(\{x\}) \cap M \neq \emptyset$ and $\mathcal{T}(\{x\}) \cap M \cap (X \setminus U) = \emptyset$. Let $p \in \mathcal{T}(\{x\}) \cap M$. Then $\mathcal{T}(\{p\}) \subset \text{Cl}(M) \cap \mathcal{T}(\{x\})$ ($\mathcal{T}$ is idempotent by Theorem 5.1.6); in particular:

$$\mathcal{T}(\{p\}) \cap (X \setminus U) \subset \text{Cl}(M) \cap \mathcal{T}(\{x\}) \cap (X \setminus U) = \emptyset,$$

since $\text{Cl}(M) \setminus M \subset U$. Thus, $\mathcal{T}(\{p\}) \subset U$. □

5.1.13 Theorem. *If X is a continuum for which $\mathcal{T}$ is continuous and W is a continuum domain of X, then $\mathcal{T}(W) = W$.*

Proof. Let $L = \{p \in X \mid \mathcal{T}(\{p\}) \subset W\}$. Let $x \in W$, and let M be a subcontinuum of X such that $x \in \text{Int}(M)$. Then $\text{Int}(M) \cap \text{Int}(W) \neq \emptyset$. Let $y \in \text{Int}(M) \cap \text{Int}(W)$. Then, since $\mathcal{T}$ is idempotent (Theorem 5.1.6),

$$y \in X \setminus (\mathcal{T}(X \setminus \text{Int}(M)) \cup \mathcal{T}(X \setminus \text{Int}(W))),$$

Theorem 2.3.4. By additivity (Theorem 5.1.10):

$$y \in X \setminus \mathcal{T}((X \setminus \text{Int}(M)) \cup (X \setminus \text{Int}(W)))$$

and

$$y \in X \setminus \mathcal{T}(X \setminus (\text{Int}(M) \cap \text{Int}(W))).$$

Hence, there exists a subcontinuum N of X such that $y \in \text{Int}(N)$ and $N \subset \text{Int}(M) \cap \text{Int}(W)$. Then, by Theorems 5.1.12 and 2.3.4, there exists $p \in N$ such that

$p \in \text{Int}(N)$ and $\mathcal{T}(\{p\}) \subset N$. Thus, $\mathcal{T}(\{p\}) \subset W$ and $p \in L$. Hence, $M \cap L \neq \emptyset$ and $x \in \mathcal{T}(L)$. Since x is an arbitrary point of W, $W \subset \mathcal{T}(L)$. By the definition of L and additivity, $\mathcal{T}(L) \subset W$. Thus, $\mathcal{T}(W) = \mathcal{T}^2(L) = \mathcal{T}(L) = W$. □

The next theorem says that for continua for which $\mathcal{T}$ is continuous $\mathcal{T}$-additivity is equivalent to $\mathcal{T}$-symmetry.

5.1.14 Theorem. *If X is a continuum for which $\mathcal{T}$ is continuous, then X is $\mathcal{T}$-symmetric if and only if X is $\mathcal{T}$-additive.*

Proof. If X is $\mathcal{T}$-symmetric, by Theorem 2.2.11, X is $\mathcal{T}$-additive.

Suppose X is $\mathcal{T}$-additive. Let A and B be two closed subsets of X such that $A \cap \mathcal{T}(B) = \emptyset$. Then, by the definition of $\mathcal{T}$, compactness and Corollary 2.3.7, there exists a finite collection $\{W_j\}_{j=1}^n$ such that W_j is a continuum domain, for each $j \in \{1, \ldots, n\}$, $A \subset \bigcup_{j=1}^n \text{Int}(W_j)$ and $B \cap \left(\bigcup_{j=1}^n W_j\right) = \emptyset$. Then, by additivity and Theorem 5.1.13, $\mathcal{T}\left(\bigcup_{j=1}^n W_j\right) = \bigcup_{j=1}^n W_j$. Hence, $\mathcal{T}(A) \subset \bigcup_{j=1}^n W_j$. Therefore, $\mathcal{T}(A) \cap B = \emptyset$. □

5.1.15 Remark. Observe that the reverse implication of Theorem 5.1.14 also follows from Theorem 2.2.17, because, by Theorem 3.2.9, $\mathcal{G} = \{\mathcal{T}(\{x\}) \mid x \in X\}$ is a decomposition. Hence, X is point $\mathcal{T}$-symmetric.

As a consequence of Corollary 5.1.11 and Theorem 5.1.14, we have the following.

5.1.16 Corollary. *If X is a continuum for which $\mathcal{T}$ is continuous, then the following hold and are equivalent:*

(1) *X is point $\mathcal{T}$-symmetric,*
(2) *X is $\mathcal{T}$-symmetric,*
(3) *X is $\mathcal{T}$-additive.*

5.1.17 Corollary. *If X is an aposyndetic continuum for which $\mathcal{T}$ is continuous, then X is locally connected.*

Proof. By Theorem 5.1.10, X is $\mathcal{T}$-additive. Thus, by Theorem 2.2.14, X is locally connected. □

5.1.18 Theorem. *If X is a continuum for which $\mathcal{T}$ is continuous and X is almost connected im kleinen at $p \in X$, then X is semi-locally connected at p.*

Proof. Let

$$\mathcal{L} = \{A \in 2^X \mid p \in \text{Int}(A)\}.$$

Note that $X \in \mathcal{L}$. Hence, $\mathcal{L} \neq \emptyset$. By the almost connectedness im kleinen of X at p and Theorem 5.1.12, the set $B(A) = \{x \in X \mid \mathcal{T}(\{x\}) \subset A\}$ is nonempty for each $A \in \mathcal{L}$. By continuity of $\mathcal{T}$, $B(A)$ is closed ($\mathcal{G} = \{\mathcal{T}(\{x\}) \mid x \in X\}$ is a continuous decomposition, Theorem 3.2.9) for each $A \in \mathcal{L}$. Hence, $\{B(A) \mid A \in \mathcal{L}\}$ is a filterbase of closed subsets, and $\bigcap_{A \in \mathcal{L}} B(A) \neq \emptyset$. But,

$$\bigcap_{A\in\mathcal{L}} B(A) \subset \bigcap_{A\in\mathcal{L}} A = \{p\}.$$

Thus, $\mathcal{T}(\{p\}) \subset \bigcap_{A\in\mathcal{L}} A = \{p\}$. Therefore, by Theorem 2.1.35, X is semi-locally connected at p. □

5.1.19 Theorem. *Let X be a continuum. If $\mathcal{T}$ is continuous for X and $p \in X$, then the following are equivalent:*

(1) *X is connected im kleinen at p.*
(2) *X is almost connected im kleinen at p.*
(3) *X is semi-locally connected at p.*

Proof. By Theorem 3.2.9, $\mathcal{G} = \{\mathcal{T}(\{x\}) \mid x \in X\}$ is a decomposition. Hence, X is point $\mathcal{T}$-symmetric. Note that, by Corollary 5.1.16, X is $\mathcal{T}$-symmetric. Thus, by Theorem 2.2.5, X is connected im kleinen at p if and only if X is semi-locally connected at p.

Clearly, if X is connected im kleinen at p, then X is almost connected im kleinen at p.

By Theorem 5.1.18, if X is almost connected im kleinen at p, then X is semi-locally connected at p. □

5.1.20 Theorem. *Let X be a continuum for which $\mathcal{T}$ is continuous. Then X is semi-locally connected if and only if X is locally connected.*

Proof. If X is semi-locally connected, then X is aposyndetic (Corollary 2.1.36). Hence, by Corollary 5.1.17, X is locally connected.

If X is locally connected, by Theorem 2.2.14, X is aposyndetic and $\mathcal{T}$-additive. Thus, by Corollary 2.1.36, X is semi-locally connected. □

As a consequence of Theorems 5.1.18 and 5.1.20, we have the following.

5.1.21 Corollary. *Let X be a continuum for which $\mathcal{T}$ is continuous. If X is almost connected im kleinen at each of its points, then X is locally connected.*

5.1.22 Lemma. *Let X be a continuum, let A be a nonempty closed subset of X and let $\mathcal{A} = \{\mathcal{T}(\{a\}) \mid a \in A\}$. If $\mathcal{T}|_{\mathcal{F}_1(X)} \colon \mathcal{F}_1(X) \to 2^X$ is continuous, then $\mathcal{A}$ is closed in 2^X.*

Proof. Let $B \in \mathrm{Cl}_{2^X}(\mathcal{A})$. Then there exists a net $\{a_\lambda\}_{\lambda\in\Lambda}$ of elements of A such that $\{\mathcal{T}(\{a_\lambda\})\}_{\lambda\in\Lambda}$ converges to B ([40, 1.6.3], [104, Theorem 4] and [97, Proposition 3.38]). Since A is compact, without loss of generality, the net $\{a_\lambda\}_{\lambda\in\Lambda}$ converges to a point a of A [40, 3.1.23 and 1.6.1]. Since $\mathcal{T}|_{\mathcal{F}_1(X)}$ is continuous, we obtain that $\mathcal{T}(\{a\}) = B$. Therefore, $B \in \mathcal{A}$ and $\mathcal{A}$ is closed in 2^X. □

5.1.23 Theorem. *Let X be a continuum such that $\mathcal{T}(A) = \bigcup\{\mathcal{T}(\{a\}) \mid a \in A\}$ for all $A \in 2^X$. If $\mathcal{T}|_{\mathcal{F}_1(X)} \colon \mathcal{F}_1(X) \to 2^X$ is continuous, then $\mathcal{T}$ is continuous.*

Proof. Let $U \in \mathfrak{U}_X$. Since $\mathcal{T}|_{\mathcal{F}_1(X)}$ is uniformly continuous, Theorem 1.3.15, there exists $V \in \mathfrak{U}_X$ such that if a and b are elements of X with $\rho_{2^X}(\{a\},\{b\}) < 2^V$, then $\rho_{2^X}(\mathcal{T}(\{a\}),\mathcal{T}(\{b\})) < 2^U$.

Let A and B be points of 2^X such that $\rho_{2^X}(A,B) < 2^V$. Let $\mathcal{A} = \{\mathcal{T}(\{a\}) \mid a \in A\}$, and let $\mathcal{B} = \{\mathcal{T}(\{b\}) \mid b \in B\}$. By Lemma 5.1.22, $\mathcal{A}$ and $\mathcal{B}$ are closed subsets of 2^X. Let $\mathcal{T}(\{a\}) \in \mathcal{A}$. Since $\rho_{2^X}(A,B) < 2^V$, there exists $b \in B$ such that $\rho_{2^X}(\{a\},\{b\}) < 2^V$. Thus, $\rho_{2^X}(\mathcal{T}(\{a\}),\mathcal{T}(\{b\})) < 2^U$. This implies that $\mathcal{A} \subset B(\mathcal{B}, 2^{2^U})$. Similarly, $\mathcal{B} \subset B(\mathcal{A}, 2^{2^U})$. Then, by Lemma 1.6.13, $\rho_{2^X}(\bigcup\mathcal{A},\bigcup\mathcal{B}) < 2^U$. By hypothesis, $\mathcal{T}(A) = \bigcup\mathcal{A}$ and $\mathcal{T}(B) = \bigcup\mathcal{B}$. Therefore, $\mathcal{T}$ is continuous, Theorem 1.3.15. □

By Lemma 1.6.13 and Theorem 5.1.23, we obtain the following.

5.1.24 Corollary. *If X is a $\mathcal{T}$-additive continuum such that $\mathcal{T}|_{\mathcal{F}_1(X)} \colon \mathcal{F}_1(X) \to 2^X$ is continuous, then $\mathcal{T}$ is continuous.*

5.1.25 Theorem. *Let X be a continuum for which $\mathcal{T}_X$ is continuous such that $\mathcal{T}_X(\{x\})$ is a terminal subcontinuum of X for all points x of X. If Z is a continuum and $f \colon X \twoheadrightarrow Z$ is a monotone map, then $\mathcal{T}_Z$ is continuous.*

Proof. Since $\mathcal{T}_X$ is continuous, by Theorem 3.2.9, $\mathcal{G} = \{\mathcal{T}_X(\{x\}) \mid x \in X\}$ is a continuous decomposition; also, by Theorem 5.1.10, X is $\mathcal{T}_X$-additive. Hence, by Theorem 2.2.20, Z is $\mathcal{T}_Z$-additive. By Corollary 5.1.24, we only need to show that $\mathcal{T}_Z|_{\mathcal{F}_1(Z)} \colon \mathcal{F}_1(Z) \to 2^Z$ is continuous.

Let $\{z_\lambda\}_{\lambda\in\Lambda}$ be a net of elements of Z converging to $z \in Z$. We prove that $\{\mathcal{T}_Z(\{z_\lambda\})\}_{\lambda\in\Lambda}$ converges to $\mathcal{T}(\{z\})$. Recall that for each $z' \in Z$, $\mathcal{T}_Z(\{z'\})$ is a continuum (Theorem 2.1.27). Since $\mathcal{C}_1(X)$ and $\mathcal{C}_1(Z)$ are compact (Theorem 1.6.7) and f is monotone, we may assume that $\{f^{-1}(z_\lambda)\}_{\lambda\in\Lambda}$ converges to an element K of $\mathcal{C}_1(X)$ ([40, 3.1.23 and 1.6.1] and [104, Theorem 4]). Observe that $K \subset f^{-1}(z)$.

We prove that $f\mathcal{T}_X(K) = f\mathcal{T}_X f^{-1}(z) = \mathcal{T}_Z(\{z\})$. To this end, let $x_0 \in f^{-1}(z)$. Then $x_0 \in \mathcal{T}_X(\{x_0\}) \cap f^{-1}(z)$. Since $\mathcal{T}_X(\{x_0\})$ is a terminal subcontinuum of X, either $\mathcal{T}_X(\{x_0\}) \subset f^{-1}(z)$ or $f^{-1}(z) \subset \mathcal{T}_X(\{x_0\})$. First, suppose that $\mathcal{T}_X(\{x_0\}) \subset f^{-1}(z)$. Note that this implies that for each $w \in f^{-1}(z)$, $\mathcal{T}_X(\{w\}) \subset f^{-1}(z)$. To see this, let $w \in f^{-1}(z)$. Then, since $\mathcal{T}_X(\{w\})$ is a terminal subcontinuum of X, either $\mathcal{T}_X(\{w\}) \subset f^{-1}(z)$ or $f^{-1}(z) \subset \mathcal{T}_X(\{w\})$. If we have the second case, then $\mathcal{T}_X(\{x\}) \subset \mathcal{T}_X(\{w\})$. Hence, $\mathcal{T}_X(\{x\}) = \mathcal{T}_X(\{w\})$, because $\mathcal{G}$ is a decomposition, and $\mathcal{T}_X(\{w\}) = f^{-1}(z)$. By Corollary 2.2.13, $\mathcal{T}_X(K) = \bigcup\{\mathcal{T}_X(\{k\}) \mid k \in K\} \subset f^{-1}(z)$ and $\mathcal{T}_X f^{-1}(z) = \bigcup\{\mathcal{T}_X(\{w\}) \mid w \in f^{-1}(z)\} \subset f^{-1}(z)$. Then $\mathcal{T}_X(K) \subset \mathcal{T}_X f^{-1}(z) \subset f^{-1}(z)$. Thus, $f\mathcal{T}_X(K) \subset f\mathcal{T}_X f^{-1}(z) \subset \{z\}$. Therefore, $f\mathcal{T}_X(K) = f\mathcal{T}_X f^{-1}(z)$.

Next, assume that $f^{-1}(z) \subset \mathcal{T}_X(\{x_0\})$. With a similar argument to the one given in the previous paragraph, we have that for each $w \in f^{-1}(z)$, $f^{-1}(z) \subset \mathcal{T}_X(\{w\})$. Let $k \in K$. Then $\{k\} \subset K \subset f^{-1}(z) \subset \mathcal{T}_X(\{k\})$. Since $\mathcal{T}_X$ is continuous, $\mathcal{T}_X$ is idempotent (Theorem 5.1.6). Hence, $\mathcal{T}_X(\{k\}) \subset \mathcal{T}_X(K) \subset \mathcal{T}_X f^{-1}(z) \subset \mathcal{T}_X\mathcal{T}_X(\{k\}) = \mathcal{T}_X(\{k\})$. Then $f\mathcal{T}_X(K) = f\mathcal{T}_X f^{-1}(z)$.

In both cases, we obtain that $f\mathcal{T}_X(K) = f\mathcal{T}_X f^{-1}(z)$. Since, by Theorem 2.1.51 (c), $\mathcal{T}_Z(\{z\}) = f\mathcal{T}_X f^{-1}(z)$, we have that $f\mathcal{T}_X(K) = f\mathcal{T}_X f^{-1}(z) = \mathcal{T}_Z(\{z\})$.

Recall that $\{z_\lambda\}_{\lambda\in\Lambda}$ is a net of elements of Z converging to z. Since f is monotone, for each $\lambda \in \Lambda$, $\mathcal{T}_Z(\{z_\lambda\}) = f\mathcal{T}_X f^{-1}(z_\lambda)$ (Theorem 2.1.51 (c)). Since $\{f^{-1}(z_\lambda)\}_{\lambda\in\Lambda}$ converges to K and $\mathcal{T}_X$ and f are continuous, we have that $\{f\mathcal{T}_X f^{-1}(z_\lambda)\}_{\lambda\in\Lambda}$ converges to $f\mathcal{T}_X(K) = \mathcal{T}_Z(\{z\})$). Therefore, $\{\mathcal{T}_Z(\{z_\lambda\})\}_{\lambda\in\Lambda}$ converges to $\mathcal{T}_Z(\{z\})$ and $\mathcal{T}_Z$ is continuous [97, Propositon 3.38]. □

By Theorems 3.2.9, 3.1.10 and 5.1.25, we obtain the following.

5.1.26 Corollary. *Let X be a continuum for which $\mathcal{T}_X$ is continuous and has the property of Kelley. If Z is a continuum and $f: X \twoheadrightarrow Z$ is a monotone map, then $\mathcal{T}_Z$ is continuous.*

5.1.27 Theorem. *Let X be a decomposable continuum for which $\mathcal{T}_X$ is continuous. Suppose there exists a subcontinuum L of X such that $\mathcal{T}_X(\{x\}) \cap L \neq \emptyset$ for all $x \in X$. If $q: X \twoheadrightarrow X/L$ is the quotient map. Then q is a monotone map and $\mathcal{T}_{X/L}$ is not continuous.*

Proof. Since $\mathcal{T}_X$ is continuous, by Theorem 3.2.9, $\mathcal{G} = \{\mathcal{T}_X(\{x\}) \mid x \in X\}$ is a continuous decomposition. Let χ_0 represent the element of X/L such that $q(L) = \{\chi_0\}$. We show that $\mathcal{T}_{X/L}|_{\mathcal{F}_1(X/L)}$ is not continuous at $\{\chi_0\}$.

Let Γ be subcontinuum of X/L with nonempty interior. Since q is monotone, $q^{-1}(\Gamma)$ is a subcontinuum of X (Lemma 1.4.46) with nonempty interior. Let $x_0 \in \text{Int}(q^{-1}(\Gamma))$, by Lemma 3.2.3, $\mathcal{T}_X(\{x_0\}) \subset q^{-1}(\Gamma)$. Since $L \cap \mathcal{T}_X(\{x_0\}) \neq \emptyset$, $\chi_0 \in q\mathcal{T}_X(\{x_0\}) \subset qq^{-1}(\Gamma) = \Gamma$. Thus, $\mathcal{T}_{X/L}(\{\chi_0\}) = X/L$.

Let $x \in X \setminus L$. Then, by Theorem 2.1.51 (c), $\mathcal{T}_{X/L}(\{q(x)\}) = q\mathcal{T}_X q^{-1}(\{q(x)\}) = q\mathcal{T}_X(\{x\})$. Since $\mathcal{T}_X(\{x\}) \setminus L \neq \emptyset$ and $\mathcal{T}_X(\{x\}) \cap L \neq \emptyset$, there exists a net $\{z_\lambda\}_{\lambda\in\Lambda}$ of elements of $\mathcal{T}_X(\{x\}) \setminus L$ converging to an element z of $\mathcal{T}_X(\{x\}) \cap L$ [40, 1.6.3]. Since $\mathcal{G}$ is a decomposition, for all $\lambda \in \Lambda$, $\mathcal{T}_X(\{z_\lambda\}) = \mathcal{T}_X(\{x\})$. Since $z \in L$, $q(z) = \chi_0$. Let $\chi_\lambda = q(z_\lambda)$. Since $\{z_\lambda\}_{\lambda\in\Lambda}$ converges to z and q is continuous, we have that $\{\chi_\lambda\}_{\lambda\in\Lambda}$ converges to χ_0 [97, Proposition 3.38]. Hence, as before, for each $\lambda \in \Lambda$, $\mathcal{T}_{X/L}(\{\chi_\lambda\}) = q\mathcal{T}_X(\{z_\lambda\}) = q\mathcal{T}_X(\{x\})$.

Since X is decomposable and $\mathcal{G}$ is a decomposition, $\mathcal{G}$ has at least two elements. Thus, $\mathcal{T}_X(\{x\}) \neq X$. Let $p \in X \setminus \mathcal{T}_X(\{x\})$. Then $\mathcal{T}_X(\{p\}) \cap \mathcal{T}_X(\{x\}) = \emptyset$ ($\mathcal{G}$ is a decomposition). Since L intersects all the elements of $\mathcal{G}$, we can find a net $\{r_\lambda\}_{\lambda\in\Lambda}$ of elements of $L \setminus \mathcal{T}_X(\{x\})$ converging to an element $r \in L \cap \mathcal{T}_X(\{x\})$ [40, 1.6.3]. Hence, $\mathcal{T}_X(\{r\}) = \mathcal{T}_X(\{x\})$ ($\mathcal{G}$ is a decomposition). Since $\{r_\lambda\}_{\lambda\in\Lambda}$ converges to r and $\mathcal{T}_X$ is continuous, $\{\mathcal{T}_X(\{r_\lambda\})\}_{\lambda\in\Lambda}$ converges to $\mathcal{T}_X(\{r\})$ ([97, Proposition 3.38] and [104, Theorem 4]). For each $\lambda \in \Lambda$, there exists $x_\lambda \in \mathcal{T}_X(\{r_\lambda\})$ such that the net $\{x_\lambda\}_{\lambda\in\Lambda}$ converges to x [40, 3.1.23 and 1.6.1] . Since $x \in X \setminus L$, we may assume that $x_\lambda \in X \setminus L$ for all $\lambda \in \Lambda$ [40, 1.6.3] . Let $\lambda_0 \in \Lambda$. Then $\mathcal{T}_X(\{x_{\lambda_0}\}) = \mathcal{T}_X(\{r_{\lambda_0}\})$. Thus, $\mathcal{T}_X(\{x_{\lambda_0}\}) \cap \mathcal{T}_X(\{x\}) = \emptyset$, and $q(x_{\lambda_0}) \in q\mathcal{T}_X(\{x_{\lambda_0}\}) \setminus q\mathcal{T}_X(\{x\})$. This implies that $q\mathcal{T}_X(\{x\}) \neq X/L$.

Since $\{\chi_\lambda\}_{\lambda\in\Lambda}$ converges to χ_0 and $\mathcal{T}_{X/L}(\{\chi_\lambda\}) \subset q\mathcal{T}_X(\{x\}) \subsetneq X/L = \mathcal{T}_{X/L}(\{\chi_0\})$, the net $\{\mathcal{T}_{X/L}(\{\chi_\lambda\})\}_{\lambda\in\Lambda}$ does not converge to $\mathcal{T}_{X/L}(\{\chi_0\})$. Therefore, $\mathcal{T}_{X/L}$ is not continuous [97, Proposition 3.38]. □

5.1.28 Remark. In [64, pp. 139-158] two metric continua X and Z and a monotone map $f\colon X \twoheadrightarrow Z$ are given such that $\mathcal{T}_X$ is continuous but $\mathcal{T}_Z$ is not.

We end this section with a characterization of continua with the uniform property of Effros for which the set function $\mathcal{T}$ is continuous.

5.1.29 Theorem. *Let X be a continuum with the uniform property of Effros. Then $\mathcal{T}_X$ is continuous if and only if one of the following conditions holds:*

(1) *X is indecomposable.*
(2) *X is not aposyndetic and $X/\mathcal{G}$ is locally connected, where $\mathcal{G} = \{\mathcal{T}_X(\{x\}) \mid x \in X\}$. If X is metric, then $X/\mathcal{G}$ is homeomorphic to either the simple closed curve or the Menger universal curve.*
(3) *X is locally connected.*

Proof. By Theorem 1.4.59, X is a homogeneous continuum. Suppose $\mathcal{T}_X$ is continuous. If X is indecomposable, then (1) holds. Assume X is decomposable. If X is not aposyndetic, $\mathcal{G} = \{\mathcal{T}_X(\{x\}) \mid x \in X\}$ is a continuous decomposition (Theorem 3.2.7) and, by Theorem 3.2.9, $X/\mathcal{G}$ is locally connected. Hence, (2) is satisfied (the metric part follows from [92, Theorem 5.1.25]). If X is aposyndetic, then, by Corollary 5.1.17, X is locally connected. Thus, (3) holds.

Conversely, if (1) holds, by Theorem 2.1.44, $\mathcal{T}_X$ is a constant map. Hence, $\mathcal{T}_X$ is continuous. If (2) is satisfied, by Theorem 3.3.12, $\mathcal{T}_X$ is continuous. If (3) holds, then, by Theorem 2.1.37, $\mathcal{T}_X$ is the identity map on 2^X. Thus, $\mathcal{T}_X$ is continuous. □

5.2 Examples of Metric Continua for Which $\mathcal{T}$ Is Continuous

In 1970, David P. Bellamy asked in [3] if any nonlocally connected continuum for which $\mathcal{T}$ is continuous had to be indecomposable. In 1983, Bellamy [5, Remark 2, p. 10] answered the question in the negative by showing that the circle of pseudo-arcs (Definition 5.2.3) is a decomposable nonlocally connected continuum for which $\mathcal{T}$ is continuous.

We give several families of decomposable nonlocally connected metric continua for which $\mathcal{T}$ is continuous.

A proof of the next theorem is due to Wayne Lewis and may be found in [78].

5.2.1 Theorem. *If X is a one-dimensional metric continuum, then there exists a one-dimensional metric continuum $\widehat{X}$ such that $\widehat{X}$ has a terminal (Definition 1.4.50) continuous decomposition into pseudo-arcs (Definition 4.4.25) such that the decomposition space is homeomorphic to X.*

5.2.2 Remark. Observe that the metric continua constructed by Wayne Lewis do not contain arcs. Hence, they are not locally connected, and, by construction, the continuum $\widehat{X}$ is decomposable if and only if X is decomposable.

5.2.3 Definition. The *circle of pseudo-arcs* is the space $\widehat{X}$, obtained from Theorem 5.2.1 when X is the circle.

Now we are ready to present our first family of examples.

5.2.4 Theorem. *If X is a locally connected one-dimensional metric continuum, then the set function $\mathcal{T}_{\widehat{X}}$ is continuous for the continuum $\widehat{X}$ given in Theorem 5.2.1.*

Proof. Let X be a one-dimensional locally metric connected metric continuum, and let $\widehat{X}$ be the one-dimensional metric continuum given in Theorem 5.2.1. Since $\widehat{X}$ admits a continuous decomposition into pseudo-arcs, the quotient map $q\colon \widehat{X} \twoheadrightarrow X$ is monotone, open and surjective. Let W be a proper subcontinuum of $\widehat{X}$, let $\hat{x} \in \widehat{X} \setminus W$ and let $P_{\hat{x}}$ be the pseudo-arc of the decomposition that contains $\hat{x}$. Since $P_{\hat{x}}$ is a terminal subcontinuum of $\widehat{X}$, either $W \subset P_{\hat{x}}$ or $W \cap P_{\hat{x}} = \emptyset$. In any case, $q(W)$ is a proper subcontinuum of X. Therefore, by Theorem 5.1.2, $\mathcal{T}_{\widehat{X}}$ is continuous. □

5.2.5 Remark. What Bellamy does to show the continuity of $\mathcal{T}_X$ in Theorem 5.1.2 is to prove that $\mathcal{T}_X = \Im(f) \circ 2^f$; that is, $\mathcal{T}_X(A) = f^{-1} \circ f(A)$ for each $A \in 2^X$. Thus, for the continuum $\widehat{X}$ of Theorem 5.2.4, $\mathcal{T}_{\widehat{X}}(\{\hat{x}\}) = P_{\hat{x}}$. This implies that $\widehat{X}$ is point $\mathcal{T}$-symmetric. Hence, $\widehat{X}$ is $\mathcal{T}$-additive by Theorem 5.1.10.

The next example shows that the requirement in Theorem 5.1.2, that the images of proper subcontinua of the domain are proper subcontinua of the range, is essential. The existence of the continuum $\widetilde{Z}$ is due to Karen Villarreal [121].

5.2.6 Example. Let Z be the circle of pseudo-arcs. Then there exists a two-dimensional aposyndetic nonlocally connected homogeneous metric continuum $\widetilde{Z} \subset Z \times Z$ such that if π_1 is the projection map of $Z \times Z$ onto the first factor, then $\pi_1|_{\widetilde{Z}}\colon \widetilde{Z} \twoheadrightarrow Z$ is open, monotone and surjective. Let $\tilde{q}\colon Z \twoheadrightarrow \mathcal{S}^1$ be the quotient map from Z onto the unit circle. Then $\tilde{q} \circ \pi_1|_{\widetilde{Z}}\colon \widetilde{Z} \twoheadrightarrow \mathcal{S}^1$ is open, monotone and surjective. From the construction of $\widetilde{Z}$, we know that $\Delta_Z = \{(z,z) \mid z \in Z\}$ is a proper subcontinuum of $\widetilde{Z}$ such that $\left(\tilde{q} \circ \pi_1|_{\widetilde{Z}}\right)(\Delta_Z) = \mathcal{S}^1$. Observe that, since $\widetilde{Z}$ is aposyndetic and nonlocally connected, $\mathcal{T}_{\widetilde{Z}}$ is not continuous, Corollary 5.1.17.

Let us note the following question:

5.2.7 Question. Let X be a nonlocally connected decomposable metric continuum for which the set function $\mathcal{T}$ is continuous. Does X admit a terminal continuous decomposition into pseudo-arcs such that the decomposition space is locally connected?

The answer to this question is negative as can be seen in the following:

5.2.8 Example. Let Z be the circle of pseudo-arcs, and let $q\colon Z \twoheadrightarrow \mathcal{S}^1$ be the quotient map. Let $A = \{(a,b) \in \mathcal{S}^1 \mid a \geq 0 \text{ and } b \geq 0\}$. Let $\mathcal{L} =$

$\{q^{-1}((a,b)) \mid (a,b) \in A\} \cup \{\{\hat{x}\} \mid \hat{x} \in Z$ and $q(\hat{x}) \in \mathcal{S}^1 \setminus A\}$. Then $\mathcal{L}$ is an upper semicontinuous decomposition of Z such that $Z/\mathcal{L}$ is a continuum containing an arc. Let $q_{\mathcal{L}} \colon Z \twoheadrightarrow Z/\mathcal{L}$ be the quotient map. Note that for each $\chi \in Z/\mathcal{L}$, there exists $z \in \mathcal{S}^1$ such that $q_{\mathcal{L}}^{-1}(\chi) \subset q^{-1}(z)$. Hence, by [36, 3.2, p. 123], the function $f \colon Z/\mathcal{L} \twoheadrightarrow \mathcal{S}^1$ given by $f = q \circ q_{\mathcal{L}}^{-1}$ is well defined and continuous. Since q is monotone and open, f is monotone and open. Let $\mathcal{W}$ be a proper subcontinuum of $Z/\mathcal{L}$. Then $q_{\mathcal{L}}^{-1}(\mathcal{W})$ is a proper subcontinuum of Z (Lemma 1.4.46). Thus, $f(\mathcal{W}) = qq_{\mathcal{L}}^{-1}(\mathcal{W})$ is a proper subcontinuum of $\mathcal{S}^1$. Therefore, by Theorem 5.1.2, $\mathcal{T}_{Z/\mathcal{L}}$ is continuous for $Z/\mathcal{L}$. Since $Z/\mathcal{L}$ contains an arc, there does not exist a continuous decomposition of $Z/\mathcal{L}$ into pseudo-arcs.

The next theorem characterizes the class of irreducible metric continua for which $\mathcal{T}$ is continuous.

5.2.9 Theorem. *Let X be an irreducible metric continuum. Then $\mathcal{T}_X$ is continuous for X if and only if X is continuously irreducible. Moreover, $\mathcal{G} = \{\mathcal{T}_X(\{x\}) \mid x \in X\}$ is the finest continuous monotone decomposition of X such that $X/\mathcal{G}$ is an arc.*

Proof. Suppose X is an irreducible metric continuum for which $\mathcal{T}_X$ is continuous. Hence,

$$\mathcal{G} = \{\mathcal{T}_X(\{x\}) \mid x \in X\}$$

is a monotone continuous decomposition of X such that $X/\mathcal{G}$ is a locally connected continuum, and the elements of $\mathcal{G}$ are nowhere dense, Theorem 3.2.9. Since X is irreducible and the quotient map, $q \colon X \twoheadrightarrow X/\mathcal{G}$, is monotone, $X/\mathcal{G}$ is an irreducible continuum [75, Theorem 3, p. 192]. Hence, $X/\mathcal{G}$ is an arc, Theorem 2.2.18. Thus, X is a continuum of type λ, Remark 1.4.64. We identify $X/\mathcal{G}$ with $[0,1]$.

Let $\mathbb{G}$ be the finest upper semicontinuous decomposition of X each element of which is nowhere dense and $X/\mathbb{G}$ is an arc. Let $q' \colon X \twoheadrightarrow X/\mathbb{G} = [0,1]$ be the quotient map. Let x and z be two points of X such that $q'(x) \neq q'(z)$. Then there exists a subinterval A of $[0,1]$ such that $q'(z) \in \mathrm{Int}_{[0,1]}(A)$ and $q'(x) \in [0,1] \setminus A$. Hence, $z \in \mathrm{Int}_X((q')^{-1}(A))$ and $x \in X \setminus (q')^{-1}(A)$. Thus, $z \in X \setminus \mathcal{T}_X(\{x\})$. Therefore, $\mathcal{T}_X(\{x\}) \subset (q')^{-1}q'(x)$. Let $w \in (q')^{-1}q'(x)$, and suppose W is a subcontinuum of X such that $w \in \mathrm{Int}_X(W)$. Then $(q')^{-1}q'(x) \subset W$ [119, Theorem 5, p. 10]. Hence, $w \in \mathcal{T}_X(\{x\})$ and $\mathcal{T}_X(\{x\}) = (q')^{-1}q'(x)$. Thus, $\mathbb{G} = \{\mathcal{T}_X(\{x\}) \mid x \in X\} = \mathcal{G}$ is the finest upper semicontinuous monotone decomposition of X such that $X/\mathbb{G}$ is an arc. Since $\mathcal{T}_X$ is continuous, $\mathcal{G}$ is continuous. Therefore, X is a continuously irreducible continuum.

Suppose X is a continuously irreducible continuum. Let $\mathbb{G}$ be the finest continuous monotone decomposition of X such that $X/\mathbb{G}$ is an arc, and let $q \colon X \twoheadrightarrow [0,1]$ be the quotient map. Then q is monotone and open. Since X is continuously irreducible, if Z is a proper subcontinuum of X, then $q(Z)$ is a proper subcontinuum of $[0,1]$. Hence, by Theorem 5.1.2, $\mathcal{T}_X$ is continuous. In fact, $\mathcal{T}_X(A) = q^{-1}q(A)$

for every nonempty closed subset A of X. In particular, $\mathcal{T}_X(\{x\}) = q^{-1}q(x)$ for each $x \in X$, and $\mathbb{G} = \{\mathcal{T}_X(\{x\}) \mid x \in X\}$. □

The next result characterizes the class of metric θ-continua of type A for which the set function $\mathcal{T}$ is continuous.

5.2.10 Theorem. *Let X be a metric θ-continuum of type A. Then X is a continuously type A' θ-continuum if and only if $\mathcal{T}_X$ is continuous. Moreover, $\mathcal{G} = \{\mathcal{T}_X(\{x\}) \mid x \in X\}$ is the finest monotone continuous decomposition of X such that $X/\mathcal{G}$ is a graph.*

Proof. Since X is a metric θ-continuum of type A, there exists a monotone upper semicontinuous decomposition $\mathcal{D}$ of X such that $X/\mathcal{D}$ is a graph. Let $q \colon X \twoheadrightarrow X/\mathcal{D}$ be the quotient map.

Suppose X is a continuously type A' θ-continuum. Observe that q is monotone and open. Also, by Lemma 1.4.69, for each subcontinuum K of X, $q(K) \neq D$. Hence, by Theorem 5.1.2, $\mathcal{T}_X$ is continuous.

Assume that $\mathcal{T}_X$ is continuous. We have that $\mathcal{G} = \{\mathcal{T}_X(\{x\}) \mid x \in X\}$ is a monotone continuous decomposition of X such that each element of $\mathcal{G}$ has empty interior and $X/\mathcal{G}$ is a locally connected continuum, Theorem 3.2.9. Let $x \in X$. Then, by Lemma 2.1.56, $\mathcal{T}_X(\{x\}) \subset q^{-1}q(x)$. Let $z \in q^{-1}q(x)$ and let W be a subcontinuum of X such that $z \in \mathrm{Int}_X(W)$. Since $X/\mathcal{D}$ is a graph and $z \in \mathrm{Int}_X(W)$, by [54, Lemma 1], we have that $q^{-1}q(z) = q^{-1}q(x) \subset W$. Thus, $x \in W$ and $z \in \mathcal{T}_X(\{x\})$. Therefore, $\mathcal{T}_X(\{x\}) = q^{-1}q(x)$, and $\mathcal{G} = \{q^{-1}q(x) \mid x \in X\}$. As a consequence of this, we obtain that X is a continuously type A' θ-continuum. The fact that $\mathcal{G}$ is the finest decomposition follows from Lemma 2.1.56 □

Let us observe that all examples presented, so far, of decomposable nonlocally connected metric continua X for which the set function $\mathcal{T}$ is continuous have the property that there exist many points $x \in X$ such that $\mathcal{T}(\{x\})$ is a pseudo-arc (Theorem 5.2.4, Example 5.2.8 and Theorem 5.2.9). In Theorem 5.2.13 we present a new family of one-dimensional nonlocally connected metric continua for which the set function $\mathcal{T}$ is continuous and no member of the family contains pseudo-arcs.

The following theorem is due to L. Mohler and L. G. Oversteegen [100, Corollary 1.1 and Theorem 2.1].

5.2.11 Theorem. *Let X be any continuously irreducible one-dimensional metric continuum. Then there exist a continuously irreducible one-dimensional metric continuum $\widehat{X}$ such that every nondegenerate subcontinuum of $\widehat{X}$ contains an arc, and an atomic map $g \colon \widehat{X} \twoheadrightarrow X$ (Definition 2.1.54).*

As a consequence of Theorems 5.2.11 and 5.2.9, we have the following.

5.2.12 Theorem. *For each one-dimensional continuously irreducible metric continuum X, there exist a one-dimensional continuously irreducible metric continuum $\widehat{X}$ such that $\mathcal{T}_{\widehat{X}}(\{\hat{x}\})$ does not contain a pseudo-arc for any $\hat{x} \in \widehat{X}$, and an atomic map $g \colon \widehat{X} \twoheadrightarrow X$.*

Let $\mathcal{Z}$ be the class of one-dimensional continuously irreducible metric continua, and let $\widehat{\mathcal{Z}} = \left\{\widehat{X} \mid X \in \mathcal{Z}\right\}$, where $\widehat{X}$ is given in Theorem 5.2.12. Hence, we have the following.

5.2.13 Theorem. *The class $\widehat{\mathcal{Z}}$, defined above, is a class of one-dimensional non-locally connected metric continua $\widehat{X}$ for which the set function $\mathcal{T}_{\widehat{X}}$ is continuous such that $\mathcal{T}_{\widehat{X}}(\{\hat{x}\})$ does not contain a pseudo-arc for any $\hat{x} \in \widehat{X}$. In particular, no element of $\widehat{\mathcal{Z}}$ contains a pseudo-arc.*

5.2.14 Theorem. *Let X be a continuum, and let $\mathcal{G}$ be a monotone, terminal and continuous decomposition of X (Definition 1.4.50). Then there exists an embedding of $\mathcal{C}_1(X/\mathcal{G})$ into $\mathcal{C}_1(X)$.*

Proof. Let $q\colon X \twoheadrightarrow X/\mathcal{G}$ be the quotient map. Note that q is open (Corollary 1.1.24) and monotone. Let $\Im(q)\colon \mathcal{C}_1(X/\mathcal{G}) \to \mathcal{C}_1(X)$ be given by $\Im(q)(\Gamma) = q^{-1}(\Gamma)$. Since q is monotone, $\Im(q)$ is well defined. Since q is open, $\Im(q)$ is continuous, Theorem 1.6.16. We show that $\Im(q)$ is one-to-one. To this end, let Γ_1 and Γ_2 be two distinct elements of $\mathcal{C}_1(X/\mathcal{G})$. Without loss of generality, we assume that there exists $\chi \in \Gamma_1 \setminus \Gamma_2$. Then, since $\mathcal{G}$ is a monotone terminal decomposition, we have that $q^{-1}(\chi) \subset q^{-1}(\Gamma_1) \setminus q^{-1}(\Gamma_2)$. Thus, $q^{-1}(\Gamma_1) \neq q^{-1}(\Gamma_2)$. Hence, $\Im(q)(\Gamma_1) \neq \Im(q)(\Gamma_2)$. Therefore, $\Im(q)$ is an embedding. □

As a consequence of Theorem 5.2.14, we have the following.

5.2.15 Corollary. *Let X be a metric continuum, and let $\mathcal{G}$ be a monotone, terminal and continuous decomposition of X. Then $\dim(\mathcal{C}_1(X/\mathcal{G})) \leq \dim(\mathcal{C}_1(X))$.*

5.2.16 Theorem. *Let X be a metric continuum, and let $\mathcal{G}$ be a monotone, terminal and continuous decomposition of X. If $X/\mathcal{G}$ is an aposyndetic continuum and $\dim(\mathcal{C}_1(X)) < \infty$, then $\mathcal{T}_X$ is continuous for X.*

Proof. Since $\dim(\mathcal{C}_1(X)) < \infty$, by Corollary 5.2.15, $\dim(\mathcal{C}_1(X/\mathcal{G})) < \infty$. Hence, since $X/\mathcal{G}$ is aposyndetic, by Corollary 2.1.40, $X/\mathcal{G}$ is locally connected. Let $q\colon X \twoheadrightarrow X/\mathcal{G}$ be the quotient map. Since $\mathcal{G}$ is a monotone, terminal and continuous decomposition of X, q is monotone and open (Corollary 1.1.24), and for every proper subcontinuum W of X, $q(W) \neq X$. Therefore, by Theorem 5.1.2, $\mathcal{T}_X$ is continuous for X. □

5.3 Continuity of $\mathcal{T}$ on Continua

We study continua for which the restriction of the set function $\mathcal{T}$ to continua is continuous. We present a couple of properties and show the equivalence of the continuity of $\mathcal{T}$ and the continuity of the restriction of $\mathcal{T}$ to continua for the product of two continua. We start with the following.

5.3.1 Theorem. *Let X be a continuum for which $\mathcal{T}_X$ is continuous on continua, and let Z be a continuum. If $f: X \twoheadrightarrow Z$ is a surjective, monotone and open map, then $\mathcal{T}_Z$ is continuous on continua.*

Proof. Since f is surjective and monotone, $\mathcal{T}_Z(B) = f\mathcal{T}_X f^{-1}(B)$ for all subsets B of Z, Theorem 2.1.51 (c). In particular, the equality is true for all subcontinua B of Z. Let $\zeta: \mathcal{C}_1(Z) \to \mathcal{C}_1(X)$ be given by $\zeta(B) = f^{-1}(B)$. Since f is monotone, ζ is well defined. Since f is open, ζ is continuous, Theorem 1.6.16. Also, by Theorem 1.6.15, $\mathcal{C}_1(f)$ is continuous. Hence, $\mathcal{T}_Z|_{\mathcal{C}_1(Z)} = \mathcal{C}_1(f) \circ \mathcal{T}_X|_{\mathcal{C}_1(X)} \circ \zeta$, and $\mathcal{T}_Z|_{\mathcal{C}_1(Z)}$ is the composition of three maps. Therefore, $\mathcal{T}_Z$ is continuous on continua for Z. □

5.3.2 Corollary. *Let X and Y be continua. If $\mathcal{T}_{X\times Y}$ is continuous on continua for $X \times Y$, then $\mathcal{T}_X$ and $\mathcal{T}_Y$ are continuous on continua for X and Y, respectively.*

Proof. The projection maps $\pi_X: X \times Y \twoheadrightarrow X$ and $\pi_Y: X \times Y \twoheadrightarrow Y$ are surjective, monotone and open. The corollary now follows from Theorem 5.3.1. □

5.3.3 Corollary. *Let X be a decomposable continuum for which $\mathcal{T}_X$ is idempotent on singletons and $\mathcal{T}_X$ is continuous on continua for X. If $\mathcal{G} = \{\mathcal{T}_X(\{x\}) \mid x \in X\}$, then $\mathcal{T}_{X/\mathcal{G}}$ is continuous on continua for $X/\mathcal{G}$.*

Proof. By Theorem 3.2.7, $\mathcal{G}$ is a continuous decomposition. Let $q: X \twoheadrightarrow X/\mathcal{G}$ be the quotient map. Then q is monotone and open. The corollary now follows from Theorem 5.3.1. □

5.3.4 Notation. If X is a continuum, then $\mathfrak{T}_{\mathcal{C}}(X) = \mathfrak{T}(X) \cap \mathcal{C}_1(X)$. Recall that $\mathfrak{T}(X)$ denotes the family of $\mathcal{T}$-closed sets of X (Definition 4.1.1).

Next, we consider the set function $\mathcal{T}$ defined on the product of two continua X and Y.

5.3.5 Lemma. *Let $Z = X \times Y$, where X and Y are continua. Suppose that we have one of the following:*

(1) *$\mathfrak{T}(Z)$ is compact,*

or

(2) *$\mathfrak{T}_{\mathcal{C}}(Z)$ is compact.*

Then Z is locally connected.

Proof. We prove that X and Y are locally connected continua. We see that X is locally connected. To this end, we show that $\mathcal{T}_X(A) = A$ for each subcontinuum A of X (Theorem 2.1.38). Let A be a subcontinuum of X. It is clear that $\mathcal{T}_X(X) = X$. Hence, we assume that $A \neq X$. Let $\{B_\lambda\}_{\lambda\in\Lambda}$ be a net of proper subcontinua of Y such that if λ and λ' are elements of Λ such that $\lambda' \leq \lambda$, then $B_{\lambda'} \subset B_\lambda$. Assume that $\{B_\lambda\}_{\lambda\in\Lambda}$ converges to Y [104, Theorem 4]. Note that $\mathcal{T}_Z(A \times B_\lambda) = A \times B_\lambda$ for each $\lambda \in \Lambda$, Corollary 2.3.26. Let $\mathfrak{D} \in \{\mathfrak{T}(X), \mathfrak{T}_{\mathcal{C}}(X)\}$. Then $A \times B_\lambda \in \mathfrak{D}$, for each $\lambda \in \Lambda$. Since $\mathfrak{D}$ is compact, $A \times Y \in \mathfrak{D}$ and $\mathcal{T}_Z(A \times Y) = A \times Y$.

Let $x \in X \setminus A$, and let y_0 be an arbitrary point of Y. Since $\mathcal{T}_Z(A \times Y) = A \times Y$, $(x, y_0) \in Z \setminus \mathcal{T}_Z(A \times Y)$. Hence, there exists a continuum W such that $(x, y_0) \in \mathrm{Int}_Z(W)$ and $W \cap (A \times Y) = \emptyset$. Since $\pi_1 \colon Z \to X$ is open, we have that $x \in \mathrm{Int}_X(\pi_1(W))$, $\pi_1(W)$ is a continuum and $\pi_1(W) \cap A = \emptyset$. Thus, $x \notin \mathcal{T}_X(A)$ and $\mathcal{T}_X(A) \subseteq A$. Therefore, $\mathcal{T}_X(A) = A$ and X is locally connected. Similarly we prove that Y is locally connected. Therefore, Z is locally connected. □

The next result extends Corollary 5.3.2.

5.3.6 Theorem. *Let X and Y be continua, and let $Z = X \times Y$. Then the following are equivalent:*

(1) *$\mathcal{T}_Z$ is continuous,*
(2) *$\mathcal{T}_Z|_{\mathcal{C}_1(Z)}$ is continuous,*
(3) *$\mathfrak{T}(Z)$ is a continuum,*
(4) *$\mathfrak{T}(Z)$ is compact,*
(5) *$\mathfrak{T}_C(Z)$ is a continuum,*
(6) *$\mathfrak{T}_C(Z)$ is compact,*
(7) *Z is locally connected.*

Proof. Observe that, from the definitions, (1) implies (2), (3) implies (4), and (5) implies (6). If $\mathcal{T}_Z$ is continuous, then $\mathcal{T}_Z$ is idempotent, Theorem 5.1.6. Hence, $\mathcal{T}_Z(2^Z) = \mathfrak{T}(Z)$ and $\mathfrak{T}(Z)$ is a continuum. Thus, (1) implies (3). Similarly, we may show that (1) implies (5).

We see that (2) implies (6). Let $\{A_\lambda\}_{\lambda \in \Lambda}$ be a net in $\mathfrak{T}_C(Z)$ converging to A, for some $A \in \mathcal{C}_1(Z)$. Since $\mathcal{T}_Z|_{\mathcal{C}_1(Z)}$ is continuous, $\{\mathcal{T}_Z(A_\lambda)\}_{\lambda \in \Lambda}$ converges to $\mathcal{T}_Z(A)$ [97, Proposition 3.38]. But $\mathcal{T}_Z(A_\lambda) = A_\lambda$ for each $\lambda \in \Lambda$. Thus, $\mathcal{T}_Z(A) = A$ and $A \in \mathfrak{T}_C(Z)$. Therefore, $\mathfrak{T}_C(Z)$ is compact, Theorem 1.6.7.

Note that (4) implies (7) and (6) implies (7), follow from Lemma 5.3.5. Now, if Z is locally connected, $\mathcal{T}_Z$ is the identity on 2^Z, Theorem 2.1.37. Therefore, $\mathcal{T}_Z$ is continuous and (7) implies (1). □

5.3.7 Corollary. *Let X and Y be continua, and let $Z = X \times Y$. If $\mathcal{T}_Z|_{\mathcal{C}_1(Z)}$ is continuous, then $\mathcal{T}_Z^2(A) = \mathcal{T}_Z(A)$ for each $A \in \mathcal{C}_1(Z)$.*

Proof. Since $\mathcal{T}_Z|_{\mathcal{C}_1(Z)}$ is continuous, $\mathcal{T}_Z$ is continuous, by Theorem 5.3.6. By Theorem 5.1.6, $\mathcal{T}_Z$ is idempotent. In particular, $\mathcal{T}_Z$ is idempotent on continua. □

5.3.8 Theorem. *Let X and Y be continua, and let $Z = X \times Y$. If $\mathcal{T}_Z(2^Z)$ is compact, then $\mathcal{T}_Z$ is surjective.*

Proof. Observe that $\mathcal{F}(Z) \subseteq \mathcal{T}_Z(2^Z)$, Theorem 4.1.7. Since $\mathcal{F}(Z)$ is dense in 2^Z (Lemma 1.6.9), $\mathcal{T}_Z(2^Z) = 2^Z$. Therefore, $\mathcal{T}_Z$ is surjective. □

5.3.9 Lemma. *Let $Z = X \times Y$, where X and Y are continua. If $\mathcal{T}_Z(2^Z)$ is closed, then $X \times \{y_0\} \in \mathfrak{T}_C(Z)$, for each $y_0 \in Y$.*

Proof. Let $y_0 \in Y$. Observe that $\mathcal{T}_Z(D) = D$, for each proper closed subset D of $X \times \{y_0\}$, Corollary 2.3.26. Since $\mathcal{T}_Z(2^Z)$ is closed, $X \times \{y_0\} \in \mathcal{T}_Z(2^Z)$. Hence, there exists a closed subset E of $X \times \{y_0\}$ such that $\mathcal{T}_Z(E) = X \times \{y_0\}$. Then $E = X \times \{y_0\}$, Corollary 2.3.26. Therefore, $X \times \{y_0\} \in \mathfrak{T}_C(Z)$. □

5.3.10 Theorem. *Let $Z = X \times Y$, where X and Y are continua. If $\mathcal{T}_Z(2^Z)$ is compact, then Z is locally connected.*

Proof. We see that both X and Y are locally connected. We prove that X is locally connected. The proof for Y is similar. Let A be a subcontinuum of X. By Theorem 2.1.38, it is enough to show that $\mathcal{T}_X(A) = A$.

Observe that $\mathcal{T}_X(X) = X$. Hence, we may assume that $X \setminus A \neq \emptyset$. Note that there exists a net $\{x_\lambda\}_{\lambda\in\Lambda}$ of elements of $X \setminus A$ such that $\mathrm{Cl}_X(\{x_\lambda\}_{\lambda\in\Lambda}) = \{x_\lambda\}_{\lambda\in\Lambda} \cup \mathrm{Bd}_X(A)$. Let $B = \{x_\lambda\}_{\lambda\in\Lambda} \cup \mathrm{Bd}_X(A)$. Since $\mathcal{T}_Z$ is surjective (Theorem 5.3.8), there exists $L \in 2^Z$ such that $\mathcal{T}_Z(L) = B\times Y$. Note that $\{x_\lambda\}\times Y$ is a component of $B\times Y$, for each $\lambda \in \Lambda$. Hence, $\mathcal{T}_Z(L\cap(\{x_\lambda\}\times Y)) = \{x_\lambda\}\times Y$ for every $\lambda \in \Lambda$, Corollary 2.1.48. We know that $\mathcal{T}_Z(R) = R$ for each closed proper subset R of $\{x_\lambda\} \times Y$, Corollary 2.3.26. Thus, $L \cap (\{x_\lambda\} \times Y) = \{x_\lambda\} \times Y$ and $\{x_\lambda\} \times Y \subseteq L$. Since x_λ is an arbitrary point, $\bigcup_{\lambda\in\Lambda}(\{x_\lambda\} \times Y) \subseteq L$. Therefore, $B \times Y \subseteq L$ and $\mathcal{T}_Z(B \times Y) = B \times Y$.

Since the natural projection onto the first factor, $\pi\colon Z \twoheadrightarrow X$, is monotone, we have that $\mathcal{T}_X(B) = B$, Theorem 2.1.51 (c).

Now, we see that $\mathcal{T}_X(A) = A$. Let $x \in X \setminus A$. We consider two cases:

Case 1. $x \in X \setminus B$.

Then there exists a subcontinuum S of X such that $x \in \mathrm{Int}_X(S)$ and $S\cap B = \emptyset$. Since $x \notin A$ and $S\cap\mathrm{Bd}_X(A) = \emptyset$, we have that $S\cap A = \emptyset$. Thus, $x \in X\setminus\mathcal{T}_X(A)$.

Case 2. $x = x_\lambda$, for some $\lambda \in \Lambda$.

Let $B_0 = B\setminus\{x_k\}$. The same argument that is used above shows that $\mathcal{T}_X(B_0) = B_0$. Thus, $x \in X \setminus \mathcal{T}_X(A)$ by Case 1.

Therefore, X is locally connected. □

By Theorems 5.3.6 and 5.3.10, we have the next theorem.

5.3.11 Theorem. *Let X and Y be continua. Let $Z = X \times Y$. The following are equivalent:*

(1) *$\mathcal{T}_Z$ is continuous,*
(2) *$\mathcal{T}_Z(2^Z)$ is a continuum,*
(3) *$\mathcal{T}_Z(2^Z)$ is compact,*
(4) *Z is locally connected.*

5.4 Continuity of $\mathcal{K}$

We consider F. Burton Jones' set function $\mathcal{K}$ and prove its continuity assuming the continuity of $\mathcal{T}$.

F. B. Jones defined the set function $\mathcal{K}$ in [68] to study aposyndetic continua. In fact, he defined this function only for points (rather than subsets). Since then many properties related to the set function $\mathcal{K}$ have been studied. Also, this function has been applied to study continua. For example, C. L. Hagopian uses the set function $\mathcal{K}$ to study plane continua ([57, 58] and [59]). E. Vought also uses $\mathcal{K}$ to study monotone decompositions of continua [123], and continua such that for each pair of its points there exists an irreducible continuum between those two points that is decomposable [124].

We present results regarding the continuity of the set function $\mathcal{K}$ using the set function $\mathcal{T}$. In Chapter 7, we give more relationships between $\mathcal{T}$ and $\mathcal{K}$.

5.4.1 Definition. Let X be a compactum. Define

$$\mathcal{K}\colon \mathcal{P}(X) \to \mathcal{P}(X)$$

by

$$\mathcal{K}(A) = \bigcap\{W \in \mathcal{C}_1(X) \mid A \subset \mathrm{Int}(W)\},$$

for each $A \in \mathcal{P}(X)$. The function $\mathcal{K}$ is called *Jones' set function* $\mathcal{K}$.

5.4.2 Remark. Let X be a compactum. If A is a subset of X, then $A \subset \mathcal{K}(A)$ and $\mathcal{K}(A)$ is closed. Hence, the range of $\mathcal{K}$ is $2^X \cup \{\emptyset\}$.

The following example shows that the image of a continuum under the set function $\mathcal{K}$ is not necessarily a continuum.

5.4.3 Example. Let X be the cone over the Cantor set (Example 2.1.15). With centre at $(\frac{1}{2}, 0)$, consider all semicircles passing through all the points of the Cantor set and having ordinates at most 0. If x is a point of X different from the vertex ν_X, then $\mathcal{K}(\{x\}) = \{\nu_X, x\}$.

5.4.4 Theorem. *Let X be continuum for which $\mathcal{T}$ is continuous, and let A be a subcontinuum of X. If $x \in X \setminus \mathcal{K}(A)$, then there exists a strong continuum domain W of X such that $x \in X \setminus W$ and $A \subset \mathrm{Int}(W)$.*

Proof. Let $x \in X \setminus \mathcal{K}(A)$. Then there exists a subcontinuum W' of X such that $x \in X \setminus W'$ and $A \subset \mathrm{Int}(W')$. This implies that $A \cap \mathcal{T}(\{x\}) = \emptyset$. Note that since $\mathcal{T}$ is continuous, then $\mathcal{T}$ is idempotent, Theorem 5.1.6. Hence, by Theorem 2.3.9, for each $a \in A$, there exists a strong continuum domain W_a of X such that $a \in \mathrm{Int}(W_a) \subset W_a \subset X \setminus \{x\}$. Note that $\{\mathrm{Int}(W_a) \mid a \in A\}$ is an open cover of A. Since A is compact, there exist $a_1, \dots, a_n$ in A such that $A \subset \bigcup_{j=1}^n \mathrm{Int}(W_{a_j})$. Observe that $U = \bigcup_{j=1}^n \mathrm{Int}(W_{a_j})$ is an open connected subset of X. Let $W = \mathrm{Cl}(U)$. Then W is a strong continuum domain of X, $x \in X \setminus W$ and $A \subset \mathrm{Int}(W)$. □

5.4.5 Theorem. *Let X be a continuum for which $\mathcal{T}_X$ is continuous. If $\mathcal{G} = \{\mathcal{T}_X(\{x\}) \mid x \in X\}$, then $\mathcal{K}_X(A) = q^{-1}\mathcal{K}_{X/\mathcal{G}}q(A)$ for every nonempty closed subset A of X, where $q\colon X \twoheadrightarrow X/\mathcal{G}$ is the quotient map.*

Proof. By Theorem 3.2.9, $\mathcal{G}$ is a continuous decomposition of X. Hence, it follows from Theorem 1.1.22 that $X/\mathcal{G}$ is a continuum. Let A be a nonempty closed subset of X, and let $x \in X \setminus q^{-1}\mathcal{K}_{X/\mathcal{G}}q(A)$. Then $q(x) \in X/\mathcal{G} \setminus \mathcal{K}_{X/\mathcal{G}}q(A)$. Thus, there exists a subcontinuum $\mathcal{W}$ of $X/\mathcal{G}$ such that $q(x) \in X/\mathcal{G} \setminus \mathcal{W}$ and $q(A) \subset \mathrm{Int}_{X/\mathcal{G}}(\mathcal{W})$. Hence, since q is monotone, $q^{-1}(\mathcal{W})$ is a subcontinuum of X (Lemma 1.4.46) such that $x \in q^{-1}q(x) \subset X \setminus q^{-1}(\mathcal{W})$ and $A \subset q^{-1}q(A) \subset \mathrm{Int}_X(q^{-1}(\mathcal{W}))$. Therefore, $x \in X \setminus \mathcal{K}_X(A)$.

Now, let $x \in X \setminus \mathcal{K}_X(A)$. Then, by Theorem 5.4.4, there exists a strong continuum domain W of X such that $x \in X \setminus W$ and $A \subset \mathrm{Int}_X(W)$. By Theorem 5.1.10, X is $\mathcal{T}_X$-additive. Hence, by Theorem 5.1.13, $\mathcal{T}_X(W) = W$. Also, since X is $\mathcal{T}_X$ additive, $W = \bigcup\{\mathcal{T}_X(\{w\}) \mid w \in W\}$, Theorem 2.2.10. Thus, $\mathcal{T}_X(\{x\}) \cap W = \emptyset$. Hence, since q is continuous and open, $q(W)$ is a subcontinuum of $X/\mathcal{G}$ such that $q(x) \in X/\mathcal{G} \setminus q(W)$ and $q(A) \subset \mathrm{Int}_{X/\mathcal{G}}(q(W))$. This implies that $q(x) \in X/\mathcal{G} \setminus \mathcal{K}_{X/\mathcal{G}}q(A)$. Then, $x \in q^{-1}q(x) \subset X \setminus q^{-1}\mathcal{K}_{X/\mathcal{G}}q(A)$. Therefore, $\mathcal{K}_X(A) = q^{-1}\mathcal{K}_{X/\mathcal{G}}q(A)$. □

5.4.6 Corollary. *Let X be a continuum for which $\mathcal{T}_X$ is continuous. Then $\mathcal{K}_X$ is continuous.*

Proof. Let $\mathcal{G} = \{\mathcal{T}_X(\{x\}) \mid x \in X\}$. By Theorem 3.2.9, $\mathcal{G}$ is a continuous decomposition. Then it follows from Theorem 1.1.22 that $X/\mathcal{G}$ is a continuum and X is a point $\mathcal{T}_X$-symmetric continuum. Hence, by Theorem 3.2.9, $X/\mathcal{G}$ is locally connected. Thus, by [53, Theorem 36], $\mathcal{K}_{X/\mathcal{G}}$ is continuous. Let $q\colon X \twoheadrightarrow X/\mathcal{G}$ be the quotient map, and let $\Im(q)\colon 2^{X/\mathcal{G}} \to 2^X$ be given by $\Im(q)(\Gamma) = q^{-1}(\Gamma)$. Since q is open, by Theorem 1.6.16, $\Im(q)$ is continuous. By Theorem 1.6.15, 2^q is continuous. Since $\mathcal{K}_X(A) = q^{-1}\mathcal{K}_{X/\mathcal{G}}q(A)$ for each nonempty closed subset A of X, we have that $\mathcal{K}_X = \Im(q) \circ \mathcal{K}_{X/\mathcal{G}} \circ 2^q$. Therefore, $\mathcal{K}_X$ is continuous. □

References for Chapter 5

Section 5.1: [3, 20, 40, 64, 83, 90–92, 97, 104].
Section 5.2: [3, 5, 13, 14, 36, 54, 75, 78, 82, 87, 88, 92, 94, 100, 119, 121].
Section 5.3: [20, 87].
Section 5.4: [53, 57–59, 68, 87, 91, 123, 124].

Chapter 6
Images of $\mathcal{T}$

For a metric continuum X, we study possible images of the set function $\mathcal{T}$. We begin by showing that $\mathcal{T}(2^X)$ is an analytic set for every metric continuum X. We are interested when either $\mathcal{T}(\mathcal{F}_1(X))$ or $\mathcal{T}(2^X)$ is finite or countable. The notion of ω-indecomposable continuum is given as a generalization of the well known concept of n-indecomposable continuum. We also present results about the connectivity and compactness of $\mathcal{T}(2^X)$. We show that if X is a metric continuum such that $\mathcal{T}(2^X)$ is compact, then $\mathcal{T}(2^X)$ is either finite or uncountable. We give an example of a continuum X such that $\mathcal{T}(2^X)$ is countable.

6.1 A Topological Property

We show that $\mathcal{T}(2^X)$ is an analytic set for every metric continuum X.

6.1.1 Definition. A topological space X is called *completely metrizable* if it admits a compatible metric d such that (X, d) is complete. A separable completely metrizable space is called a *Polish space*.

6.1.2 Definition. Let Z be a topological space. A subset A of Z is an *analytic set*, provided that there exist a polish space Y and a map $f\colon Y \to Z$ such that $f(Y) = A$.

6.1.3 Theorem. *Let X be a metric continuum. Then the set $\mathcal{T}(2^X)$ is an analytic set.*

Proof. Let $\mathcal{F} = \{(A, B) \in 2^X \times 2^X \mid \mathcal{T}(A) = B\}$. We prove that $\mathcal{F}$ is a G_δ-set. Let $\mathcal{M} = \{(A, B) \in 2^X \times 2^X \mid B \subseteq \mathcal{T}(A)\}$ and let $\mathcal{N} = \{(A, B) \in 2^X \times 2^X \mid \mathcal{T}(A) \subseteq B\}$.

We claim that $\mathcal{M}$ is a closed subset of $2^X \times 2^X$. To see this is true, let $(C, D) \in 2^X \times 2^X \setminus \mathcal{M}$. Hence, there exists $x \in D \setminus \mathcal{T}(C)$. Since X is regular, there exist

S. Macías, *Set Function $\mathcal{T}$*, Developments in Mathematics 67,
https://doi.org/10.1007/978-3-030-65081-0_6

open sets U and V such that $x \in U$, $\mathcal{T}(C) \subseteq V$ and $U \cap V = \emptyset$. Since $\mathcal{T}$ is upper semicontinuous, Theorem 5.1.1, there exists an open set $\mathcal{U} \subseteq 2^X$ such that $C \in \mathcal{U}$ and $\mathcal{T}(E) \subseteq V$ for each $E \in \mathcal{U}$. Let $\mathcal{W} = \mathcal{U} \times \langle X, V \rangle$. Observe that $(C, D) \in \mathcal{W}$ and $\mathcal{W}$ is an open subset of $2^X \times 2^X$. Furthermore, if $(E, F) \in \mathcal{W}$, then $\mathcal{T}(E) \subseteq V$ and $F \cap V \neq \emptyset$. Thus, $F \setminus \mathcal{T}(E) \neq \emptyset$ and $(E, F) \in 2^X \times 2^X \setminus \mathcal{M}$. Therefore, $\mathcal{W} \subseteq (2^X \times 2^X) \setminus \mathcal{M}$ and $\mathcal{M}$ is closed.

Now, we claim that $\mathcal{N}$ is a G_δ-set. To show this, for each $n \in \mathbb{N}$, let

$$\mathcal{L}_n = \mathrm{Cl}_{2^X \times 2^X}(\{(A, B) \in 2^X \times 2^X \mid \mathcal{V}_{\frac{1}{n}}(a) \cap B = \emptyset,$$

$$\text{for some } a \in \mathcal{T}(A)\}).$$

We see that $\mathcal{N} = (2^X \times 2^X) \setminus \bigcup_{n \in \mathbb{N}} \mathcal{L}_n$. Let $(A, B) \in 2^X \times 2^X \setminus \mathcal{N}$. Then there exists $x \in \mathcal{T}(A) \setminus B$. Let $n_0 \in \mathbb{N}$ be such that $\mathcal{V}_{\frac{1}{n_0}}(x) \cap B = \emptyset$. Thus, $(A, B) \in \mathcal{L}_{n_0} \subseteq \bigcup_{n \in \mathbb{N}} \mathcal{L}_n$. Therefore, $(2^X \times 2^X) \setminus \bigcup_{n \in \mathbb{N}} \mathcal{L}_n \subseteq \mathcal{N}$.

Conversely, suppose that $(A, B) \in \mathcal{L}_n$, for some $n \in \mathbb{N}$. Let $\{(A_m, B_m)\}_{m=1}^\infty$ be a sequence in $\{(A, B) \mid \mathcal{V}_{\frac{1}{n}}(a) \cap B = \emptyset$, for some $a \in \mathcal{T}(A)\}$, such that $\lim_{m \to \infty}(A_m, B_m) = (A, B)$. For each $m \in \mathbb{N}$, let $a_m \in \mathcal{T}(A_m)$ be such that $\mathcal{V}_{\frac{1}{n}}(a_m) \cap B_m = \emptyset$. Since X is compact, there exists a subsequence $\{a_{m_k}\}_{k=1}^\infty$ of $\{a_m\}_{m=1}^\infty$ such that $\lim_{k \to \infty} a_{m_k} = a_0$, for some $a_0 \in X$. Since $\mathcal{T}$ is upper semicontinuous, Theorem 5.1.1, $\limsup \mathcal{T}(A_{m_k}) \subseteq \mathcal{T}(A)$. Thus, $a_0 \in \mathcal{T}(A)$. Suppose that $\mathcal{V}_{\frac{1}{2n}}(a_0) \cap B \neq \emptyset$. Since $\lim_{k \to \infty} a_{m_k} = a_0$ and $\lim_{k \to \infty} B_{m_k} = B$, we have that there exists $l \in \mathbb{N}$ such that $a_{m_l} \in \mathcal{V}_{\frac{1}{2n}}(a_0)$ and $B_{m_l} \cap \mathcal{V}_{\frac{1}{2n}}(a_0) \neq \emptyset$. Let $z \in B_{m_l} \cap \mathcal{V}_{\frac{1}{2n}}(a_0)$. Note that $d(a_{m_l}, z) \leq d(a_{m_l}, a_0) + d(a_0, z) < \frac{1}{2n} + \frac{1}{2n} = \frac{1}{n}$. Hence, $\mathcal{V}_{\frac{1}{n}}(a_{m_l}) \cap B_{m_l} \neq \emptyset$, a contradiction. Thus, $\mathcal{V}_{\frac{1}{2n}}(a_0) \cap B = \emptyset$ and $a_0 \in \mathcal{T}(A) \setminus B$. Therefore, $(A, B) \in 2^X \times 2^X \setminus \mathcal{N}$, and $\mathcal{N}$ is a G_δ-set.

Note that $\mathcal{F} = \mathcal{M} \cap \mathcal{N}$. Thus, $\mathcal{F}$ is a G_δ-set. By [69, Theorem 3.11], $\mathcal{F}$ is a Polish space. Furthermore, $\mathcal{T}(2^X) = \pi_2(\mathcal{F})$, where $\pi_2 \colon 2^X \times 2^X \to 2^X$ is defined by $\pi_2(A, B) = B$, for each $(A, B) \in 2^X \times 2^X$. Therefore, $\mathcal{T}(2^X)$ is an analytic set. □

6.1.4 Remark. Note that if $\mathcal{T}$ is idempotent on closed sets, then $\mathcal{T}(2^X) = \{A \in 2^X \mid \mathcal{T}(A) = A\}$.

Next corollary follows from Theorem 4.1.11.

6.1.5 Corollary. *Let X be a metric continuum. If $\mathcal{T}$ is idempotent on closed sets, then $\mathcal{T}(2^X)$ is a G_δ subset of 2^X.*

6.2 Finite and Countable Images of $\mathcal{T}$

We study metric continua X for which either $\mathcal{T}(\mathcal{F}_1(X))$ or $\mathcal{T}(2^X)$ is finite or countable. The notion of ω-indecomposable continuum [20, p. 8] is given as a generalization of the well known concept of n-indecomposable continuum. We begin with a technical proposition. First, we need the following.

6.2.1 Notation. Given X a continuum and a point x of X, we define $L_x = \{z \in X \mid \mathcal{T}(\{z\}) = \mathcal{T}(\{x\})\}$.

6.2.2 Proposition. *Let X be a metric continuum. Let L_x be as in Notation 6.2.1, for each $x \in X$. If $\operatorname{Int}(\operatorname{Cl}(L_x)) \neq \emptyset$, for some $x \in X$, then there exists an indecomposable continuum $M \subseteq \mathcal{T}(\{x\})$ such that M is irreducible about L_x.*

Proof. Let $x \in X$ be such that $\operatorname{Int}(\operatorname{Cl}(L_x)) \neq \emptyset$. Note that $\mathcal{T}(\{x\})$ is a continuum, Theorem 2.1.27, such that $L_x \subset \mathcal{T}(\{x\})$. Let M be a subcontinuum of $\mathcal{T}(\{x\})$ irreducible about L_x [129, Theorem 28.4]. We show that M is indecomposable. Suppose that $M = A \cup B$, where A and B are proper subcontinua of M. Since $\operatorname{Int}(\operatorname{Cl}(L_x)) \subset M$, either $\operatorname{Int}(\operatorname{Cl}(L_x)) \cap \operatorname{Int}(A) \neq \emptyset$ or $\operatorname{Int}(\operatorname{Cl}(L_x)) \cap \operatorname{Int}(B) \neq \emptyset$. Suppose that $\operatorname{Int}(\operatorname{Cl}(L_x)) \cap \operatorname{Int}(B) \neq \emptyset$. Since M is irreducible about L_x, $L_x \cap (A \setminus B) \neq \emptyset$. Let $a \in L_x \cap (A \setminus B)$, and let $b \in L_x \cap \operatorname{Int}(\operatorname{Cl}(L_x)) \cap \operatorname{Int}(B) \subset B$. Hence, $b \in \operatorname{Int}(B)$ and $a \notin B$, i.e., $b \notin \mathcal{T}(\{a\})$, but this contradicts the fact that $\mathcal{T}(\{a\}) = \mathcal{T}(\{b\}) = \mathcal{T}(\{x\})$. Therefore, M is indecomposable. □

6.2.3 Theorem. *Let X be a metric continuum, and let L_x be as in Notation 6.2.1, for each $x \in X$. If $\mathcal{T}(\mathcal{F}_1(X))$ is a countable set, then there exists a family of indecomposable continua $\{M_n \mid n \in S\}$, where $S \subset \mathbb{N}$, such that:*

(1) $X = \operatorname{Cl}(\bigcup_{n \in S} M_n)$,
(2) *for every $n \in S$, there exists $x_n \in M_n$ satisfying:*

  (a) $\bigcup_{n \in S} \operatorname{Cl}(L_{x_n})$ *is dense in X,*
  (b) $\operatorname{Int}(\operatorname{Cl}(L_{x_n})) \neq \emptyset$, *for all $n \in S$,*
  (c) M_n *is irreducible about L_{x_n}, for each $n \in S$,*
  (d) $L_{x_n} \subset \operatorname{Cl}(L_{x_n}) \subset M_n \subset \mathcal{T}(\{x_n\})$, *for every $n \in S$, and*
  (e) $\operatorname{Int}(\operatorname{Cl}(L_{x_n})) \subset X \setminus \operatorname{Cl}(\bigcup_{k \in S \setminus \{n\}} M_k)$, *for all $n \in S$.*

Proof. We assume that $\mathcal{T}(\mathcal{F}_1(X))$ is infinite; if $\mathcal{T}(\mathcal{F}_1(X))$ is finite, the argument is similar. Let $\{x_n\}_{n=1}^{\infty}$ be a sequence of elements of X such that $\mathcal{T}(\mathcal{F}_1(X)) = \{\mathcal{T}(\{x_n\})\}_{n=1}^{\infty}$. Note that $X = \bigcup\{\operatorname{Cl}(L_{x_n}) \mid n \in \mathbb{N}\}$. Since X is a Baire space, $\operatorname{Int}(\operatorname{Cl}(L_{x_n})) \neq \emptyset$, for some $n \in \mathbb{N}$ [129, Theorem 25.3]. Let

$$S = \{n \in \mathbb{N} \mid \operatorname{Int}(\operatorname{Cl}(L_{x_n})) \neq \emptyset\}.$$

Let $D = \bigcup\{\operatorname{Cl}(L_{x_n}) \mid n \in S\}$. We prove that D is dense in X. Suppose that $U = X \setminus \operatorname{Cl}(D) \neq \emptyset$. Note that $U \cap \operatorname{Cl}(L_{x_n}) = \emptyset$, for each $n \in S$. Thus, $U = \bigcup\{U \cap \operatorname{Cl}(L_{x_n}) \mid n \in \mathbb{N} \setminus S\}$. Since U is a Baire space [129, Theorem 25.3], $U \cap \operatorname{Int}(\operatorname{Cl}(L_{x_n}) \neq \emptyset$ and $\operatorname{Int}(\operatorname{Cl}(L_{x_n})) \neq \emptyset$, for some $n \in \mathbb{N} \setminus S$, a contradiction. Therefore, D is dense in X.

Now, by Proposition 6.2.2, for each $n \in S$, there exists an indecomposable continuum M_n irreducible about L_{x_n}. Since $L_{x_n} \subset M_n$, we have that $\operatorname{Cl}(L_{x_n}) \subset M_n$, for every $n \in S$. Therefore, by the previous paragraph, $X = \operatorname{Cl}(\bigcup_{n \in S} M_n)$.

Let k and l be elements of $\mathbb{N}$. Suppose that $\text{Int}(\text{Cl}(L_{x_k})) \cap M_l \neq \emptyset$. Since $\text{Int}(\text{Cl}(L_{x_k})) \subset M_k$ and $M_l \subset \mathcal{T}(\{x_l\})$, we have that $\text{Int}(M_k) \cap \mathcal{T}(\{x_l\}) \neq \emptyset$. Hence, $L_{x_l} \subset M_k$ and $\text{Cl}(L_{x_l}) \subset M_k$. Observe that $\text{Int}(\text{Cl}(L_{x_l})) \cap M_k \neq \emptyset$ and, with a similar argument, we have that $\text{Cl}(L_{x_k}) \subset M_l$. Therefore, $\text{Cl}(L_{x_l}) \cup \text{Cl}(L_{x_k}) \subset M_l \cap M_k$. We see that $\mathcal{T}(\{x_l\}) = \mathcal{T}(\{x_k\})$. Let $z \in \mathcal{T}(\{x_l\})$, and suppose that $z \notin \mathcal{T}(\{x_k\})$. Then there exists a continuum K such that $z \in \text{Int}(K)$ and $x_k \notin K$. Since $z \in \mathcal{T}(\{x_l\})$, we have that $L_{x_l} \subset K$ and $\text{Cl}(L_{x_l}) \subset K$. Note that $\text{Int}(\text{Cl}(L_{x_l})) \subset K \cap \mathcal{T}(\{x_k\})$. If $y \in \text{Int}(\text{Cl}(L_{x_l}))$, then $y \in \mathcal{T}(\{x_k\})$ and K is a continuum containing y in its interior. Hence, $x_k \in K$, a contradiction. Then $\mathcal{T}(\{x_l\}) \subset \mathcal{T}(\{x_k\})$. Similarly, we obtain that $\mathcal{T}(\{x_k\}) \subset \mathcal{T}(\{x_l\})$. Thus, $\mathcal{T}(\{x_l\}) = \mathcal{T}(\{x_k\})$. In this case we have that $\text{Int}(\text{Cl}(L_{x_k})) \cap M_l = \emptyset$, whenever $k \neq l$. Therefore, for all $n \in S$, $\text{Int}(\text{Cl}(L_{x_n})) \cap \text{Cl}(\bigcup_{k \in S \setminus \{n\}} M_k) = \emptyset$ and $\text{Int}(\text{Cl}(L_{x_n})) \subset X \setminus \text{Cl}(\bigcup_{k \in S \setminus \{n\}} M_k)$. □

The next result is in [45, p. 139].

6.2.4 Lemma. *Let X be a metric continuum, and let F be a finite subset of X. If X is irreducible about F, then X is a θ-continuum.*

6.2.5 Proposition. *Let X be a metric continuum. If X is an n-indecomposable continuum, for some $n \in \mathbb{N}$, then X is $\mathcal{T}$-additive.*

Proof. Note that X is irreducible about some finite set, Theorem 2.4.10. Hence, by Lemma 6.2.4, X is a θ-continuum. Thus, X is a weakly irreducible continuum, Theorem 1.4.70. Then X is $\mathcal{T}$-symmetric, Theorem 2.2.2. Therefore, X is $\mathcal{T}$-additive, Theorem 2.2.11. □

In Theorem 2.4.10, it is mentioned that a metric continuum X is n-indecomposable, for some $n \in \mathbb{N}$, if and only if $\mathcal{T}(\mathcal{F}_1(X))$ is finite. We extend this result to $\mathcal{T}(2^X)$.

6.2.6 Theorem. *Let X be a metric continuum. Then $\mathcal{T}(2^X)$ is a finite set if and only if there exists $n \in \mathbb{N}$ such that X is an n-indecomposable continuum.*

Proof. Suppose $\mathcal{T}(2^X)$ is a finite set. In particular, $\{\mathcal{T}(\{x\}) \mid x \in X\}$ is finite. Hence, by Theorem 2.4.10, there exists $n \in \mathbb{N}$ such that X is an n-indecomposable continuum.

Suppose X is an n-indecomposable continuum, for some $n \in \mathbb{N}$. Then, by Theorem 2.4.10, X is the union of n indecomposable continua $M_1, \ldots, M_n$ such that no one of them is a subset of the union of the others, and it is irreducible about some finite set. We show that $\mathcal{T}(\{x\}) = \bigcup\{M_j \mid x \in M_j\}$. By [18, Theorem 7], $\bigcup\{M_j \mid x \in M_j\} \subset \mathcal{T}(\{x\})$. Let $p \in X \setminus \bigcup\{M_j \mid x \in M_j\}$, and let $X_p = \bigcup\{M_j \mid p \in M_j\}$. Then X_p is a subcontinuum of X. Since X is a θ-continuum (Lemma 6.2.4), $p \in \text{Int}(X_p) \subset X_p \subset X \setminus \{x\}$. Hence, $p \in X \setminus \mathcal{T}(\{x\})$, $\mathcal{T}(\{x\}) \subset \bigcup\{M_j \mid x \in M_j\}$ and $\mathcal{T}(\{x\}) = \bigcup\{M_j \mid x \in M_j\}$. Note that X is $\mathcal{T}$-additive, Proposition 6.2.5. Thus, if A is a nonempty closed subset of X, then $\mathcal{T}(A) = \bigcup\{M_j \mid A \cap M_j \neq \emptyset\}$, Theorem 2.2.10. Therefore, $\mathcal{T}(2^X)$ is a finite set. □

6.2.7 Lemma. *Let X be a decomposable metric continuum. If X is a θ-continuum and $\mathcal{T}(\{x_0\}) = X$ for some $x_0 \in X$, then $X = \bigcup_{j=1}^{m} M_j$, such that:*

(1) *each M_j is a proper subcontinuum of X;*
(2) *$x_0 \in M_j$, for all $j \in \{1, \ldots, m\}$;*
(3) *M_j is a strong continuum domain, for every $j \in \{1, \ldots, m\}$;*
(4) *$\mathrm{Int}(M_j) \cap (\bigcup_{k \neq j} M_k) = \emptyset$, for each $j \in \{1, \ldots, m\}$.*

Proof. Let L be a proper subcontinuum of X such that $\mathrm{Int}(L) \neq \emptyset$, Lemma 1.4.34. Since X is a θ-continuum, $X \setminus L$ only has finitely many components. Let $U_1, \ldots, U_n$ be the components of $X \setminus L$. Observe that U_j is open, for all $j \in \{1, \ldots, n\}$. Let $M_j = \mathrm{Cl}(U_j)$, for each $j \in \{1, \ldots, n\}$. Note that M_j is a strong continuum domain, for every $j \in \{1, \ldots, n\}$. Since $\mathcal{T}(\{x_0\}) = X$, $x_0 \in M_j$, for all $j \in \{1, \ldots, n\}$. Thus, $\bigcup_{j=1}^{n} M_j$ is a continuum. Then, since X is a θ-continuum, $X \setminus (\bigcup_{j=1}^{n} M_j) = \bigcup_{k=1}^{l} V_k$, where each V_k is an open component of $X \setminus (\bigcup_{j=1}^{n} M_j)$, for all $k \in \{1, \ldots, l\}$. Let $M_{k+n} = \mathrm{Cl}(V_k)$, for every $k \in \{1, \ldots, l\}$. Therefore, $X = \bigcup_{j=1}^{m} M_j$, where $M_1, \ldots, M_m$ satisfy (1), (2), (3) and (4), for $m = n + l$. □

From Lemmas 6.2.7 and 6.2.4, we have the following.

6.2.8 Corollary. *Let X be a decomposable metric continuum. If X is irreducible about a finite set and $\mathcal{T}(\{x_0\}) = X$ for some $x_0 \in X$, then $X = \bigcup_{j=1}^{m} M_i$, such that*

(1) *each M_j is a proper subcontinuum of X;*
(2) *$x_0 \in M_j$, for all $j \in \{1, \ldots, m\}$;*
(3) *M_j is a strong continuum domain, for every $j \in \{1, \ldots, m\}$;*
(4) *$\mathrm{Int}(M_j) \cap (\bigcup_{k \neq j} M_k) = \emptyset$, for each $j \in \{1, \ldots, m\}$.*

6.2.9 Theorem. *Let X be a metric continuum. If X is irreducible about a finite set and $\mathcal{T}(\{x_0\}) = X$ for some $x_0 \in X$, then X is n-indecomposable, for some $n \in \mathbb{N}$.*

Proof. If X is indecomposable, we have the result for $n = 1$. Hence, suppose that X is decomposable. Thus, $X = \bigcup_{j=1}^{m} M_j$, where $M_1, \ldots, M_m$ satisfy (1), (2), (3) and (4) of Corollary 6.2.8. Suppose X is irreducible about a finite set F. Note that $\mathrm{Int}(M_j) \cap F \neq \emptyset$, since $\bigcup_{k \neq j} M_k$ is a continuum, for each $j \in \{1, \ldots, m\}$. Hence, m is less than or equal to the number of elements of F. If M_j is indecomposable, for every $j \in \{1, \ldots, m\}$, then X is m-indecomposable, Theorem 2.4.10.

Suppose that M_1 is decomposable. Let L be a proper subcontinuum of M_1 such that $\mathrm{Int}_{M_1}(L) \neq \emptyset$, Lemma 1.4.34. Since $M_1 = \mathrm{Cl}_X(\mathrm{Int}_X(M_1))$ and $\mathrm{Int}_X(M_1) \cap (\bigcup_{j=2}^{m} M_j) = \emptyset$, we have that $\mathrm{Int}_X(L) \neq \emptyset$. Furthermore, since $\mathcal{T}(\{x_0\}) = X$, $x_0 \in L$ and $L \cup (\bigcup_{j=2}^{m} M_j)$ is a proper subcontinuum of X. Thus, $X \setminus (L \cup (\bigcup_{j=2}^{m} M_j)) = \bigcup_{j=1}^{s} U_j$, where U_j is an open component of $X \setminus (L \cup (\bigcup_{j=2}^{m} M_j))$, for all $j \in \{1, \ldots, s\}$ (Lemma 6.2.4). Let $N_j = \mathrm{Cl}_X(U_j)$, for each $j \in \{1, \ldots, s\}$. Since $\mathcal{T}(\{x_0\}) = X$, $x_0 \in N_j$ for every $j \in \{1, \ldots, s\}$, and $(\bigcup_{j=1}^{s} N_j) \cup (\bigcup_{j=2}^{m} M_j)$ is a proper subcontinuum of X. Similarly, $X \setminus ((\bigcup_{j=1}^{s} N_j) \cup (\bigcup_{j=2}^{m} M_j)) = \bigcup_{j=1}^{l} V_j$, where V_j is an open component of $X \setminus ((\bigcup_{j=1}^{s} N_j) \cup (\bigcup_{j=2}^{m} M_j))$, for all $j = \{1, \ldots, l\}$, Lemma 6.2.4. Let $W_j = \mathrm{Cl}_X(V_j)$, for each $j \in \{1, \ldots, l\}$. Therefore, $X = (\bigcup_{j=1}^{l} W_j) \cup (\bigcup_{j=1}^{s} N_j) \cup (\bigcup_{j=2}^{m} M_j)$, where $W_1, \ldots, W_l, N_1, \ldots, N_s, M_2, \ldots, M_m$ satisfy conditions (1), (2), (3) and (4) of Corollary 6.2.8.

Let $K_1, \ldots, K_{m_1}$ be subcontinua of X such that $\{K_1, \ldots, K_{m_1}\} = \{W_1, \ldots, W_l, N_1, \ldots, N_s, M_2, \ldots, M_m\}$. Thus, $X = \bigcup_{j=1}^{m_1} K_j$, where $K_1, \ldots, K_{m_1}$ satisfy conditions (1), (2), (3) and (4) of Corollary 6.2.8. Since $x_0 \in K_j$, $\bigcup_{k \neq j} K_k$ is a proper subcontinuum of X, for all $j \in \{1, \ldots, m_1\}$. Therefore, $\mathrm{Int}_X(K_j) \cap F \neq \emptyset$, for each $j \in \{1, \ldots, n\}$, and $m < m_1 \leq |F|$, where $|F|$ denotes the cardinality of F. If K_j is indecomposable for every $j \in \{1, \ldots, m_1\}$, then X is m_1-indecomposable, Theorem 2.4.10.

Suppose that K_1 is decomposable. As we did before, there exist subcontinua $L_1, \ldots, L_{m_2}$ of X such that $X = \bigcup_{j=1}^{m_2} L_j$, where $L_1, \ldots, L_{m_2}$ satisfy conditions (1), (2), (3) and (4) of Corollary 6.2.8, and $m < m_1 < m_2 \leq |F|$. Since F is finite, without loss of generality, we assume that $L_1, \ldots, L_{m_2}$ are indecomposable and X is m_2-indecomposable, Theorem 2.4.10. □

6.2.10 Example. There exists a metric continuum X such that X is irreducible about three points, $\{x_1, x_2, x_3\}$, such that $X = \mathcal{T}(\{x_1\}) \cup \mathcal{T}(\{x_2\}) \cup \mathcal{T}(\{x_3\})$ and X is not 3-indecomposable. To see this, let X_1, X_2, X_3 and X_4 be four pairwise disjoint indecomposable continua (e.g., four copies of the pseudo-arc). Assume X_j is irreducible between the points a_j and b_j, $j \in \{1, 2, 3, 4\}$. Let $X' = X_1 \cup X_2 \cup X_3 \cup X_4$ and define an equivalence relation $\sim$ on X' by $p \sim q$ if and only if $p = q$ or $\{p, q\} = \{b_j, a_{j+1}\}$ for some $j \in \{1, 2, 3\}$. Let $X = X'/\sim$, let $q \colon X' \twoheadrightarrow X$ be the quotient map and let $x_1 = q(a_1)$, $x_2 = q(a_3)$ and $x_3 = q(b_4)$. Then X is a continuum irreducible about $\{x_1, x_2, x_3\}$, $X = \mathcal{T}(\{x_1\}) \cup \mathcal{T}(\{x_2\}) \cup \mathcal{T}(\{x_3\})$ and X is not 3-indecomposable (in fact, X is 4-indecomposable).

6.2.11 Definition. Let X be a continuum. We say that X is *ω-indecomposable* provided that $\mathcal{T}(\mathcal{F}_1(X))$ is a countably infinite set.

6.2.12 Proposition. *Let X be a metric continuum, and let L_x be as in Notation 6.2.1, for each $x \in X$. If X is an ω-indecomposable continuum, then there exists a family of indecomposable continua $\{M_n\}_{n=1}^{\infty}$ such that:*

(1) $X = \mathrm{Cl}(\bigcup_{n=1}^{\infty} M_n)$,
(2) *for every $n \in \mathbb{N}$, there exists $x_n \in M_n$ satisfying:*

  (*a*) $\bigcup_{n=1}^{\infty} \mathrm{Cl}(L_{x_n})$ *is dense in X;*
  (*b*) $\mathrm{Int}(\mathrm{Cl}(L_{x_n})) \neq \emptyset$;
  (*c*) *M_n is irreducible about L_{x_n};*
  (*d*) $L_{x_n} \subset \mathrm{Cl}(L_{x_n}) \subset M_n \subset \mathcal{T}(\{x_n\})$;
  (*e*) $\mathrm{Int}(\mathrm{Cl}(L_{x_n})) \subset X \setminus \mathrm{Cl}(\bigcup_{k \in \mathbb{N} \setminus \{n\}} M_k)$.

Proof. By Theorem 6.2.3, there exists $S \subset \mathbb{N}$ such that the proposition is satisfied for each $n \in S$. Thus, we have to show that S is countably infinite. Suppose that S is finite; that is, $X = M_1 \cup \cdots \cup M_k$. Let $x_n \in M_n$ be such that, if $L_{x_n} = \{x \in X \mid \mathcal{T}(\{x\}) = \mathcal{T}(\{x_n\})\}$, then $L_{x_n} \subset \mathrm{Cl}(L_{x_n}) \subset M_n \subset \mathcal{T}(\{x_n\})$, where $\operatorname{Int}(\operatorname{Cl}(L_{x_n})) \neq \emptyset$, for every $n \in \{1, \ldots, k\}$, and $X = \bigcup_{n=1}^{k} \operatorname{Cl}(L_{x_n})$ (Theorem 6.2.3). Suppose that $X = \bigcup_{n=1}^{k+1} R_n$, where R_n is a continuum, for each $n \in \{1, \ldots, k+1\}$, and $X \setminus (\bigcup_{n \neq j} R_n) \neq \emptyset$, for all $j \in \{1, \ldots, k+1\}$. Since $X = \bigcup_{n=1}^{k} \operatorname{Cl}(L_{x_n})$, there exists $j \in \{1, \ldots, k\}$ such that $\operatorname{Cl}(L_{x_j}) \cap (X \setminus (\bigcup_{n \neq j} R_n)) \neq \emptyset$. Let $z \in L_{x_j} \cap (X \setminus (\bigcup_{n \neq j} R_n))$. Observe that $z \in \operatorname{Int}(R_j)$. Since $\mathcal{T}(\{z\}) = \mathcal{T}(\{x_j\})$, for every $z \in L_{x_j}$, $L_{x_j} \subset R_j$. Thus, for each $l \in \{1, \ldots, k\}$, $L_{x_j} \subset R_{\ell_l}$, for some $\ell_l \in \{1, \ldots, k+1\}$. Since $X = \bigcup_{n=1}^{k} \operatorname{Cl}(L_{x_n})$, there exists $j_0 \in \{1, \ldots, k+1\}$ such that $\bigcup_{n \neq j_0} R_n = X$, a contradiction. Thus, by the definition of an n-indecomposable continuum, X is m-indecomposable for some $m \leq k$, and $\mathcal{T}(\mathcal{F}_1(X))$ is finite (Theorem 2.4.10), a contradiction. Therefore, S is countably infinite. □

Regarding part (2) (a) of Proposition 6.2.12, we have the following.

6.2.13 Lemma. *Let X be a metric continuum. If $\mathcal{T}(2^X)$ is a countably infinite set, then $\bigcup_{n=1}^{\infty} \operatorname{Int}(\operatorname{Cl}(L_{x_n}))$ (Notation 6.2.1) is dense in X.*

Proof. Let U be a nonempty open subset of X. Since $\mathcal{T}(\mathcal{F}_1(X))$ is countably infinite, there exists a countable set S of $\mathbb{N}$, such that $\mathcal{T}(\mathcal{F}_1(X)) = \{\mathcal{T}(\{z_j\}) \mid z_j \in X,\ j \in S\}$, where $\{x_n\}_{n=1}^{\infty} \subseteq \{z_j \mid j \in S\}$. Observe that by the proof of Theorem 6.2.3,

$$\operatorname{Int}(\operatorname{Cl}(L_{z_j})) \neq \emptyset \text{ if and only if } z_j = x_k, \text{ for some } k \in \mathbb{N}.$$

It is clear that $X = \bigcup_{j \in S} \operatorname{Cl}(L_{z_j})$. Thus, $U = \bigcup_{j \in S}(U \cap \operatorname{Cl}(L_{z_j}))$. Since U is a Baire space [129, Theorem 25.3], there exists $j_0 \in S$ such that $U \cap \operatorname{Cl}(L_{z_{j_0}})$ has nonempty interior. Hence, $U \cap \operatorname{Int}(\operatorname{Cl}(L_{z_{j_0}})) \neq \emptyset$. Therefore, there exists $k \in \mathbb{N}$ such that $U \cap \operatorname{Int}(\operatorname{Cl}(L_{x_k})) \neq \emptyset$ and $\bigcup_{n=1}^{\infty} \operatorname{Int}(\operatorname{Cl}(L_{x_n}))$ is dense in X. □

6.2.14 Lemma. *Let X be a metric continuum such that $\mathcal{T}(2^X)$ is countably infinite. Let $\{M_n\}_{n=1}^{\infty}$ be a family of indecomposable continua satisfying Proposition 6.2.12. If K is an indecomposable subcontinuum of X such that $L_{x_j} \subseteq K$, for some $j \in \mathbb{N}$, then $K \subseteq \mathcal{T}(\{x_j\})$.*

Proof. Suppose that K is an indecomposable continuum and $K \not\subseteq \mathcal{T}(\{x_j\})$ for any $j \in \mathbb{N}$. Let $j \in \mathbb{N}$ and let $y \in K \setminus \mathcal{T}(\{x_j\})$. Then there exists a subcontinuum H of X such that $y \in \operatorname{Int}(H)$ and $x_j \notin H$. Since $\operatorname{Int}(H) \cap K$ is an open subset of K, each composant of K intersects H.

We prove that $\operatorname{Int}(\operatorname{Cl}(L_{x_j})) \setminus H \neq \emptyset$. To this end, suppose $\operatorname{Int}(\operatorname{Cl}(L_{x_j})) \subset H$. Then $\operatorname{Int}(\operatorname{Cl}(L_{x_j})) \subset \operatorname{Int}(H)$. Hence, $\operatorname{Cl}(L_{x_j}) \cap \operatorname{Int}(H) \neq \emptyset$ and $L_{x_j} \cap \operatorname{Int}(H) \neq \emptyset$. This is a contradiction to the fact that $x_j \notin H$. Thus, $\operatorname{Int}(\operatorname{Cl}(L_{x_j})) \setminus H \neq \emptyset$.

Let $U = \operatorname{Int}(\operatorname{Cl}(L_{x_j})) \setminus H$. Since $L_{x_j} \subset K$, $U \subset K$. For each $x \in U$, let Q_x be the component of $K \setminus \operatorname{Int}(H)$ containing x. By Theorem 1.4.36, $Q_x \cap H \neq \emptyset$.

Let z and w be points of U such that z and w belong to different composants of K. Let $\varepsilon > 0$ be such that $w \notin \mathcal{V}_\varepsilon(z)$ and $\mathcal{V}_\varepsilon(z) \subseteq U$. Let $W_\varepsilon = \operatorname{Cl}\left(\bigcup\{Q_x \mid x \in \mathcal{V}_\varepsilon(z)\}\right) \cup H$. Then W_ε is a continuum and $z \in \operatorname{Int}(W_\varepsilon)$.

We claim that there exists $\delta > 0$, such that $w \notin W_\delta$. To see this, suppose that for each $n \in \mathbb{N}$, $w \in W_{\frac{1}{n}}$. Since $w \notin H$, there exists $r_n \in \mathcal{V}_{\frac{1}{n}}(z)$ such that $\min\{d(w,r) \mid r \in Q_{r_n}\} < \frac{1}{n}$. Note that $\{Q_{r_n}\}_{n=1}^{\infty}$ is a sequence of $\mathcal{C}_1(K)$ such that $Q_{r_n} \cap \operatorname{Int}(H) = \emptyset$, for each $n \in \mathbb{N}$. Since $\mathcal{C}_1(K)$ is compact, Theorem 1.6.7, $\{Q_{r_n}\}_{n=1}^{\infty}$ has a limit point Q in $\mathcal{C}_1(K)$, where $Q \cap \operatorname{Int}(H) = \emptyset$. Observe that $\{z, w\} \subseteq \liminf Q_{r_n}$. Thus, $\{z, w\} \subseteq Q$ and Q is a proper subcontinuum of K. This contradicts the fact that z and w belong to different composants of K. Therefore, there exists $\delta > 0$, such that $w \notin W_\delta$.

Now, observe that W_δ is a continuum such that $\operatorname{Int}(\operatorname{Cl}(L_{x_j})) \cap \operatorname{Int}(W_\delta) \neq \emptyset$ and $(X \setminus W_\delta) \cap \operatorname{Int}(\operatorname{Cl}(L_{x_j})) \neq \emptyset$. Thus, there exist p and q in L_{x_j} such that $p \in \operatorname{Int}(W_\delta)$ and $q \notin W_\delta$. This contradicts the fact that $\mathcal{T}(\{p\}) = \mathcal{T}(\{q\}) = \mathcal{T}(\{x_j\})$. Thus, $K \setminus \mathcal{T}(\{x_j\}) = \emptyset$, and $K \subset \mathcal{T}(\{x_j\})$. □

6.2.15 Theorem. *Let X be a metric continuum, and let $\mathcal{M}$ be an infinite collection of closed subsets of X. Suppose that:*

(1) *$\{M \mid M \cap M_0 \neq \emptyset\}$ is finite for each $M_0 \in \mathcal{M}$;*
(2) *$\mathcal{T}(\{x\}) = \bigcup\{M \in \mathcal{M} \mid x \in M\}$ and $\operatorname{Int}(\mathcal{T}(\{x\})) \neq \emptyset$, for every $x \in X$.*

Then $\mathcal{T}(\mathcal{F}_1(X))$ is an infinite discrete set. In particular, X is ω-indecomposable.

Proof. Observe that for each finite subfamily $\mathcal{M}'$ of $\mathcal{M}$, the family $\{M \in \mathcal{M} \mid M \cap (\cup\mathcal{M}') \neq \emptyset\}$ is also finite, by (1).

Note that $X = \bigcup \mathcal{M}$ by (2). If $\mathcal{T}(\mathcal{F}_1(X))$ were finite, then

$$X = \bigcup \mathcal{T}(\mathcal{F}_1(X))$$

would be a union of a finite subfamily of $\mathcal{M}$. Hence, $\mathcal{M}$ would be finite, by the first observation, a contradiction.

Let $x \in X$, and let $\mathcal{M}_x = \{M \in \mathcal{M} \mid x \in M\}$. Note that $\mathcal{M}'_x = \{M \in \mathcal{M} \setminus \mathcal{M}_x \mid M \cap (\cup\mathcal{M}_x) \neq \emptyset\}$ and $\mathcal{M}''_x = \{M \in \mathcal{M} \setminus (\mathcal{M}_x \cup \mathcal{M}'_x) \mid M \cap ((\cup\mathcal{M}_x) \cup (\cup\mathcal{M}'_x)) \neq \emptyset\}$ are finite, by finiteness of $\mathcal{M}_x$.

Let $U_x = \operatorname{Int}(\mathcal{T}(\{x\}))$. Observe that $\langle X, U_x\rangle_1 \cap \mathcal{T}(\mathcal{F}_1(X))$ is open in $\mathcal{T}(\mathcal{F}_1(X))$ and finite (it contains at most $2^{|\mathcal{M}_x|} + 2^{|\mathcal{M}'_x|} + 2^{|\mathcal{M}''_x|}$ elements). Therefore, $\mathcal{T}(\mathcal{F}_1(X))$ is discrete. $\mathcal{T}(\mathcal{F}_1(X))$ is countable, since it is a discrete subset of the compact metric space 2^X. □

The next example gives us a continuum X such that $\mathcal{T}(\mathcal{F}_1(X))$ is homeomorphic to $\mathbb{N}$.

6.2.16 Example. Let $\{M_n\}_{n=1}^{\infty} \cup \{M\}$ be a family of indecomposable metric continua such that:

(1) $M_i \cap M_j \neq \emptyset$ if and only if $|i - j| \leq 1$, and $M \cap M_j = \emptyset$, for each $j \in \mathbb{N}$;
(2) $M_n \cap M_{n+1} = \{p_n\}$, for every $n \in \mathbb{N}$;
(3) $\lim_{n\to\infty} M_n = \{p\}$, for some $p \in M$.

Then $X = (\bigcup_{n=1}^{\infty} M_n) \cup M$ is a metric continuum (Figure 6.1) such that $\mathcal{T}(\mathcal{F}_1(X))$ is an infinite discrete set, Theorem 6.2.15.

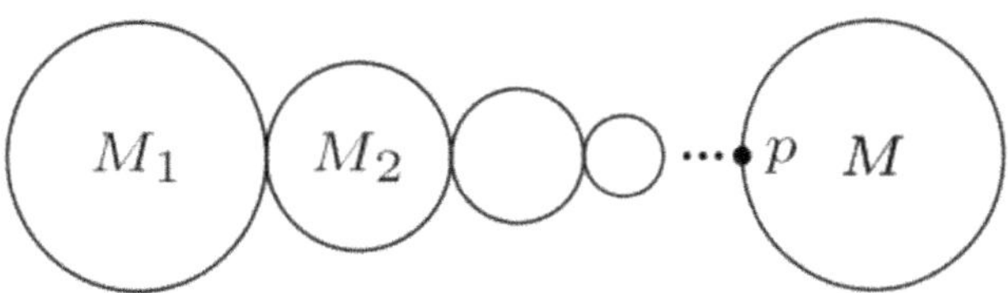

Fig. 6.1 A family of indecomposable metric continua

The following example shows an ω-indecomposable metric continuum X, such that $\mathcal{T}(\{x\}) = \bigcup\{M_j \mid x \in M_j\}$, for each $x \in X$, and $\mathcal{T}(\mathcal{F}_1(X))$ is homeomorphic to $\{\frac{1}{n}\}_{n=1}^{\infty} \cup \{0\}$.

6.2.17 Example. Let M_0 be an indecomposable metric continuum, and let $\{p_n\}_{n=1}^{\infty} \subset M_0$ be such that $\lim_{n\to\infty} p_n = p$, for some $p \in M_0$. Let $\{M_n\}_{n=1}^{\infty}$ be a collection of pairwise disjoint indecomposable metric continua such that:

1. $M_n \cap M_0 = \{p_n\}$, for each $n \in \mathbb{N}$, and
2. $\lim_{n\to\infty} \operatorname{diam}(M_n) = 0$.

If $X = \bigcup_{n=0}^{\infty} M_n$ (Figure 6.2), then X is an ω-indecomposable metric continuum such that $\mathcal{T}(\mathcal{F}_1(X))$ has exactly one limit point.

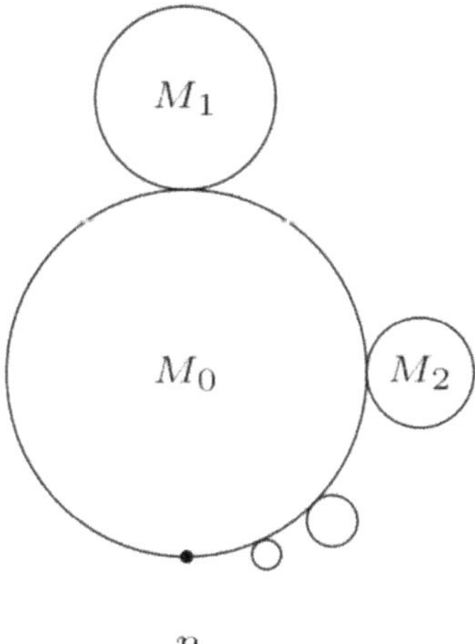

Fig. 6.2 A second family of indecomposable metric continua

The next example shows that Theorem 2.4.10 does not extend to ω-indecomposable metric continua. It also shows that the converse of Theorem 6.2.3 is not true.

6.2.18 Example. There exist a metric continuum X and a family of indecomposable metric continua $\{M_n\}_{n=1}^{\infty}$ such that:

(1) $X = \mathrm{Cl}(\bigcup_{n=1}^{\infty} M_n)$;
(2) for each $n \in \mathbb{N}$, there exists a composant C_n of M_n, such that $C_n \cap \mathrm{Cl}(\bigcup_{j \neq n} M_j) = \emptyset$;
(3) X is not ω-indecomposable.

Let P be the Pełczyński compactum (P is a metric compactification of the integers with a Cantor set as a remainder) in $\mathbb{R}^2$, Figure 6.3. Each segment in Figure 6.4 represents an indecomposable metric continuum. Let X be the continuum in Figure 6.4. Then $\mathcal{T}(\{p\}) = \{p\}$ for each p in the remainder of P. Therefore, $\mathcal{T}(\mathcal{F}_1(X))$ is not ω-indecomposable.

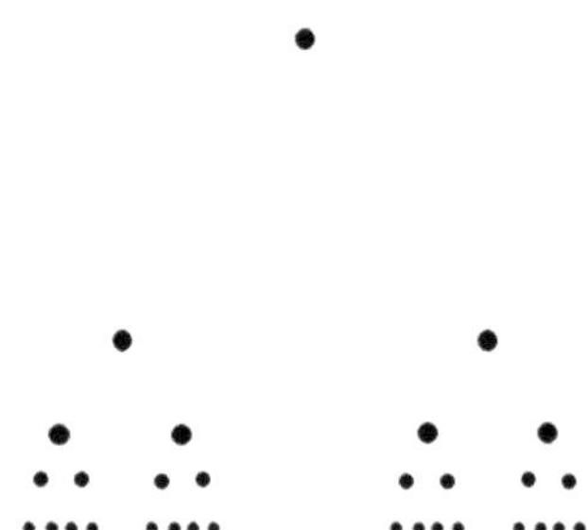

Fig. 6.3 Pełczyński compactum P

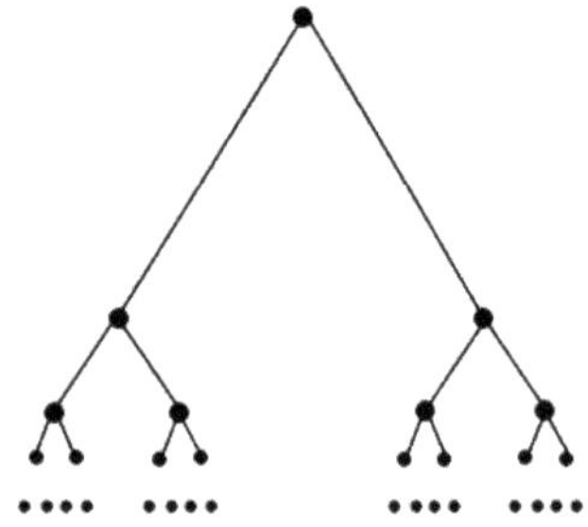

Fig. 6.4 Non-ω-indecomposable metric continuum

6.2.19 Theorem. *If X is an ω-indecomposable metric continuum, then $\mathcal{T}(2^X)$ has a limit point.*

Proof. By Proposition 6.2.12, there exists a family of indecomposable metric continua $\{M_n\}_{n=1}^{\infty}$, such that:

(1) $X = \mathrm{Cl}(\bigcup_{n=1}^{\infty} M_n)$;
(2) for every $n \in \mathbb{N}$, there exists $x_n \in M_n$ satisfying:

   (a) $\mathrm{Int}(\mathrm{Cl}(L_{x_n})) \neq \emptyset$;
   (b) $L_{x_n} \subset \mathrm{Cl}(L_{x_n}) \subset M_n \subset \mathcal{T}(\{x_n\})$;
   (c) $\mathrm{Int}(\mathrm{Cl}(L_{x_k})) \subset X \setminus \mathrm{Cl}(\bigcup_{n \in \mathbb{N}\setminus\{k\}} M_n)$.

Let $y_n \in L_{x_n} \cap \mathrm{Int}(\mathrm{Cl}(L_{x_n}))$, for all $n \in \mathbb{N}$. Let $A_n = \{y_1, \dots, y_n\}$, for each $n \in \mathbb{N}$. Note that if $j > n$, then $M_j \cap A_n = \emptyset$, by (2) (c). Hence, $\mathcal{T}(A_n) \neq X$ for any $n \in \mathbb{N}$. Since $\mathcal{T}(\{y_j\}) = \mathcal{T}(\{x_j\})$ and $M_j \subset \mathcal{T}(\{y_j\})$, by (2) (b), we obtain that $\bigcup_{j=1}^{n} M_j \subset \mathcal{T}(A_n)$. Also, we have that $\mathcal{T}(A_n) \subset \mathcal{T}(A_{n+1})$, for all $n \in \mathbb{N}$. Thus, $\bigcup_{n=1}^{\infty} M_n \subset \lim_{n\to\infty} \mathcal{T}(A_n) = \mathrm{Cl}(\bigcup_{n=1}^{\infty} \mathcal{T}(A_n))$ [74, 7., p. 339]. Since $X = \mathrm{Cl}(\bigcup_{n=1}^{\infty} M_n)$, by (1), $X = \lim_{n\to\infty} \mathcal{T}(A_n)$. Hence, $X \in \mathcal{T}(2^X)$. Therefore, X is a limit point of $\mathcal{T}(2^X)$. □

6.2.20 Corollary. *Let X be a metric continuum. Then $\mathcal{T}(2^X)$ is discrete if and only if X is n-indecomposable, for some $n \in \mathbb{N}$.*

6.2.21 Example. Let $\{M_n\}_{n=1}^{\infty}$ be a sequence of indecomposable metric continua such that:

(1) for each pair of elements i and j of $\mathbb{N}$, $M_i \cap M_j \neq \emptyset$ if and only if $|i - j| \leq 1$;
(2) $M_n \cap M_{n+1} = \{p_n\}$, for every $n \in \mathbb{N}$;
(3) $\lim_{n\to\infty} M_n = \{p\}$, for some p, and

$X = (\bigcup_{n=1}^{\infty} M_n) \cup \{p\}$ is an irreducible continuum (Figure 6.5).

Then $\mathcal{T}(2^X)$ is homeomorphic to the Pełczyński compactum P (Example 6.2.18).

Note that

$$\mathcal{T}(\{x\}) = \begin{cases} M_1, & \text{if } x \in M_1 \setminus \{p_1\}; \\ M_n, & \text{if } x \in M_n \setminus \{p_{n-1}, p_n\} \text{ and } n \geq 2; \\ M_n \cup M_{n+1}, & \text{if } x = p_n; \\ \{p\}, & \text{if } x = p. \end{cases}$$

Let $y_1 \in M_1 \setminus \{p_1\}$, and for each $n \geq 2$, let $y_n \in M_n \setminus \{p_{n-1}, p_n\}$. Let $Y = \{p\} \cup \{y_n\}_{n=1}^{\infty}$. Then Y is a closed set, and it has one limit point. Observe that $\mathcal{T}(A) = A \cup \bigcup\{M_n \mid M_n \cap A \neq \emptyset\}$, for $A \in 2^Y$. Also note that $\mathcal{T}|_{2^Y}$ is a homeomorphism onto its image $\mathcal{T}(2^Y) = \mathcal{T}(2^X)$. Since 2^Y is homeomorphic to the Pełczyński compactum P [108, Corollary 3], $\mathcal{T}(2^X)$ is homeomorphic to P.

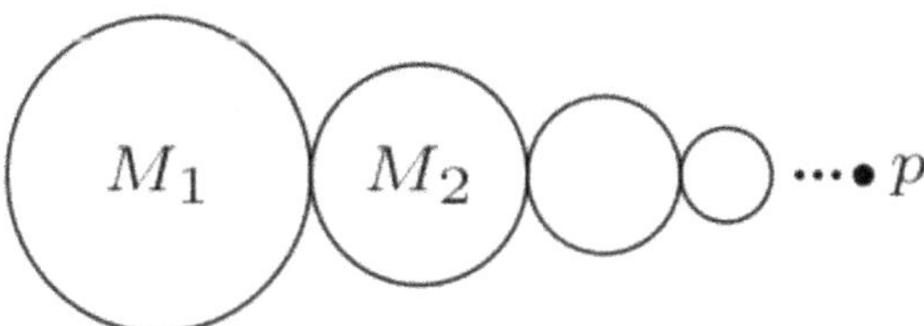

Fig. 6.5 A third family of indecomposable metric continua

6.2.22 Example. There exists a metric continuum X such that $\mathcal{T}(2^X)$ is a countably infinite set. Let $\{M_n\}_{n=1}^{\infty}$ be a sequence of indecomposable continua in $\mathbb{R}^2$ such that

(1) $M_i \cap M_j = \emptyset$ for each $i \neq j$;

(2) $\lim_{n\to\infty} M_n = \{p\}$, for some $p \in \mathbb{R}^2$.

Let $Z = \bigcup_{n=1}^{\infty} M_n \cup \{p\}$ (Figure 6.5). Note that Z is a compact subset of $\mathbb{R}^2$. Let $x_n \in M_n$, for each $n \in \mathbb{N}$, and define $S = \{x_n\}_{n=1}^{\infty} \cup \{p\}$. Since $\lim_{n\to\infty} M_n = \{p\}$, S is compact. Let $X = Z/S$. Then for each $A \in 2^X$,

$$\mathcal{T}(A) = \begin{cases} X, & \text{if } p \in A; \\ \bigcup_{i=1}^{k} M_{n_i} & \text{if } p \notin A, \text{ and } \{M_{n_1}, \dots, M_{n_k}\} = \\ & \qquad \{M_n \mid M_n \cap A \neq \emptyset\}. \end{cases}$$

Thus, $|\mathcal{T}(2^X)| = |\{\bigcup_{j\in F} M_j \mid j \in F, F \text{ is finite}\}|$. Therefore, $\mathcal{T}(2^X)$ is countably infinite.

6.3 Connected and Compact Images of $\mathcal{T}$

Since the set function $\mathcal{T}$, in general, is not continuous, one can ask about the compactness and connectedness of $\mathcal{T}(2^X)$, for some continuum X. We begin this section by proving the compactness of $\mathcal{T}(2^X)$ implies that $\mathcal{T}(2^X)$ is either finite or uncountable. Then we study conditions on a metric continuum, in order to have that $\mathcal{T}(2^X)$ is connected.

6.3.1 Theorem. *Let X be a metric continuum such that $\mathcal{T}(2^X)$ is compact. Then $\mathcal{T}(2^X)$ is a finite or an uncountable set.*

Proof. Suppose that $\mathcal{T}(2^X)$ is a countably infinite set. By Theorems 2.4.10 and 6.2.6, $\mathcal{T}(\mathcal{F}_1(X))$ is also countably infinite. Thus, there exists a sequence $\{x_n\}_{n=1}^{\infty}$ of points of X satisfying Proposition 6.2.12.

We claim that there exist an infinite subset $\mathcal{N}$ of $\mathbb{N}$ and a sequence $\{z_j\}_{j=1}^{\infty}$ of elements of X such that $z_j \in \mathrm{Int}(\mathrm{Cl}(L_{x_{n_j}})) \cap L_{x_{n_j}}$ (Notation 6.2.1), for each $n_j \in \mathcal{N}$, $\lim_{j\to\infty} z_j = z_0$ and $\mathrm{Int}(\mathrm{Cl}(L_{x_{n_j}})) \cap \mathcal{T}(\{z_0\}) = \emptyset$ for every $n_j \in \mathcal{N}$.

To prove this, note that since X is compact, without loss of generality, we may assume that there exists $y_n \in \mathrm{Int}(\mathrm{Cl}(L_{x_n})) \cap L_{x_n}$, for each $n \in \mathbb{N}$, such that $\lim_{n\to\infty} y_n = y_0$ and $\lim_{n\to\infty} \mathcal{T}(\{y_n\}) = L$, for some $L \in 2^X$. Since $\mathcal{T}$ is upper semicontinuous, Theorem 5.1.1, $L \subseteq \mathcal{T}(\{y_0\})$. If $\mathrm{Int}(\mathrm{Cl}(L_{x_n})) \cap \mathcal{T}(\{y_0\}) = \emptyset$, for infinitely many indexes n, then we complete the proof of our claim. Hence, suppose that $\mathrm{Int}(\mathrm{Cl}(L_{x_n})) \cap \mathcal{T}(\{y_0\}) \neq \emptyset$, for any $n \in \mathbb{N}$. Note that if $w \in \mathrm{Int}(\mathrm{Cl}(L_{x_n}))$ and K is a continuum such that $w \in \mathrm{Int}(K)$, then $L_{x_n} \subseteq K$. Also, $\mathrm{Int}(\mathrm{Cl}(L_{x_n})) \subseteq K$ and $y_0 \in K$. Thus, $\mathrm{Int}(\mathrm{Cl}(L_{x_n})) \subseteq \mathcal{T}(\{y_0\})$, for every $n \in \mathbb{N}$. Now, we see that $L \cap \mathrm{Int}(\mathrm{Cl}(L_{x_n})) = \emptyset$, for each $n \in \mathbb{N}$. Suppose that there exists $k \in \mathbb{N}$ such that $L \cap \mathrm{Int}(\mathrm{Cl}(L_{x_k})) \neq \emptyset$. Since $\lim_{n\to\infty} \mathcal{T}(\{y_n\}) = L$, there exists $l \neq k$ such that $\mathcal{T}(\{y_l\}) \cap \mathrm{Int}(\mathrm{Cl}(L_{x_k})) \neq \emptyset$. Since $\mathrm{Int}(\mathrm{Cl}(L_{x_k})) \subseteq M_k$, $y_l \in M_k$. This contradicts the fact that $\mathrm{Int}(\mathrm{Cl}(L_{x_l})) \cap M_k = \emptyset$, (e) of Proposition 6.2.12. Therefore, $L \cap \mathrm{Int}(\mathrm{Cl}(L_{x_n})) = \emptyset$, for each $n \in \mathbb{N}$, and $L \subsetneq \mathcal{T}(\{y_0\})$.

Since $\mathcal{T}(2^X)$ is compact, $L \in \mathcal{T}(2^X)$. Hence, there exists $A \in 2^X$, $A \subseteq L$, such that $\mathcal{T}(A) = L$.

Since $\bigcup_{n=1}^{\infty} \text{Int}(\text{Cl}(L_{x_n}))$ is dense, Lemma 6.2.13, there exists an infinite subset $\mathcal{N}$ of $\mathbb{N}$ such that, for every $n_j \in \mathcal{N}$, there exists $z_j \in \text{Int}(\text{Cl}(L_{x_{n_j}})) \cap L_{x_{n_j}}$, where $\lim_{j\to\infty} z_j = z_0$ and $z_0 \in A$. Since $L \cap \text{Int}(\text{Cl}(L_{x_n})) = \emptyset$, for each $n \in \mathbb{N}$, and $\mathcal{T}(\{z_0\}) \subseteq \mathcal{T}(A) = L$, we have that $\text{Int}(\text{Cl}(L_{x_{n_s}})) \cap \mathcal{T}(\{z_0\}) = \emptyset$, for every $n_s \in \mathcal{N}$.

Let $P = \{z_j\}_{j=1}^{\infty} \cup \{z_0\} \in 2^X$. We claim that for each $j \in \mathbb{N}$, there exists a subcontinuum E_j of X such that $z_j \in \text{Int}(E_j)$ and $E_j \cap P = \{z_j\}$.

To show this, let $j \in \mathbb{N}$. Since $z_j \in \text{Int}(\text{Cl}(L_{x_{n_j}}))$ and $\text{Int}(\text{Cl}(L_{x_{n_j}})) \cap \mathcal{T}(\{z_0\}) = \emptyset$, we have that $z_j \notin \mathcal{T}(\{z_0\})$. Thus, there exists a subcontinuum K of X such that $z_j \in \text{Int}(K)$ and $z_0 \notin K$. Since $z_j \in \mathcal{T}(\{x_{n_j}\})$, $L_{x_{n_j}} \subseteq K$. By [129, Theorem 28.4], there exists a subcontinuum E_j of K such that E_j is irreducible about $L_{x_{n_j}}$. Note that E_j is indecomposable (Proposition 6.2.2). Therefore, $E_j \subseteq \mathcal{T}(\{x_{n_j}\})$, Lemma 6.2.14, and $E_j \cap P = \{z_j\}$, for each $j \in \mathbb{N}$.

Let S and R be two different subsequences of P. If $z_j \in S \setminus R$, then $z_j \in \text{Int}(E_j)$ and $E_j \cap (R \cup \{z_0\}) = \emptyset$. Thus, $z_j \in \mathcal{T}(S \cup \{z_0\}) \setminus \mathcal{T}(R \cup \{z_0\})$. Since P has uncountably many subsequences, $\mathcal{T}(2^X)$ is uncountable, a contradiction. Therefore, $\mathcal{T}(2^X)$ is either finite or uncountable. □

6.3.2 Theorem. *Let X be a metric continuum such that $\mathcal{T}$ is idempotent on closed sets. Let $M = \{x \in X \mid \mathcal{T}(\{x\})$ has property $BL\}$ and $L = \bigcup\{\mathcal{T}(\{x\}) \mid x \in M\}$. If L is compact and $\mathcal{T}(2^X)$ is connected, then L is connected.*

Proof. Suppose that $L = A \cup B$, where A and B are disjoint and closed sets. By Theorem 3.1.6, for each $D \in 2^X$, there exists $x \in M$ such that $\mathcal{T}(\{x\}) \subset \mathcal{T}(D)$. Therefore, we have that

$$\mathcal{T}(2^X) = (\langle X, A\rangle \cap \mathcal{T}(2^X)) \cup (\langle X, B\rangle \cap \mathcal{T}(2^X)).$$

Since A is closed, $\langle X, A\rangle \cap \mathcal{T}(2^X)$ is closed in $\mathcal{T}(2^X)$ [51, p. 2]. We prove that there exists an open subset U of X such that $A \subset U$ and $\langle X, A\rangle \cap \mathcal{T}(2^X) = \langle X, U\rangle \cap \mathcal{T}(2^X)$. Suppose that there exists a sequence of open sets $\{U_n\}_{n=1}^{\infty}$ such that $A = \bigcap_{n=1}^{\infty} U_n$ and $\langle X, U_n\rangle \cap (\mathcal{T}(2^X) \setminus \langle X, A\rangle) \neq \emptyset$, for any $n \in \mathbb{N}$. Let $E_n \in 2^X$ be such that $\mathcal{T}(E_n) \in \langle X, U_n\rangle \cap (\mathcal{T}(2^X) \setminus \langle X, A\rangle)$, for each $n \in \mathbb{N}$. Hence, $\mathcal{T}(E_n) \cap U_n \neq \emptyset$ and $\mathcal{T}(E_n) \cap A = \emptyset$, for every $n \in \mathbb{N}$. Note that $\mathcal{T}(E_n) \in \langle X, B\rangle \cap \mathcal{T}(2^X)$, for all $n \in \mathbb{N}$. Let $x_n \in \mathcal{T}(E_n) \cap U_n$ be such that $\lim_{n\to\infty} x_n = x_0$, for some $x_0 \in X$. Since $A = \bigcap_{n=1}^{\infty} U_n$, $x_0 \in A$. Also, since $\mathcal{T}$ is idempotent on closed sets, $\mathcal{T}(\{x_n\}) \subset \mathcal{T}(E_n)$, for all $n \in \mathbb{N}$. Thus, $\mathcal{T}(\{x_n\}) \cap A = \emptyset$ and $\mathcal{T}(\{x_n\}) \in \langle X, B\rangle$. Let W be an open set such that $A \subset W$ and $W \cap B = \emptyset$. Since $x_0 \in A$, $\mathcal{T}(\{x_0\})$ is connected (Theorem 2.1.27) and $\mathcal{T}(\{x_0\}) \subset L$, we have that $\mathcal{T}(\{x_0\}) \subset A \subset W$. Since $\mathcal{T}$ is upper semicontinuous, Theorem 5.1.1, there exists an open subset $\mathcal{V}$ of 2^X such that $\{x_0\} \in \mathcal{V}$ and $\mathcal{T}(E) \subset W$ for each $E \in \mathcal{V}$, but this contradicts the facts that $\lim_{n\to\infty}\{x_n\} = \{x_0\}$, $\mathcal{T}(\{x_n\}) \in \langle X, B\rangle$ and $W \cap B = \emptyset$. Therefore, there exists an open set U such that $A \subset U$ and

$\langle X, U\rangle \cap \mathcal{T}(2^X) \subset \langle X, A\rangle \cap \mathcal{T}(2^X)$. Since $\langle X, U\rangle$ is open in 2^X, Theorem 1.6.5, $\langle X, A\rangle \cap \mathcal{T}(2^X)$ is open in $\mathcal{T}(2^X)$. Note that $\langle X, A\rangle \cap \mathcal{T}(2^X) \subset \langle X, U\rangle \cap \mathcal{T}(2^X)$ and $\langle X, A\rangle \cap \mathcal{T}(2^X) = \langle X, U\rangle \cap \mathcal{T}(2^X)$. Since $\mathcal{T}(2^X)$ is connected, $\mathcal{T}(2^X) = \langle X, A\rangle \cap \mathcal{T}(2^X)$ and $B = \emptyset$. Therefore, L is connected. □

6.3.3 Remark. Let X be the Cantor fan, Example 2.1.15, and let M be as in Theorem 6.3.2. Then $L = \bigcup\{\mathcal{T}(\{x\}) \mid x \in M\}$ is homeomorphic to $\mathcal{C}$. Therefore, $\mathcal{T}(2^X)$ is compact but not connected, Theorem 6.3.2.

The proof of the next lemma is due to David P. Bellamy, which simplifies the original proof.

6.3.4 Lemma. *Let X be a metric continuum. If U is an open set such that X is locally connected at x, for each $x \in U$, then $\mathcal{T}(A) = A$, for every nonempty closed subset A of X that is contained in U.*

Proof. Let V be an arbitrary open set such that A is a subset of V and $\mathrm{Cl}(V)$ is a subset of U. Since X is locally connected at every point of U, there exists a finite cover of the boundary of V by the interiors of continua $K_1, \ldots, K_n$ such that each K_j is a subset of U and is disjoint from A. Then the union of $X \setminus V$ and the K_j's only has finitely many components, since each such component intersects some K_j, Theorem 1.4.36. Hence, every point p of $X \setminus V$ belongs to the interior of one of these components, which is disjoint from A. Thus, p does not belong to $\mathcal{T}(A)$. Therefore, $\mathcal{T}(A)$ is a subset of V, and thus, $\mathcal{T}(A) = A$, since no other closed set is contained in every such open set V. □

6.3.5 Example. There exists a metric continuum X such that

$$X = \bigcup\{\mathcal{T}(\{x\}) \mid \mathcal{T}(\{x\}) \text{ has property } BL\},$$

but $\mathcal{T}(2^X)$ is not connected. Let $X = U \cup (\{0\} \times [-2, 2]) \cup V$, where $U = \{(t, 2\sin(\frac{1}{t}) \mid 0 < t \leq 1)\}$ and $V = \{(-t, \sin(\frac{1}{t})) \mid 0 < t \leq 1\}$. Note that $\mathcal{T}(\{z\}) = \{z\}$, for each $z \in U \cup V$, and $\mathcal{T}(\{w\}) = \{0\} \times [-2, 2]$, for every $w \in \{0\} \times [-2, 2]$. Thus, for each point x of X, $\mathcal{T}(\{x\})$ has property BL. Let $W = X \cap \{(x, y) \mid x > 0 \text{ or } |y| > 1\}$. Note that $\mathcal{T}(2^X) = (\langle V\rangle \cap \mathcal{T}(2^X)) \cup (\langle X, W\rangle \cap \mathcal{T}(2^X))$. Therefore, $\mathcal{T}(2^X)$ is not connected.

6.3.6 Theorem. *Let X be a metric continuum. If there exists an open, connected and dense subset U of X such that X is locally connected at every point of U, then $\mathcal{T}(2^X)$ is connected.*

Proof. Let $\mathcal{L} = \{\mathcal{T}(A) \mid A \in 2^X \text{ and } A \subset U\}$. By Lemma 6.3.4, $\mathcal{L} = \langle U\rangle$. Hence, $\mathcal{L}$ is open and connected. Since U is dense in X, $\mathrm{Cl}_{2^X}(\langle U\rangle) = 2^X$. Hence, $\mathcal{L} \subset \mathcal{T}(2^X) \subset 2^X = \mathrm{Cl}_{2^X}(\mathcal{L})$. Therefore, $\mathcal{T}(2^X)$ is connected. □

6.3.7 Remark. If X is a metric compactification of a ray $[0, \infty)$, then $\mathcal{T}(2^X)$ is connected, Theorem 6.3.6. Note that in Example 6.3.5, $U \cup V$ is dense, X is locally connected in every point of $U \cup V$, but $\mathcal{T}(2^X)$ is not connected. Thus, the connectedness of U in Theorem 6.3.6 is essential.

6.3.8 Example. There exists a metric continuum X, such that $\mathcal{T}_X(2^X)$ is homeomorphic to the Cantor set $\mathcal{C}$. Let $Z = \mathcal{C} \times [0,1]$, where $\mathcal{C}$ is the Cantor set in $[0,1]$. Let X be the "V-Λ" continuum defined in [75, Fig. 3, p. 191], and let $q\colon Z \twoheadrightarrow X$ be the quotient map. We see that $\mathcal{T}_X(2^X)$ is homeomorphic to the Cantor set $\mathcal{C}$. If $f\colon 2^{\mathcal{C}} \to 2^X$ is given by $f(A) = q(A \times [0,1])$, then f is a homeomorphism from $2^{\mathcal{C}}$ onto its image $f(2^{\mathcal{C}}) = \mathcal{T}_X(2^X)$. Since $2^{\mathcal{C}}$ is homeomorphic to $\mathcal{C}$ [105, (0.69.8)], $\mathcal{T}_X(2^X)$ is homeomorphic to $\mathcal{C}$.

6.3.9 Theorem. *Let X be a one-dimensional metric continuum. If $\mathcal{T}(2^X) = 2^X$, then X is locally connected.*

Proof. By Theorem 2.1.37, we only need to show that $\mathcal{T}(A) = A$ for all $A \in 2^X$. Suppose there exists $A \in 2^X$ such that $\mathcal{T}(A) \neq A$. Let $p \in \mathcal{T}(A) \setminus A$. Since X is one-dimensional and A is closed, there exists an open subset U of X such that $p \in U$, $\text{Cl}(U) \cap A = \emptyset$ and $\dim(\text{Bd}(U)) = 0$. This implies that $\text{Bd}(U)$ is totally disconnected [63, A), p. 20]. By Corollary 4.1.4, we have that $\mathcal{T}(\text{Bd}(U)) = \text{Bd}(U)$. Let W be a subcontinuum of X such that $p \in \text{Int}(W)$. Since $p \in \mathcal{T}(A)$, $W \cap A \neq \emptyset$. Thus, $W \cap \text{Bd}(U) \neq \emptyset$. Hence, $p \in \mathcal{T}(\text{Bd}(U)) = \text{Bd}(U)$, a contradiction. Therefore, $\mathcal{T}(A) = A$ for all $A \in 2^X$. □

6.3.10 Theorem. *Let X be a continuum with the uniform property of Effros. Then $\mathcal{T}(\mathcal{F}_1(X))$ is a closed subset of 2^X. Moreover $\mathcal{T}(2^X)$ is a closed subset of 2^X if and only if $\mathcal{T}$ is continuous.*

Proof. If X is indecomposable, then $\mathcal{T}(\mathcal{F}_1(X)) = \{X\}$, Theorem 2.1.44. Hence, $\mathcal{T}(\mathcal{F}_1(X))$ is a closed subset of 2^X. Observe that, in this case, $\mathcal{T}$ is continuous.

Suppose X is a decomposable continuum. By Theorem 1.4.59, X is a homogeneous continuum. By Theorem 3.3.8,

$$\mathcal{G} = \{\mathcal{T}(\{x\}) \mid x \in X\}$$

is a continuous decomposition of X. In particular, we have that $\mathcal{T}|_{\mathcal{F}_1(X)}\colon \mathcal{F}_1(X) \to 2^X$ is continuous. Thus, $\mathcal{T}(\mathcal{F}_1(X))$ is a closed subset of 2^X. Also, by Theorems 1.6.22 and 2.3.18, $\mathcal{T}$ is idempotent on closed sets. Therefore, by Theorem 5.1.7, $\mathcal{T}(2^X)$ is a closed subset of 2^X if and only if $\mathcal{T}$ is continuous. □

6.3.11 Theorem. *There exists a one-dimensional homogeneous metric continuum X such that $\mathcal{T}(\mathcal{F}_1(X)) = \mathcal{F}_1(X)$ and $\mathcal{T}(2^X)$ is not a closed subset of 2^X.*

Proof. There exists a one-dimensional arcwise connected homogeneous continuum X that is not locally connected [109]. By [92, Corollary 5.1.21], each arcwise connected homogeneous metric continuum is aposyndetic. Hence, by Theorem 2.1.34, for each $x \in X$, $\mathcal{T}(\{x\}) = \{x\}$. In particular, $\mathcal{T}(\mathcal{F}_1(X)) = \mathcal{F}_1(X)$. Since aposyndetic continua for which $\mathcal{T}$ is continuous are locally connected, Corollary 5.1.17, we have that $\mathcal{T}$ is not continuous on X. Therefore, by Theorem 6.3.10, $\mathcal{T}(2^X)$ is not a closed subset of 2^X. □

References for Chapter 6

Section 6.1: [19, 69].
Section 6.2: [16–20, 45, 74, 108, 129].
Section 6.3: [16–18, 53, 63, 75, 92, 105, 109].

Chapter 7
Applications

We present several applications of the set function $\mathcal{T}$. We start with properties of continuously irreducible metric continua and their hyperspace of subcontinua and continuously type A' metric θ-continua. We give sufficient conditions for the noncontractibility of continua. We study strict point $\mathcal{T}$-asymmetry of arc-smooth metric continua and dendroids. We consider R-continua in dendroids. We give a couple of characterizations of local connectedness. We present sufficient conditions for a closed subset of a continuum to have its image under $\mathcal{T}$ to be a shore set in the continuum. We consider generalized inverse limits of nonaposyndetic metric homogeneous continua X using the set function $\mathcal{T}|_{\mathcal{F}_1(X)}$ as an upper semicontinuous bonding function. We present more relationships between the set functions $\mathcal{T}$ and $\mathcal{K}$.

7.1 Continuously Irreducible and Type A' Continua

We characterize the finest continuous monotone decomposition of continuously irreducible and continuously type A' metric continua such that the quotient space is an arc and a graph, respectively. We give properties of the hyperspaces of these classes of continua.

7.1.1 Lemma. *Let X be a continuously irreducible continuum, and let $\mathcal{G}$ be the finest continuous monotone decomposition of X such that $X/\mathcal{G}$ is an arc. If $q\colon X \twoheadrightarrow [0,1]$ is the quotient map, then q is atomic, and $q^{-1}(t)$ is a terminal subcontinuum of X for every $t \in [0,1]$.*

Proof. Let X be a continuously irreducible continuum, and let $q\colon X \twoheadrightarrow [0,1]$ be the quotient map obtained from the finest continuous monotone decomposition of X. Let K be a subcontinuum of X such that $q(K)$ is nondegenerate. It is always true that $K \subset q^{-1}q(K)$. Let $x \in q^{-1}q(K)$. Then there exists $y \in K$ such that

S. Macías, *Set Function $\mathcal{T}$*, Developments in Mathematics 67,
https://doi.org/10.1007/978-3-030-65081-0_7

$q(y) = q(x)$. Hence, by [119, Theorem 5, p. 10], $q^{-1}q(x) \subset K$. Therefore, $K = q^{-1}q(K)$, and q is atomic.

Let $t \in [0,1]$. Since q is an atomic map, by [92, Theorem 8.1.25], $q^{-1}(t)$ is a terminal subcontinuum of X. □

As a consequence of Theorem 5.2.9 and Lemma 7.1.1, we obtain the following.

7.1.2 Corollary. *Let X be a continuously irreducible continuum. If $\mathcal{G} = \{\mathcal{T}_X(\{x\}) \mid x \in X\}$, then $\mathcal{G}$ is the finest continuous decomposition of X such that $X/\mathcal{G}$ is an arc, and $\mathcal{T}_X(\{x\})$ is a terminal subcontinuum of X for every $x \in X$.*

7.1.3 Theorem. *Let X be a continuously irreducible continuum. If $f\colon X \to X$ is a map and $x \in f(X)$, then either $f(X) \subset \mathcal{T}_X(\{x\})$ or $\mathcal{T}_X(\{x\}) \subset f(X)$. Moreover, in the second case,*

$$f(X) = \bigcup \{\mathcal{T}_X(\{x\}) \mid x \in f(X)\}.$$

Proof. Let X be a continuously irreducible continuum, let $f\colon X \to X$ be a map and let $x \in f(X)$. Note that $f(X) \cap \mathcal{T}_X(\{x\}) \neq \emptyset$. Since $f(X)$ is a continuum and $\mathcal{T}_X(\{x\})$ is a terminal subcontinuum of X (Corollary 7.1.2), either $f(X) \subset \mathcal{T}_X(\{x\})$ or $\mathcal{T}_X(\{x\}) \subset f(X)$.

Suppose $\mathcal{T}_X(\{x\}) \subset f(X)$. If $\mathcal{T}_X(\{x\}) = f(X)$, then since $\mathcal{T}_X$ is continuous (Theorem 5.2.9), $\mathcal{T}_X$ is idempotent (Theorem 5.1.6). Hence, $f(X) = \bigcup \{\mathcal{T}_X(\{z\}) \mid z \in f(X)\}$, Proposition 2.3.3. Next assume that $\mathcal{T}_X(\{x\}) \neq f(X)$. Then for every $z \in f(X) \setminus \mathcal{T}_X(\{x\})$, we also have that $\mathcal{T}_X(\{z\}) \subset f(X)$. Therefore,

$$f(X) = \bigcup \{\mathcal{T}_X(\{x\}) \mid x \in f(X)\}.$$

□

As a consequence of Theorem 7.1.3, we have the following.

7.1.4 Corollary. *Let X be a continuously irreducible continuum. If $f\colon X \to X$ is a map such that either $f(X) \not\subset \mathcal{T}_X(\{x\})$ or $f(X) \neq \mathcal{T}_X(\{x\})$ for any $x \in X$, then $\mathcal{T}_X(\{x\}) = \mathcal{T}_{f(X)}(\{x\})$ for each $x \in f(X)$.*

7.1.5 Theorem. *Let X be a continuously irreducible continuum. If $f\colon X \to X$ is a map, then there exists $x \in X$ such that $f(x) \in \mathcal{T}_X(\{x\})$.*

Proof. Let X be a continuously irreducible continuum, and let $q\colon X \twoheadrightarrow [0,1]$ be the quotient map given by the finest continuous monotone decomposition of X. Let $f\colon X \to X$ be a map, and consider the following sets:

$$A = \{x \in X \mid q(x) < qf(x)\};$$

$$B = \{x \in X \mid q(x) = qf(x)\};$$

$$C = \{x \in X \mid q(x) > qf(x)\}.$$

We show that $B \neq \emptyset$. To this end, suppose $B = \emptyset$. Note that if $x_0 \in q^{-1}(0)$ and $x_1 \in q^{-1}(1)$, then $x_0 \in A$ and $x_1 \in C$. Also observe that $A \cap C = \emptyset$ and $X = A \cup C$.

Let $x \in \mathrm{Cl}_X(A)$. Then there exists a sequence $\{a_n\}_{n=1}^{\infty}$ of points of A converging to x. Since both q and f are continuous, $\{q(a_n)\}_{n=1}^{\infty}$ converges to $q(x)$ and $\{qf(a_n)\}_{n=1}^{\infty}$ converges to $qf(x)$. Since $\{a_n\}_{n=1}^{\infty} \subset A$, $q(a_n) < qf(a_n)$ for every positive integer n. Hence, $q(x) \leq qf(x)$. Since $B = \emptyset$, $q(x) < qf(x)$. Thus, $x \in A$. Therefore, A is closed in X. Similarly, C is also closed in X. Thus, X is not connected, a contradiction. Therefore, $B \neq \emptyset$.

Let $x \in B$. Then, by Theorem 5.2.9, $q^{-1}q(x) = \mathcal{T}_X(\{x\})$. Therefore, $f(x) \in \mathcal{T}_X(\{x\})$. □

7.1.6 Theorem. *Let X be a continuously irreducible continuum such that $\mathcal{T}_X(\{x\})$ has the fixed-point property for each $x \in X$. If $f\colon X \to X$ is a map such that f is monotone onto $f(X)$, then f has a fixed point.*

Proof. Let X be a continuously irreducible continuum. Suppose $\mathcal{T}_X(\{x\})$ has the fixed-point property for every $x \in X$. Let $f\colon X \to X$ be a map such that f is monotone onto $f(X)$. If there exists $x \in X$ such that either $f(X) \subset \mathcal{T}_X(\{x\})$ or $f(X) = \mathcal{T}_X(\{x\})$, then there is nothing to prove. Assume that $f(X) \not\subset \mathcal{T}_X(\{x\})$ and $f(X) \neq \mathcal{T}_X(\{x\})$ for any $x \in X$. By Theorem 7.1.5, there exists $x_0 \in X$ such that $f(x_0) \in \mathcal{T}_X(\{x_0\})$. Hence, $\mathcal{T}_X(\{f(x_0)\}) = \mathcal{T}_X(\{x_0\})$. Since f is monotone onto $f(X)$, $f\mathcal{T}_X(\{x_0\}) \subset \mathcal{T}_{f(X)}(\{f(x_0)\})$, Theorem 2.1.51 (b). Since $\mathcal{T}_{f(X)}(\{f(x_0)\}) = \mathcal{T}_X(\{f(x_0)\})$ (Corollary 7.1.4), we have $f\mathcal{T}_X(\{x_0\}) \subset \mathcal{T}_X(\{f(x_0)\})$. Thus, $f|_{\mathcal{T}_X(\{x_0\})}\colon \mathcal{T}_X(\{x_0\}) \to \mathcal{T}_X(\{x_0\})$ is well defined. Since $\mathcal{T}_X(\{x_0\})$ has the fixed-point property, there exists $z \in \mathcal{T}_X(\{x_0\})$ such that $f(z) = z$. □

7.1.7 Theorem. *If X is a continuously type A' θ-continuum, which is not a graph, then X is not hereditarily decomposable.*

Proof. Let X be a continuously type A' θ-continuum that is not a graph. By Theorem 5.2.10, $\mathcal{T}_X$ is continuous, and $\mathcal{G} = \{\mathcal{T}_X(\{x\}) \mid x \in X\}$ is a monotone decomposition such that $X/\mathcal{G}$ is a graph. By Theorem 3.2.9, there are many indecomposable elements of $\mathcal{G}$. □

7.1.8 Theorem. *Let X be a continuously type A' θ-continuum, which is not a graph. If $q\colon X \twoheadrightarrow D$ is the quotient map, where D is a graph, then q is atomic.*

Proof. Let K be a subcontinuum of X such that $q(K)$ is nondegenerate. Without loss of generality, we assume that $K \neq X$. Clearly, $K \subset q^{-1}q(K)$. Let $x_0 \in q^{-1}q(K)$. By Theorem 5.2.10, for each $x \in X$, $q^{-1}q(x) = \mathcal{T}_X(\{x\})$. Then, by Lemma 3.2.3, we have that $\mathcal{T}_X(\{x\}) = q^{-1}q(x) \subset K$ for all $x \in \mathrm{Int}_X(K)$. Since D is a graph, there exists a sequence $\{t_m\}_{m=1}^{\infty}$ of points of $q(\mathrm{Int}_X(K))$ converging to $q(x_0)$. Since q is open, $\Im(q)\colon 2^D \to 2^X$ given by $\Im(q)(B) = q^{-1}(B)$ is continuous, Theorem 1.6.16. Hence, for each $m \in \mathbb{N}$, there exists

$x_m \in q^{-1}(t_m)$ such that the sequence $\{x_m\}_{m=1}^{\infty}$ converges to x_0. Since K is closed and $\{x_m\}_{m=1}^{\infty} \subset K$, we obtain that $x_0 \in K$. Thus, $q^{-1}q(K) \subset K$. Therefore, q is an atomic map. □

7.1.9 Corollary. *Let X be a continuously type A' θ-cotinuum, which is not a graph. If $\mathcal{G} = \{\mathcal{T}_X(\{x\}) \mid x \in X\}$, then $\mathcal{G}$ is the finest monotone continuous decomposition of X such that each element of $\mathcal{G}$ is nowhere dense and $X/\mathcal{G}$ is a graph, and $\mathcal{T}_X(\{x\})$ is a terminal subcontinuum of X for all $x \in X$.*

Proof. Since X is a continuously type A' θ-continuum, there exists a monotone continuous decomposition $\mathcal{D}$ of X, whose elements have empty interior, such that $X/\mathcal{D}$ is a graph. Let $q\colon X \twoheadrightarrow X/\mathcal{D}$ be the quotient map. By Theorem 7.1.8, q is an atomic map. Hence, by [92, Theorem 8.1.25], the fibres of q are terminal subcontinua of X. By Theorem 5.2.10, for each $x \in X$, $q^{-1}q(x) = \mathcal{T}_X(\{x\})$. Thus, $\mathcal{D} = \mathcal{G}$, and each $\mathcal{T}_X(\{x\})$ is a terminal subcontinuum of X. By Theorem 5.2.10, $\mathcal{G}$ is the finest monotone decomposition X such that each element of $\mathcal{G}$ is nowhere dense and $X/\mathcal{G}$ is a graph. □

7.1.10 Corollary. *Let X be a continuously type A' θ-continuum, which is not a graph. If K is a subcontinuum of X such that $K = \mathcal{T}(\{x_0\})$, for some $x_0 \in X$, or $K \not\subset \mathcal{T}(\{x\})$ for any $x \in X$, then $\mathcal{T}(K) = K$.*

Proof. Let K be a subcontinuum of X. Suppose that $K = \mathcal{T}(\{x_0\})$ for some $x_0 \in X$. Since $\mathcal{T}$ is continuous (Theorem 5.2.10), $\mathcal{T}$ is idempotent, Theorem 5.1.6. Hence, $\mathcal{T}(K) = \mathcal{T}\mathcal{T}(\{x_0\}) = \mathcal{T}(\{x_0\}) = K$.

Assume that $K \not\subset \mathcal{T}(\{x\})$ for any $x \in X$. Then, by Corollary 7.1.9, $\mathcal{T}(\{x\}) \subset K$ for all $x \in K$. This implies that $K = \bigcup\{\mathcal{T}(\{x\}) \mid x \in K\}$. Also, since, by Corollary 2.2.4 and Theorem 2.2.11, X is $\mathcal{T}$-additive, we have that $\mathcal{T}(K) = \bigcup\{\mathcal{T}(\{x\}) \mid x \in K\} = K$, Corollary 2.2.13. □

A slight modification of the proof of Corollary 7.1.10 gives the following.

7.1.11 Corollary. *Let X be a continuously irreducible continuum. If K is a subcontinuum of X such that $K = \mathcal{T}(\{x_0\})$, for some $x_0 \in X$, or $K \not\subset \mathcal{T}(\{x\})$ for any $x \in X$, then $\mathcal{T}(K) = K$.*

7.1.12 Corollary. *If X is a continuously type A' θ-continuum, then there exists $n \in \mathbb{N}$ such that X is a θ_n-continuum.*

Proof. Since X is a continuously type A' θ-continuum, there exists a monotone continuous decomposition $\mathcal{D}$ of X, where the elements of $\mathcal{D}$ are nowhere dense, such that $X/\mathcal{D}$ is a graph. By Theorem 5.2.10, we know that $\mathcal{D} = \{\mathcal{T}_X(\{x\}) \mid x \in X\}$. Let $q\colon X \twoheadrightarrow X/\mathcal{D}$ be the quotient map. By [45, Corollary 4.8], there exists $n \in \mathbb{N}$ such that $X/\mathcal{D}$ is a θ_n-continuum.

We show that X is a θ_n-continuum. Let K be a subcontinuum of X. If $q(K)$ is nondegenerate, since q is an atomic map, we have that $K = q^{-1}q(K)$. Hence, $X \setminus K$ and $X/\mathcal{D} \setminus q(K)$ have the same number of components. Thus, $X \setminus K$ has at most n components. Suppose that $K \subset \mathcal{T}_X(\{x_0\})$, for some $x_0 \in X$. If $K \subsetneq \mathcal{T}_X(\{x_0\})$, then we have that $X \setminus K$ is connected. Assume that $K = \mathcal{T}_X(\{x_0\})$. Hence, $X \setminus K$ and $X \setminus \mathcal{T}_X(\{x_0\})$ have the same number of components. Since $\mathcal{T}_X(\{x_0\})$ is nowhere dense, $\mathcal{T}_X(\{x_0\}) = q^{-1}q(x_0)$, and $\mathcal{D}$ is a continuous decomposition, we obtain that $X \setminus \mathcal{T}_X(\{x_0\})$ and $X/\mathcal{D} \setminus \{q(x_0)\}$ have the same number of components. Thus, $X \setminus K$ has at most n components. Therefore, X is a θ_n-continuum. □

7.1.13 Definition. Let X be a metric continuum. Then a map $\mu \colon \mathcal{C}_1(X) \twoheadrightarrow [0,1]$ is a *Whitney map* provided that

(1) $\mu(X) = 1$;
(2) if A and B are elements of $\mathcal{C}_1(X)$ and $A \subsetneq B$, then $\mu(A) < \mu(B)$;
(3) $\mu(\{x\}) = 0$ for all points x in X.

7.1.14 Remark. Whitney maps have been a very useful tool to study hyperspaces of metric continua. There are several constructions of them. Three constructions of Whitney maps may be found in [105, (0.50.1), (0.50.2) and (0.50.3)]. Also, many Whitney properties may be found in [105, Chapter XIV]

7.1.15 Theorem. *Let X be a metric continuum such that $\mathcal{G} = \{\mathcal{T}_X(\{x\}) \mid x \in X\}$ is a continuous terminal decomposition of X, and let $\mu \colon \mathcal{C}_1(X) \twoheadrightarrow [0,1]$ be a Whitney map such that $\mu\mathcal{T}_X(\{x\}) = t_0$ for all $x \in X$ and some $t_0 \in (0,1)$. Let $t \in (0, t_0)$, and for each $x \in X$, let $\mathcal{G}_x(t) = \{A \in \mu^{-1}(t) \mid A \subset \mathcal{T}_X(\{x\})\}$. Then $\mathfrak{G}_t = \{\mathcal{G}_x(t) \mid x \in X\}$ is a monotone continuous decomposition of $\mu^{-1}(t)$ such that $\mu^{-1}(t)/\mathfrak{G}_t$ is homeomorphic to $X/\mathcal{G}$.*

Proof. Let $x \in X$. Observe that $\mathcal{G}_x(t) = \mu^{-1}(t) \cap \mathcal{C}_1(\mathcal{T}_X(\{x\}))$. Hence, $\mathcal{G}_x(t)$ is a continuum [38, (1.4)]. Since $\mathcal{G}$ is a continuous decomposition of X and each $\mathcal{G}_x(t)$ is a continuum, we obtain that $\mathfrak{G}_t$ is a decomposition of $\mu^{-1}(t)$. We show that $\mathfrak{G}_t$ is continuous. Let $q \colon X \twoheadrightarrow X/\mathcal{G}$ be the quotient map, and let $h_t \colon \mu^{-1}(t) \twoheadrightarrow X/\mathcal{G}$ be given by $h_t = \xi \circ \mathcal{C}_1(q)|_{\mu^{-1}(t)}$, where $\xi \colon \mathcal{F}_1(X/\mathcal{G}) \twoheadrightarrow X/\mathcal{G}$ is the natural isometry given by $\xi(\{\chi\}) = \chi$ (Definition 3.1.8). Note that h_t is continuous and surjective. Also, for each $\chi \in X/\mathcal{G}$, there exists $x \in X$ such that $h_t^{-1}(\chi) = \mathcal{G}_x(t)$. This implies that h_t is monotone. To show that h_t is open, it suffices to prove that $\mathcal{C}_1(q)|_{\mu^{-1}(t)}$ is open. Observe that $\mathcal{C}_1(q) \circ \Im(q) = 1_{\mathcal{C}_1(X/\mathcal{G})}$, where $\Im(q) \colon \mathcal{C}_1(X/\mathcal{G}) \to \mathcal{C}_1(X)$ is given by $\Im(q)(A) = q^{-1}(A)$. Since q is open, $\Im(q)$ is continuous (Theorem 1.6.16). Also, open maps have the composition factor property [92, Theorem 8.1.13]. Hence, we have that $\mathcal{C}_1(q)$ is open. Let $\langle U_1, \ldots, U_n \rangle_1$ be an open subset of $\mathcal{C}_1(X)$ such that $\langle U_1, \ldots, U_n \rangle_1 \subset \mu^{-1}([0, t_0))$ and $\langle U_1, \ldots, U_n \rangle_1 \cap \mu^{-1}(t) \neq \emptyset$. We see that $\mathcal{C}_1(q)(\langle U_1, \ldots, U_n \rangle_1) = \mathcal{C}_1(q)(\langle U_1, \ldots, U_n \rangle_1 \cap \mu^{-1}(t))$. Clearly, $\mathcal{C}_1(q)(\langle U_1, \ldots, U_n \rangle_1 \cap \mu^{-1}(t)) \subset \mathcal{C}_1(q)(\langle U_1, \ldots, U_n \rangle_1)$. Let $A \in \langle U_1, \ldots, U_n \rangle_1$. Since A is a subcontinuum of X, $\mu(A) < t_0$ and $\mathcal{G}$ is a terminal decomposition of X, there exists $x \in X$ such that $A \subset \mathcal{T}_X(\{x\})$. Let $B \in \mathcal{G}_x(t)$. Then $\mathcal{C}_1(q)(A) = \mathcal{C}_1(q)(B) = \mathcal{C}_1(q)(\mathcal{T}_X(\{x\}))$. Hence,

$$\mathcal{C}_1(q)(\langle U_1, \ldots, U_n \rangle_1) \subset \mathcal{C}_1(q)(\langle U_1, \ldots, U_n \rangle_1 \cap \mu^{-1}(t)).$$

Thus,

$$\mathcal{C}_1(q)(\langle U_1, \ldots, U_n \rangle_1) = \mathcal{C}_1(q)(\langle U_1, \ldots, U_n \rangle_1 \cap \mu^{-1}(t)).$$

As a consequence of this, $\mathcal{C}_1(q)|_{\mu^{-1}(t)}$ is open, and h_t is also open. Thus, $\mathfrak{G}_t$ is a continuous decomposition of $\mu^{-1}(t)$. The fact that $\mu^{-1}(t)/\mathfrak{G}_t$ is homeomorphic to $X/\mathcal{G}$ follows from Theorem 1.1.11. □

7.1.16 Definition. Let X be a metric space, with metric d, and let A be a closed subset of X. Then A is a *Z-set* provided that for each $\varepsilon > 0$, there exists a map $f\colon X \to X \setminus A$ such that $d(x, f(x)) < \varepsilon$ for all $x \in X$.

7.1.17 Theorem. *If X is a continuously type A' θ-continuum such that for each $x \in X$, $\mathcal{T}_X(\{x\})$ is nondegenerate, then $\mathcal{F}_n(X)$ is Z-set in both 2^X and $\mathcal{C}_n(X)$ for every $n \in \mathbb{N}$.*

Proof. By hypothesis, for each $x \in X$, $\mathcal{T}_X(\{x\})$ is nondegenerate. Let $t_0 \in (0,1)$. By [126, Theorem 3.1], there exists a Whitney map $\mu\colon \mathcal{C}_1(X) \twoheadrightarrow [0,1]$ such that $\mu\mathcal{T}_X(\{x\}) = t_0$ for all $x \in X$. Let $t \in (0, t_0)$. Define $g_t\colon \mathcal{F}_1(X) \to \mathcal{C}_1(\mathcal{C}_1(X))$ by $g_t(\{x\}) = \mathcal{G}_x(t) = \{A \in \mu^{-1}(t) \mid A \subset \mathcal{T}_X(\{x\})\}$ (Theorem 7.1.15). Since, by Theorem 7.1.15, $\mathfrak{G}_t = \{\mathcal{G}_x(t) \mid x \in X\}$ is a monotone continuous decomposition of $\mu^{-1}(t)$, we have that g_t is continuous. Let $\sigma\colon 2^{2^X} \twoheadrightarrow 2^X$ be given by $\sigma(\mathcal{A}) = \cup\mathcal{A}$, σ is continuous, Lemma 1.6.13.

Let $\varepsilon > 0$. Then there exists $t \in (0, t_0)$ such that $\mathrm{diam}(A) < \frac{\varepsilon}{3}$ for each $A \in \mu^{-1}(t)$ [85, Lemma 6.3]. Since for every $x \in X$, $\lim_{t\to 0} \mathrm{diam}(\mathcal{G}_x(t)) = 0$ and each $\mu^{-1}(t)$ is compact, we also assume that $\mathrm{diam}(\mathcal{G}_x(t)) < \frac{\varepsilon}{3}$ for all $\mathcal{G}_x(t) \in \mathfrak{G}_t$. Let $f_\varepsilon\colon \mathcal{F}_1(X) \to \mathcal{C}_1(X) \setminus \mathcal{F}_1(X)$ be given by

$$f_\varepsilon(\{x\}) = \sigma \circ g_t(\{x\}) = \bigcup \mathcal{G}_x(t).$$

Note that f_ε is well defined; that is, $f_\varepsilon(\{x\}) \in \mathcal{C}_1(X)$ (Lemma 1.6.8) and it is nondegenerate. Let $A_x \in \mathcal{G}_x(t)$ be such that $x \in A_x$. Let $y \in f_\varepsilon(\{x\})$. Then there exists $A_y \in \mathcal{G}_x(t)$ such that $y \in A_y$. Let $z_x \in A_x$ and $z_y \in A_y$ be such that $d(A_x, A_y) = d(z_x, z_y)$. Observe that $d(z_x, z_y) \le \mathcal{H}(A_x, A_y)$ [105, (0.4)]. Thus, $d(x,y) \le d(x, z_x) + d(z_x, z_y) + d(z_y, y) < \frac{\varepsilon}{3} + \frac{\varepsilon}{3} + \frac{\varepsilon}{3} = \varepsilon$. Hence, $\mathcal{H}(\{x\}, f_\varepsilon(\{x\})) < \varepsilon$. Therefore, $\mathcal{F}_1(X)$ is a Z-set in $\mathcal{C}_1(X)$ [93, Theorem 2.1].

Since $\mathcal{F}_1(X)$ is a Z-set in $\mathcal{C}_1(X)$, by [93, Theorem 2.2], $\mathcal{F}_n(X)$ is a Z-set in both 2^X and $\mathcal{C}_n(X)$ for every $n \in \mathbb{N}$. □

A very similar proof to the one given for Theorem 7.1.17 shows the following.

7.1.18 Theorem. *Let X be a continuously irreducible continuum such that for each $x \in X$, $\mathcal{T}_X(\{x\})$ is nondegenerate, then $\mathcal{F}_n(X)$ is Z-set in both 2^X and $\mathcal{C}_n(X)$ for every positive integer n.*

7.1.19 Theorem. *If X is a continuously irreducible continuum, then X has a 2-cell neighbourhood in $\mathcal{C}_1(X)$.*

Proof. Let X be a continuously irreducible continuum. Then $\mathcal{T}_X$ is continuous for X, Theorem 5.2.9. Let $q\colon X \twoheadrightarrow [0,1]$ be the quotient map given by the finest continuous monotone decomposition of X. Let $\Im(q)\colon 2^{[0,1]} \to 2^X$ be given by $\Im(q)(B) = q^{-1}(B)$. By Theorem 1.6.16, $\Im(q)$ is continuous. In fact, $2^q \circ \Im(q) =$

$1_{2^{[0,1]}}$. In particular, $\Im(q)\colon 2^{[0,1]} \to \Im(q)\left(2^{[0,1]}\right)$ is a homeomorphism. By the proof of Theorem 3.2.9, we have that $\mathcal{T}_X\left(2^X\right) = \Im(q)\left(2^{[0,1]}\right)$. Note that $\mathcal{T}_X\left(\mathcal{C}_1(X)\right) = \Im(q)\left(\mathcal{C}_1[0,1]\right)$, $\Im(q)([0,1]) = X$ and $\Im(q)\left(\mathcal{F}_1([0,1])\right) = \{\mathcal{T}_X(\{x\}) \mid x \in X\}$. It is well known that $\mathcal{C}_1([0,1])$ is a 2-cell [105, (0.54)]. Therefore, $\mathcal{T}_X\left(\mathcal{C}_1(X)\right)$ is a 2-cell.

Let us consider a Whitney map $\mu\colon \mathcal{C}_1(X) \twoheadrightarrow [0,1]$ [105, (0.50)]. Let $t_0 = \max\{\mu\mathcal{T}_X(\{x\}) \mid x \in X\}$. Note that $0 < t_0 < 1$. Also observe that if $D \in \mu^{-1}([t_0,1])$, then $\mathcal{T}_X(D) = D$, Corollary 7.1.11. Thus, $\mu^{-1}([t_0,1]) \subset \mathcal{T}_X\left(\mathcal{C}_1(X)\right)$. Hence, $\mathcal{T}_X\left(\mathcal{C}_1(X)\right)$ is a neighbourhood of X in $\mathcal{C}_1(X)$. Therefore, X has a 2-cell neighbourhood in $\mathcal{C}_1(X)$. □

7.1.20 Theorem. *Let X be a continuously irreducible continuum such that $\mathcal{T}_X(\{x\})$ is nondegenerate for any $x \in X$. Then there exist a Whitney map $\mu\colon \mathcal{C}_1(X) \twoheadrightarrow [0,1]$ and $t_0 \in [0,1]$ such that $\mu\mathcal{T}_X(\{x\}) = t_0$, for each $t \in [0,1]$, $\mu^{-1}(t)$ admits a continuous decomposition into continua such that the quotient space is $[0,1]$.*

Proof. Let X be a continuously irreducible continuum such that $\mathcal{T}_X(\{x\})$ is nondegenerate for any $x \in X$, and let $t_0 \in (0,1)$. Then, by [126, Theorem 3.1], there exists a Whitney map $\mu\colon \mathcal{C}_1(X) \twoheadrightarrow [0,1]$ such that $\mu\mathcal{T}_X(\{x\}) = t_0$ for each $x \in X$. We show that $\mathcal{T}_X\left(\mathcal{C}_1(X)\right) = \mu^{-1}([t_0,1])$. Let $A \in \mathcal{C}_1(X)$. Then for each $a \in A$, $\mathcal{T}_X(\{a\}) \subset \mathcal{T}_X(A)$. Since μ is a Whitney map, $\mu\mathcal{T}_X(A) \geq t_0$. Thus, $\mathcal{T}_X(A) \in \mu^{-1}([t_0,1])$, and $\mathcal{T}_X\left(\mathcal{C}_1(X)\right) \subset \mu^{-1}([t_0,1])$. Next, let $D \in \mu^{-1}([t_0,1])$. Then, by Corollary 7.1.11, $D \in \mathcal{T}_X\left(\mathcal{C}_1(X)\right)$, and $\mu^{-1}([t_0,1]) \subset \mathcal{T}_X\left(\mathcal{C}_1(X)\right)$. Thus, $\mathcal{T}_X\left(\mathcal{C}_1(X)\right) = \mu^{-1}([t_0,1])$.

Let $q\colon X \twoheadrightarrow [0,1]$ be the quotient map given by the finest continuous monotone decomposition of X. As we saw in the proof of Theorem 7.1.19, $\mathcal{T}_X\left(\mathcal{C}_1(X)\right) = \Im(q)\left(\mathcal{C}_1([0,1])\right)$, and $\Im(q)|_{\mathcal{C}_1([0,1])}$ is a homeomorphism. Let $\omega\colon \mathcal{C}_1([0,1]) \twoheadrightarrow [0,1]$ be given by $\omega(B) = \dfrac{\mu\Im(q)(B) - t_0}{1 - t_0}$. Then ω is a Whitney map for $\mathcal{C}_1([0,1])$. By [105, (14.6)], $\omega^{-1}(s)$ is an arc for each $s \in [0,1)$. Let $t \in [t_0,1]$. We show that $\mu^{-1}(t) = \Im(q)\omega^{-1}\left(\dfrac{t-t_0}{1-t_0}\right)$. To this end, let $A \subset \mu^{-1}(t)$. Then $\omega\Im(q)^{-1}(A) = \dfrac{\mu\Im(q)\Im(q)^{-1}(A) - t_0}{1-t_0} = \dfrac{\mu(A) - t_0}{1-t_0} = \dfrac{t-t_0}{1-t_0}$. Thus, $A \in \Im(q)\omega^{-1}\left(\dfrac{t-t_0}{1-t_0}\right)$, and $\mu^{-1}(t) \subset \Im(q)\omega^{-1}\left(\dfrac{t-t_0}{1-t_0}\right)$. Now, let $D \in \Im(q)\omega^{-1}\left(\dfrac{t-t_0}{1-t_0}\right)$. Then, $\dfrac{t-t_0}{1-t_0} = \omega\Im(q)^{-1}(D) = \dfrac{\mu\Im(q)\Im(q)^{-1}(D) - t_0}{1-t_0} = \dfrac{\mu(D) - t_0}{1-t_0}$. This implies that $\mu(D) = t$. Hence, $D \in \mu^{-1}(t)$, and $\mu^{-1}(t) \subset \Im(q)\omega^{-1}\left(\dfrac{t-t_0}{1-t_0}\right)$. Thus, $\mu^{-1}(t) = \Im(q)\omega^{-1}\left(\dfrac{t-t_0}{1-t_0}\right)$. Therefore, for every $t \in [t_0,1]$, $\mu^{-1}(t)$ is an arc.

Next, suppose $t \in (0, t_0)$. For each $x \in X$, let $\mathcal{G}_x(t) = \{A \in \mu^{-1}(t) \mid A \subset \mathcal{T}_X(\{x\})\} = \mu^{-1}(t) \cap \mathcal{C}_1(\mathcal{T}_X(\{x\}))$. Then $\mathcal{G}_x(t)$ is a continuum [38, (1.4)], and $\mathfrak{G}_t = \{\mathcal{G}_x(t) \mid x \in X\}$ is a monotone decomposition of $\mu^{-1}(t)$.

Let $h_t\colon \mu^{-1}(t) \to [0,1]$ be given by $h_t = \xi \circ \mathcal{C}_1(q)|_{\mu^{-1}(t)}$ (Definition 3.1.8). Hence, h_t is continuous, monotone and surjective ($\mathcal{C}_1(q)$ is monotone by [92, Theorem 8.2.4]). In fact, for each $s \in [0,1]$, there exists $x \in X$ such that $h_t^{-1}(s) = \mathcal{G}_x(t)$. To see that h_t is open, it suffices to show that $\mathcal{C}_1(q)|_{\mu^{-1}(t)}$ is open. To this end, note that $\mathcal{C}_1(q) \circ \Im(q) = 1_{\mathcal{C}_1([0,1])}$. Since open maps have the composition factor property [92, Theorem 8.1.13], $\mathcal{C}_1(q)$ is open. Let $\langle U_1, \ldots, U_n\rangle_1$ be an open set in $\mathcal{C}_1(X)$ such that $\langle U_1, \ldots, U_n\rangle_1 \subset \mu^{-1}([0,t_0))$ and $\langle U_1, \ldots, U_n\rangle_1 \cap \mu^{-1}(t) \neq \emptyset$. We show that $\mathcal{C}_1(q)(\langle U_1, \ldots, U_n\rangle_1) = \mathcal{C}_1(q)\left(\langle U_1, \ldots, U_n\rangle_1 \cap \mu^{-1}(t)\right)$. It is clear that $\mathcal{C}_1(q)(\langle U_1, \ldots, U_n\rangle_1 \cap \mu^{-1}(t)) \subset \mathcal{C}_1(q)(\langle U_1, \ldots, U_n\rangle_1)$. Let $A \in \langle U_1, \ldots, U_n\rangle_1$. Since A is a subcontinuum of X and $\mu(A) < t_0$, there exists $x \in X$ such that $A \subset \mathcal{T}_X(\{x\})$. Let $B \in \mathcal{G}_x(t)$. Then $\mathcal{C}_1(q)(A) = \mathcal{C}_1(q)(B) = \mathcal{C}_1(q)(\mathcal{T}_X(\{x\}))$. Hence,

$$\mathcal{C}_1(q)(\langle U_1, \ldots, U_n\rangle_1) \subset \mathcal{C}_1(q)\left(\langle U_1, \ldots, U_n\rangle_1 \cap \mu^{-1}(t)\right),$$

and

$$\mathcal{C}_1(q)(\langle U_1, \ldots, U_n\rangle_1) = \mathcal{C}_1(q)\left(\langle U_1, \ldots, U_n\rangle_1 \cap \mu^{-1}(t)\right).$$

Therefore, $\mathcal{C}_1(q)|_{\mu^{-1}(t)}$ is open, and h_t is also open. Therefore, $\mathfrak{G}_t = \{\mathcal{G}_x(t) \mid x \in X\}$ is a continuous decomposition of $\mu^{-1}(t)$ such that $\mu^{-1}(t)/\mathfrak{G}_t$ is $[0,1]$. □

We end this section with the next result about metric irreducible continua.

7.1.21 Theorem. *Let X be a metric irreducible continuum, and let $x_0 \in X$. If $\mathcal{T}(\{x_0\}) = X$, then the collection $\mathcal{E} = \{\mathcal{T}(\{x\}) \mid x \in X\}$ is finite.*

Proof. First, note that if X is indecomposable, then $\mathcal{E} = \{X\}$, Theorem 2.1.44. Hence, $\mathcal{E}$ is finite.

Suppose X is decomposable and irreducible between the points a and b. Let x_0 be a point of X such that $\mathcal{T}(\{x_0\}) = X$. Since X is irreducible, X is $\mathcal{T}$-symmetric, Corollary 2.2.3. Thus, X is point $\mathcal{T}$-symmetric. Then, since $\mathcal{T}(\{x_0\}) = X$, $x_0 \in \mathcal{T}(\{a\}) \cap \mathcal{T}(\{b\})$. Hence, by the irreducibility of X, $X = \mathcal{T}(\{a\}) \cup \mathcal{T}(\{b\})$.

Now, we show that $\mathrm{Cl}(X \setminus \mathcal{T}(\{b\})) = \mathcal{T}(\{a\})$. Suppose there exists $x \in \mathcal{T}(\{a\}) \setminus \mathrm{Cl}(X \setminus \mathcal{T}(\{b\}))$. Then $\mathrm{Cl}(X \setminus \mathcal{T}(\{b\}))$ is a subcontinuum of X (Theorem 1.4.40) such that $a \in \mathrm{Int}(\mathrm{Cl}(X \setminus \mathcal{T}(\{b\}))) \subset \mathrm{Cl}(X \setminus \mathcal{T}(\{b\})) \subset X \setminus \{x\}$. Hence, $a \notin \mathcal{T}(\{x\})$. Since X is point $\mathcal{T}$-symmetric (Corollary 2.2.3), $x \notin \mathcal{T}(\{a\})$, a contradiction. Therefore, $\mathrm{Cl}(X \setminus \mathcal{T}(\{b\})) = \mathcal{T}(\{a\})$. Thus, $\mathcal{T}(\{a\})$ is irreducible between a and each point of $\mathrm{Bd}(\mathcal{T}(\{a\}))$ [75, Theorem 7, p. 194]. Since, with a similar argument, we can show that $x_0 \in \mathrm{Cl}(X \setminus \mathcal{T}(\{a\}))$, we have that $x_0 \in \mathrm{Bd}(\mathcal{T}(\{a\}))$. Therefore, $\mathcal{T}(\{a\})$ is irreducible between a and x_0. Similarly, $\mathcal{T}(\{b\})$ is irreducible between b and x_0.

Let $x \in X$. If $x \in \mathcal{T}(\{a\}) \cap \mathcal{T}(\{b\})$, then $\mathcal{T}(\{x\}) = X$. Suppose $x \in \mathcal{T}(\{a\}) \setminus \mathcal{T}(\{b\})$. Since X is point $\mathcal{T}$-symmetric (Corollary 2.2.3), $a \in \mathcal{T}(\{x\})$. Hence, $\mathcal{T}(\{x\}) = \mathcal{T}(\{a\})$, since $x \notin \mathcal{T}(\{b\})$. Similarly, if $x \in \mathcal{T}(\{b\}) \setminus \mathcal{T}(\{a\})$, $\mathcal{T}(\{x\}) = \mathcal{T}(\{b\})$. Thus, $\mathcal{E} = \{X, \mathcal{T}(\{a\}), \mathcal{T}(\{b\})\}$. Therefore, $\mathcal{E}$ has three elements. □

7.2 Noncontractibility

The set function $\mathcal{T}$ may be used to show that continua are not contractible. First, we show the next lemma.

7.2.1 Lemma. *Suppose X is a continuum, A is a closed subset of X and $r \in \mathcal{T}_X(A)$. If $H\colon X \times [0,1] \twoheadrightarrow X$ is a homotopy such that $H((x,0)) = x$ for each $x \in X$ and such that $H((r,1)) \in X \setminus \mathcal{T}_X(A)$, then $H((r,t)) \in A$ for some $t \in [0,1]$.*

Proof. Let $f\colon X \to X$ be given by $f(x) = H((x,1))$. Suppose the result is not true. Then $H^{-1}(A) \cap (\{r\} \times [0,1]) = \emptyset$. Thus, there exists an open subset U of X such that $r \in U$ and $H^{-1}(A) \cap (\mathrm{Cl}(U) \times [0,1]) = \emptyset$. Since $f(r) = H((r,1)) \in X \setminus \mathcal{T}_X(A)$, there exists a subcontinuum W of X such that $f(r) \in \mathrm{Int}(W) \subset W \subset X \setminus A$. Let $V = U \cap f^{-1}(\mathrm{Int}(W))$. Then $r \in V$, $H^{-1}(A) \cap (\mathrm{Cl}(V) \times [0,1]) = \emptyset$ and $f(\mathrm{Cl}(V)) \subset W$. Let $L = (X \times \{0\}) \cup (\mathrm{Cl}(V) \times [0,1])$. Then L is a subcontinuum of $X \times [0,1]$. Consider the disjoint union $L \cup W$. Let K be the quotient space of $L \cup W$ obtained by identifying each $(v,1) \in \mathrm{Cl}(V) \times \{1\}$ with $f(v)$, and let $q\colon L \cup W \twoheadrightarrow K$ be the quotient map.

Define $G\colon K \twoheadrightarrow X$ by

$$G(q(\omega)) = \begin{cases} \omega, & \text{if } \omega \in W; \\ H(\omega), & \text{if } \omega \in L. \end{cases}$$

Note that G is, in fact, surjective since $G(q((x,0))) = x$ for each $x \in X$. By Theorem 2.1.51 (a), $\mathcal{T}_X(A) \subset G\mathcal{T}_K G^{-1}(A)$. In particular, $r \in G\mathcal{T}_K G^{-1}(A)$. Hence, $G^{-1}(r) \cap \mathcal{T}_K G^{-1}(A) \neq \emptyset$. However, since $A \cap W = \emptyset$ and $H^{-1}(A) \cap (\mathrm{Cl}(V) \times [0,1]) = \emptyset$, it follows that $G^{-1}(A) = q(A \times \{0\})$. Thus, $G^{-1}(r) \cap \mathcal{T}_K q(A \times \{0\}) \neq \emptyset$.

Let $J = q(W \cup (\mathrm{Cl}(V) \times [0,1]))$. Then J is a subcontinuum of K, and

$$M = q(W \cup (\mathrm{Cl}(V) \times (0,1]) \cup (V \times [0,1])) \subset \mathrm{Int}_K(J).$$

Since $G^{-1}(r) \cap q(X \times \{0\}) = q(\{(r,0)\})$, $G^{-1}(r) \subset M \subset \mathrm{Int}_K(J)$. Since $q(A \times \{0\}) \cap J = \emptyset$, we have that $G^{-1}(r) \cap \mathcal{T}_K q(A \times \{0\}) = \emptyset$, a contradiction. Therefore, there exists $t \in [0,1]$ such that $H((r,t)) \in A$. □

The next theorem gives a sufficient condition for a continuum not to be contractible.

7.2.2 Theorem. *If X is a continuum and A and B are closed subsets of X such that $A \cap \mathcal{T}(B) = \emptyset$, $\mathcal{T}(A) \cap B = \emptyset$ and $\mathcal{T}(A) \cap \mathcal{T}(B) \neq \emptyset$, then X is not contractible.*

Proof. Suppose X is contractible. Then there exists a homotopy $H\colon X \times [0,1] \twoheadrightarrow X$ such that $H((x,0)) = x$ and $H((x,1)) = p$ for each $x \in X$ and some $p \in X$. Without loss of generality, we assume that $p \in A$. Let $r \in \mathcal{T}(A) \cap \mathcal{T}(B)$. Since $H((r,1)) \in A$ and $H((r,1)) \in X \setminus \mathcal{T}(B)$, by Lemma 7.2.1, there exists $t \in [0,1]$ such that $H((r,t)) \in B$. Let t_A and t_B be the smallest elements of $[0,1]$ such that $H((r,t_A)) \in A$ and $H((r,t_B)) \in B$. Thus, either $t_A < t_B$ or $t_A > t_B$, since $t_A = t_B$ is impossible. If $t_A < t_B$, applying Lemma 7.2.1 to $H|_{X\times[0,t_A]}$, we obtain an element $t' < t_A$ such that $H((r,t')) \in B$, a contradiction. Similarly, if $t_A > t_B$, applying Lemma 7.2.1 to $H|_{X\times[0,t_B]}$, we obtain an element $t'' < t_B$ such that $H((r,t'')) \in A$, a contradiction. Therefore, X is not contractible. □

7.2.3 Definition. Let X and Y be compacta. A map $f\colon X \to Y$ is said to be *interior at the point* x_0 of X provided that for every open subset U of X containing x_0, $f(x_0)$ is an interior point of $f(U)$.

7.2.4 Theorem. *Let X be a compactum, let Y be a continuum, let A and B be closed subsets of Y such that $A \cap \mathcal{T}(B) = \emptyset$, $\mathcal{T}(A) \cap B = \emptyset$ and $\mathcal{T}(A) \cap \mathcal{T}(B) \neq \emptyset$, and let $f\colon X \to Y$ be a map that is interior at a point $x_0 \in f^{-1}(\mathcal{T}(A) \cap \mathcal{T}(B))$. If $H\colon X \times [0,1] \to Y$ is a homotopy with $H((x,0)) = f(x)$ for every $x \in X$, then $H(\{x_0\} \times [0,1]) \subset \mathcal{T}(A) \cap \mathcal{T}(B))$.*

Proof. Suppose that the result is false. Then there exist a homotopy $H\colon X \times [0,1] \twoheadrightarrow X$, with $H((x,0)) = f(x)$, for each point x of X, and $t_0 \in [0,1]$ such that $H((x_0,t_0)) \in Y \setminus \mathcal{T}(A) \cap \mathcal{T}(B)$. Note that $t_0 > 0$. We claim that there exists $t_1 \in (0,t_0]$ such that $H((x_0,t_1)) \in Y \setminus \mathcal{T}(A) \cap \mathcal{T}(B)$ and $H(\{x_0\} \times [0,t_1]) \subset Y \setminus (A \cup B)$. To see this, we consider two cases. If $H(\{x_0\} \times [0,1]) \cap (A \cup B) = \emptyset$, then let $t_1 = t_0$. If $H(\{x_0\} \times [0,1]) \cap (A \cup B) \neq \emptyset$, then let $t' = \min\{t \in [0,1] \mid H((x_0,t)) \in A \cup B\}$. Without loss of generality, we assume that $H((x_0,t')) \in A$. It follows from the continuity of H and the definition of t' that there exists $\delta > 0$ such that $t_1 = t' - \delta$ satisfies the required conditions. Observe that, in this case, we have that $H((x_0,t_1)) \in (Y \setminus \mathcal{T}(A)) \cup (Y \setminus \mathcal{T}(B))$. We assume that $H((x_0,t_1)) \in Y \setminus \mathcal{T}(A)$.

Note that $H(\{x_0\} \times [0,t_1])$ is a nondegenerate path in Y whose first point is $f(x_0)$. Hence, $\{x_0\} \times [0,t_1] \subset H^{-1}(Y \setminus (A \cup B))$. This set is an open subset of $X \times [0,1]$. Since $[0,t_1]$ is compact, there exists an open subset U of X such that $x_0 \in U$ and $U \times [0,t_1] \subset H^{-1}(Y \setminus (A \cup B))$. Since $H((x_0,t_1)) \in Y \setminus \mathcal{T}(A)$, there exists a subcontinuum W of Y such that $H((x_0,t_1)) \in \operatorname{Int}(W) \subset W \subset Y \setminus A$. By the continuity of H, there exists an open subset V of X such that $x_0 \in V$ and $H(V \times \{t_1\}) \subset \operatorname{Int}(W)$.

Let $K = H(\operatorname{Cl}(V) \times [0,t_1]) \cup W$. Then K is a subcontinuum of Y disjoint from A. Since $x_0 \in V$ and f is an interior map at x_0, we obtain that $H((x_0,0)) =$

$f(x_0) \in \operatorname{Int}(f(V)) = \operatorname{Int}(H(\operatorname{Cl}(V) \times \{0\})) \subset K$. Thus, K is a subcontinuum of Y having $f(x_0)$ in its interior and missing A. Hence, $f(x_0)$ does not belong to $\mathcal{T}(A)$, a contradiction. □

As a consequence of Theorem 7.2.4 and the fact that a map is open if and only if it is interior at each of the points of its domain, we have the following.

7.2.5 Theorem. *Let X be a compactum, let Y be a continuum, let A and B be closed subsets of Y such that $A \cap \mathcal{T}(B) = \emptyset$, $\mathcal{T}(A) \cap B = \emptyset$ and $\mathcal{T}(A) \cap \mathcal{T}(B) \neq \emptyset$ and let $f\colon X \to Y$ be an open map. If $H\colon X \times [0,1] \to Y$ is a homotopy with $H((x,0)) = f(x)$ for every $x \in X$, then $H(\{x_0\} \times [0,1]) \subset \mathcal{T}(A) \cap \mathcal{T}(B)$, for each $x_0 \in f^{-1}(\mathcal{T}(A) \cup \mathcal{T}(B))$.*

7.2.6 Definition. Let X be a compactum, and let A be a nonempty subset of X. Then A is *homotopically fixed* provided that for each homotopy $H\colon X \times [0,1] \twoheadrightarrow X$ such that $H((x,0)) = x$, for all x in X, we have that $H(A \times [0,1]) \subset A$.

As a consequence of Theorem 7.2.5, we obtain:

7.2.7 Theorem. *Let X be a continuum, and let A and B be closed subsets of X such that $A \cap \mathcal{T}(B) = \emptyset$, $\mathcal{T}(A) \cap B = \emptyset$ and $\mathcal{T}(A) \cap \mathcal{T}(B) \neq \emptyset$. Then $\mathcal{T}(A) \cap \mathcal{T}(B)$ is homotopically fixed.*

7.2.8 Theorem. *Let X be a continuum. If X contains a homotopically fixed subset A, then X is not contractible.*

Proof. Suppose that X is contractible. Then there exists a homotopy $H\colon X \times [0,1] \twoheadrightarrow X$ such that $H((x,0)) = x$ and $H((x,1)) = p$ for all x in X and some p in X. This implies that X is an arcwise connected space. Note that we may take $p \in X \setminus A$, a contradiction. □

From Theorems 7.2.7 and 7.2.8, we have the following.

7.2.9 Theorem. *Let X be a continuum, and let A and B be closed subsets of X such that $A \cap \mathcal{T}(B) = \emptyset$, $\mathcal{T}(A) \cap B = \emptyset$ and $\mathcal{T}(A) \cap \mathcal{T}(B) \neq \emptyset$. Then X is not contractible.*

As a consequence of Theorem 7.2.9, we obtain the following.

7.2.10 Corollary. *If a dendroid X (Definition 7.3.3) contains two points p and q such that $p \in X \setminus \mathcal{T}(\{q\})$, $q \in X \setminus \mathcal{T}(\{p\})$ and $\mathcal{T}(\{p\}) \cap \mathcal{T}(\{q\}) \neq \emptyset$, then X is not contractible.*

7.3 Arc-Smooth Continua

We begin with strictly point $\mathcal{T}$-asymmetric metric continua. We prove that arc-smooth metric continua are strictly point $\mathcal{T}$-asymmetric. Then we give sufficient conditions for the existence of R-continua in dendroids. We end with a characterization of pointwise smooth dendroids.

7.3.1 Definition. A continuum X is *strictly point $\mathcal{T}$-asymmetric* if for any two distinct points p and q of X with $p \in \mathcal{T}(\{q\})$, we have that $q \in X \setminus \mathcal{T}(\{p\})$.

7.3.2 Theorem. *If X is an arc-smooth metric continuum, then X is strictly point $\mathcal{T}$-asymmetric.*

Proof. Let X be a metric arc-smooth continuum at the point p. Let $\alpha\colon X \to \mathcal{C}_1(X)$ be the map given by the definition of arc-smoothness (Definition 1.4.48). Let x and y be two distinct points of X such that $x \in \mathcal{T}(\{y\})$. Then $y \in W$ for each subcontinuum W of X such that $x \in \mathrm{Int}(W)$. We show that $y \in \alpha(x)$. To this end, we prove first that either $\alpha(x) \subset \alpha(y)$ or $\alpha(y) \subset \alpha(x)$. Suppose that there exists a point $z \in \alpha(x) \cap \alpha(y) \setminus \{x, y\}$ such that $\alpha(z) = \alpha(x) \cap \alpha(y)$. Let $r > 0$ be such that $\mathrm{Cl}(\mathcal{V}_r^d(x)) \cap \{z, y\} = \emptyset$ and $\mathrm{Cl}(\mathcal{V}_r^d(y)) \cap \{z, x\} = \emptyset$. By Theorem 1.4.49, $W = \bigcup_{w \in \mathrm{Cl}(\mathcal{V}_r^d(x))} \alpha(w)$ is a subcontinuum of X which contains x in its interior. Since $x \in \mathcal{T}(\{y\})$, $y \in W$. Let $N \in \mathbb{N}$ be such that $\frac{1}{n} < r$ for each $n \geq N$. Then, for every $n \geq N$, there exists $x_n \in \mathcal{V}_{\frac{1}{n}}^d(x)$ such that there exists $y_n \in \mathcal{V}_{\frac{1}{n}}^d(y)$ such that $y_n \in \alpha(x_n)$, because if for every $w \in \mathcal{V}_{\frac{1}{n}}^d(x)$, there does not exist such $w' \in \mathcal{V}_{\frac{1}{n}}^d(y)$, then, by Theorem 1.4.49, $W_n = \bigcup_{w \in \mathrm{Cl}(\mathcal{V}_{\frac{1}{n}}^d(x))} \alpha(w)$ would be a subcontinuum of X that would contain x in its interior and such that $W_n \cap \mathcal{V}_{\frac{1}{n}}^d(y) = \emptyset$, a contradiction to the fact that $x \in \mathcal{T}(\{y\})$. Hence, there exist a sequence $\{x_n\}_{n=N}^{\infty}$ that converges to x and a sequence $\{y_n\}_{n=N}^{\infty}$ converging to y such that $y_n \in \alpha(x_n)$ for all $n \geq N$. Since $y \in \lim_{n\to\infty} \alpha(x_n)$, we have that $\lim_{n\to\infty} \alpha(x_n) \neq \alpha(x)$, a contradiction to the continuity of α. Thus, either $\alpha(x) \subset \alpha(y)$ or $\alpha(y) \subset \alpha(x)$.

Next, assume that $x \in \alpha(y)$. Let $\ell > 0$ be such that $y \in X \setminus \mathcal{V}_\ell^d(x)$. By Theorem 1.4.49, $W' = \bigcup_{w \in \mathrm{Cl}(\mathcal{V}_\ell^d(x))} \alpha(w)$ is a subcontinuum of X such that $x \in \mathrm{Int}(W')$. A similar argument to the one given in the previous paragraph shows that there exist a sequence $\{x_n\}_{n=1}^{\infty}$ that converges to x and a sequence $\{y_n\}_{n=1}^{\infty}$ converging to y such that $y_n \in \alpha(x_n)$ for every $n \in \mathbb{N}$. Since $y \in \lim_{n\to\infty} \alpha(x_n)$, we obtain that $\lim_{n\to\infty} \alpha(x_n) \neq \alpha(x)$, a contradiction to the continuity of α. Thus, $y \in \alpha(x)$.

Now, let $s > 0$ be such that $\mathrm{Cl}(\mathcal{V}_s^d(x)) \cap \mathrm{Cl}(\mathcal{V}_s^d(y)) = \emptyset$. Let

$$W_y = \bigcup_{w \in \mathrm{Cl}(\mathcal{V}_s^d(y))} \alpha(w).$$

Then, by Theorem 1.4.49, we have that W_y is a subcontinuum of X which contains y in its interior. Since $y \in \alpha(x)$, we obtain that $x \in X \setminus W_y$. Thus, $y \in X \setminus \mathcal{T}(\{x\})$. Therefore, X is strictly point $\mathcal{T}$-asymmetric. □

For the next results, we need the following definitions.

7.3.3 Definition. A *dendroid* is an arcwise connected and hereditarily unicoherent metric continuum. An *end point* of a dendroid is a point of it such that it is an end point of every arc that contains it. A *fan* is a dendroid with exactly one ramification point (i.e., with only one point that is the common part of three otherwise disjoint arcs). The unique ramification point of a fan F is called the *top of* F; τ always denotes the top of a fan.

7.3.4 Definition. A dendroid D is said to be *smooth at a point* p *of* D provided that whenever $\{x_n\}_{n=1}^{\infty}$ is a sequence in D converging to a point x of D, then the sequence of arcs $\{\overline{px_n}\}_{n=1}^{\infty}$ converges to the arc $\overline{px}$. A dendroid is *smooth* if it is smooth at one of its points. A fan F is *smooth* if it is smooth at the top τ.

Note that by [47, Theorem II-4-B], a one-dimensional arc-smooth continuum is a smooth dendroid (Definition 7.3.4). Hence, as a consequence of this and Theorem 7.3.2, we have the following.

7.3.5 Corollary. *If X is a smooth dendroid, then X is strictly point $\mathcal{T}$-asymmetric.*

The next example shows that the converse of Theorem 7.3.2 is not true even for dendroids.

7.3.6 Example. Let X be the following subcontinuum of $\mathbb{R}^2$ (Notation 4.2.4):

$$X = \bigcup_{n=1}^{\infty} \overline{(-1,0), \left(0, \frac{1}{n}\right)} \cup \overline{(-1,0),(1,0)} \cup \bigcup_{n=1}^{\infty} \overline{(1,0), \left(0, -\frac{1}{n}\right)}.$$

Then X is a strictly point $\mathcal{T}$-asymmetric dendroid which is not smooth.

The converse of Theorem 7.3.2 is true for the class of fans (Definition 7.3.3). In order to prove this, we need the next lemma.

7.3.7 Lemma. *Let X be a fan with top τ, and let p be a point of $X \setminus \{\tau\}$. If $\{p_n\}_{n=1}^{\infty}$ is a sequence of points of X converging to p, then the arc $\overline{\tau p}$ is contained in* $\liminf \overline{\tau p_n}$.

Proof. Suppose that there exists a point $t \in \overline{\tau p} \setminus \liminf \overline{\tau p_n}$. Then there exist an open subset U of X such that $t \in U$ and a subsequence $\{n_k\}_{k=1}^{\infty}$ of $\mathbb{N}$ such that $\overline{\tau p_{n_k}} \cap U = \emptyset$ for each $k \in \mathbb{N}$. Note that, by Lemma 1.1.28 and Theorem 1.1.30, $\limsup \overline{\tau p_{n_k}}$ is a subcontinuum of X. Hence, since for every $k \in \mathbb{N}$, $\tau \in \overline{\tau p_{n_k}}$, and the sequence $\{p_{n_k}\}_{k=1}^{\infty}$ converges to p, we obtain that $\overline{\tau p} \subset \limsup \overline{\tau p_{n_k}}$. Thus, $t \in \limsup \overline{\tau p_{n_k}}$. Then there exists a subsequence $\{n_{k_\ell}\}_{\ell=1}^{\infty}$ of $\{n_k\}_{k=1}^{\infty}$ such that $U \cap \overline{\tau p_{n_{k_\ell}}} \neq \emptyset$ for all $\ell \in \mathbb{N}$, a contradiction to our assumption. Therefore, $\overline{\tau p} \subset \limsup \overline{\tau p_n}$. □

7.3.8 Theorem. *A fan X is strictly point $\mathcal{T}$-asymmetric if and only if it is smooth.*

Proof. Let X be a fan with top τ. If X is smooth, then, by [47, Theorem II-4-B] and Theorem 7.3.2, X is strictly point $\mathcal{T}$-asymmetric.

Suppose X is not smooth. Then there exist a point $p \in X$ and a sequence $\{p_n\}_{n=1}^{\infty}$ of points of X such that $\{p_n\}_{n=1}^{\infty}$ converges to p, but $\limsup \overline{\tau p_n} \neq \liminf \overline{\tau p_n}$. Hence, by Lemma 7.3.7, $\overline{\tau p} \subset \liminf \overline{\tau p_n}$. Thus, $\limsup \overline{\tau p_n} \neq \overline{\tau p}$. Let $q \in \limsup \overline{\tau p_n} \setminus \overline{\tau p}$, and let e_p be the end point of X such that $p \in \overline{\tau e_p}$. Then there exists a subsequence $\{p_{n_k}\}_{k=1}^{\infty}$ of $\{p_n\}_{n=1}^{\infty}$ such that for every $k \in \mathbb{N}$, there exists $q_{n_k} \in \overline{\tau p_{n_k}}$ in such a way that the sequence $\{q_{n_k}\}_{k=1}^{\infty}$ converges to q, and $p_{n_k} \in X \setminus \overline{\tau e_p}$ for all $k \in \mathbb{N}$. We consider two cases.

Case (1). $p \in \tau q$.

Let W be a subcontinuum of X containing p in its interior. Since it is easy to see that subcontinua of dendroids are dendroids, W is arcwise connected. Since $p \in \mathrm{Int}(W)$, there exists $K \in \mathbb{N}$ such that $p_{n_k} \in \mathrm{Int}(W)$ for each $k \geq K$. Since W is arcwise connected and $q_{n_k} \in \overline{\tau p_{n_k}}$ for each $k \in \mathbb{N}$, we have that $q_{n_k} \in W$ for every $k \geq K$. Thus, $q \in W$. This implies that $p \in \mathcal{T}(\{q\})$. Now if W' is a subcontinuum of X containing q in its interior, there exists $K' \in \mathbb{N}$ such that $q_{n_k} \in \mathrm{Int}(W')$ for each $k \geq K'$. Since $q_{n_k} \in \overline{\tau p_{n_k}}$ and W' is arcwise connected, the arcs $\overline{\tau q}$ and $\overline{\tau q_{n_k}}$ are contained in W' for every $k \geq K'$. Since $p \in \overline{\tau q}$, $p \in W'$. Thus, $q \in \mathcal{T}(\{p\})$. Therefore, X is not strictly point $\mathcal{T}$-asymmetric.

Case (2). $p \in X \setminus \tau q$.

Let $r \in \overline{\tau q} \setminus \{\tau, q\}$. By Lemma 1.1.28 and Theorem 1.1.30, $\limsup \overline{q_{n_k} p_{n_k}}$ is a subcontinuum of X which contains q and p. Thus, $\overline{qp} \subset \limsup \overline{q_{n_k} p_{n_k}}$ ($\limsup \overline{q_{n_k} p_{n_k}}$ is arcwise connected). Hence, there exists a sequence $\{r_{n_k}\}_{k=1}^{\infty}$ converging to r such that $r_{n_k} \in \overline{q_{n_k} p_{n_k}}$ for every $k \in \mathbb{N}$. Thus, r is in the same situation as p in Case (1); that is, $q_{n_k} \in \overline{\tau r_{n_k}}$, and the sequences $\{q_{n_k}\}_{k=1}^{\infty}$ and $\{r_{n_k}\}_{k=1}^{\infty}$ converge to q and r, respectively. Hence, $q \in \mathcal{T}(\{r\})$ and $r \in \mathcal{T}(\{q\})$. Therefore, X is not strictly point $\mathcal{T}$ asymmetric. □

7.3.9 Definition. Let X be a dendroid, and let K be a proper subcontinuum of X. Then K is an *R-continuum* if there exist $\varepsilon > 0$ and two sequences $\{C_n^1\}$ and $\{C_n^2\}$ of components of $\mathcal{V}_\varepsilon^d(K)$ such that $\limsup C_n^1 \cap \limsup C_n^2 = K$.

7.3.10 Lemma. *If X is a dendroid and A and B are closed subsets of X such that $A \cap \mathcal{T}(B) = \emptyset$, $\mathcal{T}(A) \cap B = \emptyset$ and $\mathcal{T}(A) \cap \mathcal{T}(B) \neq \emptyset$, then there exist two points $p \in A$ and $q \in B$ such that $p \in X \setminus \mathcal{T}(\{q\})$, $q \in \mathcal{T}(\{p\})$ and $\mathcal{T}(\{p\}) \cap \mathcal{T}(\{q\}) \neq \emptyset$.*

Proof. For each $a \in A$ and every $b \in B$, by Proposition 2.1.7, we have that $\mathcal{T}(\{a\}) \subset \mathcal{T}(A)$ and $\mathcal{T}(\{b\}) \subset \mathcal{T}(B)$. Hence, $\{a\} \cap \mathcal{T}(\{b\}) \subset A \cap \mathcal{T}(B) = \emptyset$ and $\mathcal{T}(\{a\}) \cap \{b\} \subset \mathcal{T}(A) \cap B = \emptyset$. Thus, $a \in X \setminus \mathcal{T}(\{b\})$ and $b \in X \setminus \mathcal{T}(\{a\})$.

Suppose the lemma is false. This means, according to the previous paragraph, that for each $a \in A$ and every $b \in B$, $\mathcal{T}(\{a\}) \cap \mathcal{T}(\{b\}) = \emptyset$.

By Theorem 2.2.12, X is $\mathcal{T}$-additive. By Corollary 2.2.13, we have that $\mathcal{T}(A) = \bigcup\{\mathcal{T}(\{a\}) \mid a \in A\}$ and $\mathcal{T}(B) = \bigcup\{\mathcal{T}(\{b\}) \mid b \in B\}$. This implies that $\emptyset = \bigcup\{\mathcal{T}(\{a\}) \cap \mathcal{T}(\{b\}) \mid a \in A \text{ and } b \in B\} = (\bigcup\{\mathcal{T}(\{a\}) \mid a \in A\}) \bigcap (\bigcup\{\mathcal{T}(\{b\}) \mid b \in B\}) = \mathcal{T}(A) \cap \mathcal{T}(B)$, a contradiction to the fact that $\mathcal{T}(A) \cap \mathcal{T}(B) \neq \emptyset$. Therefore, the lemma is true. □

7.3.11 Theorem. *If X is a dendroid and A and B are subcontinua of X such that $A \cap \mathcal{T}(B) = \emptyset$, $\mathcal{T}(A) \cap B = \emptyset$ and $\mathcal{T}(A) \cap \mathcal{T}(B) \neq \emptyset$, then $\mathcal{T}(A) \cap \mathcal{T}(B)$ is an R-continuum.*

Proof. Let $\varepsilon > 0$ be such that

$$\varepsilon < \frac{1}{3} \min\{d(\mathcal{T}(A) \cap \mathcal{T}(B), A), d(\mathcal{T}(A) \cap \mathcal{T}(B), B)\}.$$

Let $P = \{p_k\}_{k=1}^{\infty}$ be a countable dense subset of $\mathcal{T}(A) \cap \mathcal{T}(B)$. For each n and k in $\mathbb{N}$, consider $\mathcal{V}^d_{\frac{\varepsilon}{nk}}(p_k)$. Note that the smallest subcontinuum of X that contains $\mathcal{V}^d_{\frac{\varepsilon}{nk}}(p_k)$ is the closure of the union of all arcs $\overline{rp_k}$ with $r \in \mathcal{V}^d_{\frac{\varepsilon}{nk}}(p_k)$. Since every subcontinuum of X that contains p_k in its interior must intersect A, for each $n \in \mathbb{N}$, there exists a point $a_n(p_k)$ of $\mathcal{V}^d_{\frac{\varepsilon}{nk}}(p_k)$ such that the arc $\overline{p_k a_n(p_k)}$ contains a point $b_n(p_k)$ at a distance less than $\frac{\varepsilon}{nk}$ from A.

Then, for each p_k, we have a pair of sequences $\{a_n(p_k)\}_{n=1}^{\infty}$ and $\{b_n(p_k)\}_{n=1}^{\infty}$ of points of X such that $\{a_n(p_k)\}_{n=1}^{\infty}$ converges to p_k, $d(b_n(p_k), A) < \frac{\varepsilon}{nk}$, and $b_n(p_k)$ belongs to the arc $\overline{a_n(p_k)p_k}$ for all $n \in \mathbb{N}$. Similarly, there exist a sequence of points $\{a'_n(p_k)\}_{n=1}^{\infty}$ converging to p_k and a sequence of points $\{b'_n(p_k)\}_{n=1}^{\infty}$ with the properties that $d(b'_n(p_k), B) < \frac{\varepsilon}{nk}$ and $b'_n(p_k)$ belongs to the arc $\overline{a'_n(p_k)p_k}$ for every $n \in \mathbb{N}$.

Let $S_n(p_k)$ be the component of $\mathcal{V}^d_{\varepsilon}(\mathcal{T}(A) \cap \mathcal{T}(B))$ that contains $a_n(p_k)$, and let $\mathcal{A} = \{S_n(p_k) \mid p_k \in P \subset \mathrm{Cl}(P) = \mathcal{T}(A) \cap \mathcal{T}(B)\}_{n=1}^{\infty}$. Then $\mathcal{A}$ is a countable family. Hence, we can write $\mathcal{A} = \{A_j\}_{j=1}^{\infty}$. Note that $P \subset \limsup A_j$. Thus, $\mathcal{T}(A) \cap \mathcal{T}(B) \subset \limsup A_j$. Since X is hereditarily unicoherent, $\mathcal{T}(A) \cap \mathcal{T}(B)$ is a subcontinuum of X (Theorem 2.1.27). Let L be the component of $\limsup A_j$ that contains $\mathcal{T}(A) \cap \mathcal{T}(B)$. By Lemma 1.1.32, there exists a subsequence $\{A_{j_l}\}_{l=1}^{\infty}$ of $\{A_j\}_{j=1}^{\infty}$ such that $\limsup A_{j_l} = L$. Note that $\limsup A_{j_l} \subset \mathcal{T}(A)$. Suppose this is not true. Then there exists a point $x \in \limsup A_{j_l} \setminus \mathcal{T}(A)$. Hence, there exists a subcontinuum K of X such that $x \in \mathrm{Int}(K) \subset K \subset X \setminus A$. Thus, $\mathrm{Int}(K) \cap A_{j_l} \neq \emptyset$ for infinitely many indexes $l \in \mathbb{N}$. Since, for all $n \in \mathbb{N}$, $d(b_n(p_k), A) < \frac{\varepsilon}{nk}$, there exists $h_0 \in \mathbb{N}$ such that if $nk > h_0$, then $b_n(p_k) \notin K$. Since $\mathrm{Int}(K) \cap A_{j_l} \neq \emptyset$ for infinitely many indexes $l \in \mathbb{N}$, we have that there exists $j_{l_0} \in \mathbb{N}$ such that $A_{j_{l_0}} = S_{n_0}(p_{k_0})$, $n_0 k_0 > h_0$ and $A_{j_{l_0}} \cap \mathrm{Int}(K) \neq \emptyset$. Let $c \in A_{j_{l_0}} \cap \mathrm{Int}(K)$. Since x and c both belong to K, the arc $\overline{xc}$ is contained in K. Thus, $b_n(p_k)$ is not in $\overline{xc}$ (recall that $b_n(p_k) \notin K$). The arc $\overline{ca_{n_0}(p_{k_0})}$ is contained in $\mathrm{Cl}(A_{j_{l_0}})$. This implies that $b_{n_0}(p_{k_0})$ is not in the arc $\overline{ca_{n_0}(p_{k_0})}$ (because $d(b_n(p_k), A) < \frac{\varepsilon}{nk}$ and $A_{j_{l_0}} = S_{n_0}(p_{k_0})$ is the component of $\mathcal{V}^d_{\varepsilon}(\mathcal{T}(A) \cap \mathcal{T}(B))$ that contains $a_{n_0}(p_{k_0})$). The arc $\overline{xp_{k_0}}$ is contained in $\limsup A_{j_{l_0}} \subset \mathrm{Cl}(\mathcal{V}^d_{\varepsilon}(\mathcal{T}(A) \cap \mathcal{T}(B)))$. Thus, as before, $b_{n_0}(p_{k_0})$ does not belong to the arc $\overline{xp_{k_0}}$. Hence, $b_{n_0}(p_{k_0})$ does not belong to $\overline{xp_{k_0}} \cup \overline{xc} \cup \overline{ca_{n_0}(p_{k_0})}$. Thus, $b_{n_0}(p_{k_0})$ does not belong to $\overline{a_{n_0}(p_{k_0})p_{k_0}} \subset \overline{xp_{k_0}} \cup \overline{xc} \cup \overline{ca_{n_0}(p_{k_0})}$, contrary to the construction of the sequence $\{b_n(p_{k_0})\}_{n=1}^{\infty}$. Therefore, $\limsup A_{j_l} \subset \mathcal{T}(A)$.

By a similar construction, we have a sequence $\{A'_{j_l}\}_{n=1}^{\infty}$ of components of $\mathcal{V}^d_{\varepsilon}(\mathcal{T}(A) \cap \mathcal{T}(B))$ such that $\mathcal{T}(A) \cap \mathcal{T}(B) \subset \limsup A'_{j_l} \subset \mathcal{T}(B)$. Now, we have that $\mathcal{T}(A) \cap \mathcal{T}(B) \subset \limsup A_{j_l} \cap \limsup A'_{j_l} \subset \mathcal{T}(A) \cap \mathcal{T}(B)$. Therefore, $\limsup A_{j_l} \cap \limsup A'_{j_l} = \mathcal{T}(A) \cap \mathcal{T}(B)$, and $\mathcal{T}(A) \cap \mathcal{T}(B)$ is an R-continuum in X. □

7.3.12 Corollary. *If X is a dendroid and A and B are closed subsets of X such that $A \cap \mathcal{T}(B) = \emptyset$, $\mathcal{T}(A) \cap B = \emptyset$ and $\mathcal{T}(A) \cap \mathcal{T}(B) \neq \emptyset$, then X contains an R-continuum.*

7.3.13 Definition. A dendroid X is said to be *pointwise smooth* if for each point x of X, there exists a point $p(x)$ of X such that for every sequence $\{x_n\}_{n=1}^{\infty}$ of points of X that converges to x, the sequence of arcs $\{\overline{p(x)x_n}\}_{n=1}^{\infty}$ converges to the arc $\overline{p(x)x}$. A point $p(x)$ is called an *initial point* for x in X.

7.3.14 Remark. Note that the dendroid defined in Example 7.3.6 is pointwise smooth but not smooth.

The following theorem gives a characterization of pointwise smooth dendroids.

7.3.15 Theorem. *Let X be a dendroid. Then X is pointwise smooth if and only if for each pair of different points x_1 and x_2 of X, we have that either $\mathcal{T}(\{x_1\}) \cap \overline{x_1x_2} = \{x_1\}$ or $\mathcal{T}(\{x_2\}) \cap \overline{x_1x_2} = \{x_2\}$ or $\mathcal{T}(\{x_1\}) \cap \mathcal{T}(\{x_2\}) = \emptyset$.*

Proof. Suppose X is a pointwise smooth dendroid, and assume that there exist two points x_1 and x_2 of X such that $\mathcal{T}(\{x_1\}) \cap \overline{x_1x_2} \neq \{x_1\}$, $\mathcal{T}(\{x_2\}) \cap \overline{x_1x_2} \neq \{x_2\}$ and $\mathcal{T}(\{x_1\}) \cap \mathcal{T}(\{x_2\}) \neq \emptyset$. Since X is hereditarily unicoherent, we obtain that $\mathcal{T}(\{x_1\}) \cap \mathcal{T}(\{x_2\}) \cap \overline{x_1x_2} \neq \emptyset$. Let $z \in \mathcal{T}(\{x_1\}) \cap \mathcal{T}(\{x_2\}) \cap \overline{x_1x_2}$. Thus, for each subcontinuum K of X such that $z \in \operatorname{Int}(K)$, we have that $\{x_1, x_2\} \subset K$. Hence, the arc $\overline{x_1x_2}$ is contained in K. Let $\{z_n\}_{n=1}^{\infty}$ be a sequence of points of X converging to z. Thus, the arc $\overline{zx_1}$ is contained in $\limsup \overline{z_nx_1}$, and the arc $\overline{zx_2}$ is contained in $\limsup \overline{z_nx_2}$. Since X is pointwise smooth, there exists a point $p(z)$ such that the sequence of arcs $\{\overline{p(z)z_n}\}_{n=1}^{\infty}$ converges to the arc $\overline{p(z)z}$.

We show that $x_1 \in \overline{p(z)z}$. Note that the smallest subcontinuum of X that contains $\mathcal{V}^d_{\frac{1}{n}}(z)$ is the closure union of all arcs $\overline{zt}$, where $t \in \mathcal{V}^d_{\frac{1}{n}}(z)$. Since x_1 belongs to every subcontinuum of X that contains z in its interior, for each $n \in \mathbb{N}$, there exists a point a_n in $\mathcal{V}^d_{\frac{1}{n}}(z)$ such that the arc $\overline{za_n}$ contains a point b_n such that $d(b_n, x_1) < \frac{1}{n}$. Observe that the sequence $\{a_n\}_{n=1}^{\infty}$ converges to z and $\overline{za_n} \subset \overline{p(z)z_n} \cup \overline{p(z)z}$. Hence, by the construction of the b_n's, we obtain that $x_1 \in \liminf \overline{za_n}$. Since $\liminf \overline{za_n} \subset \lim \overline{p(z)a_n} \cup \overline{p(z)z} = \overline{p(z)z}$, we have that $x_1 \in \overline{p(z)z}$. Similarly, $x_2 \in \overline{p(z)z}$. Thus, $\overline{x_1x_2} \subset \overline{p(z)z}$. Since $z \in \overline{x_1x_2}$, we conclude that either $z = x_1$ or $z = x_2$. Now, we have $\mathcal{T}(\{x_1\}) \cap \mathcal{T}(\{x_2\}) \cap \overline{x_1x_2} = \{x_1\}$ or $\mathcal{T}(\{x_1\}) \cap \mathcal{T}(\{x_2\}) \cap \overline{x_1x_2} = \{x_2\}$. Suppose that $\mathcal{T}(\{x_1\}) \cap \mathcal{T}(\{x_2\}) \cap \overline{x_1x_2} = \{x_1\}$. Since $\{x_1, x_2\} \subset \mathcal{T}(\{x_2\}) \cap \overline{x_1x_2}$, we conclude that $\overline{x_1x_2} \subset \mathcal{T}(\{x_2\}) \cap \overline{x_1x_2}$ and $\mathcal{T}(\{x_1\}) \cap (\mathcal{T}(\{x_2\}) \cap \overline{x_1x_2}) \supset \mathcal{T}(\{x_1\}) \cap \overline{x_1x_2} \neq \{x_1\}$. This is a contradiction.

Now, assume that X satisfies the statement of the theorem. We prove that X is pointwise smooth. Let x be a point of X. If X is connected im kleinen at x, by [27, Theorem 2], we may take $p(x) = x$. Thus, assume that X is not connected im kleinen at x. Hence, there exists a point $x' \in X$ such that $x \in \mathcal{T}(\{x'\})$ [27, Theorem 2].

Next, we prove that if z and z' are points of X such that $x \in \mathcal{T}(\{z\}) \cap \mathcal{T}(\{z'\})$, then either $z \in \overline{xz'}$ or $z' \in \overline{xz}$. Let z and z' be such points, and assume that neither z belongs to $\overline{xz'}$ nor z' belongs to $\overline{xz}$. Then $\overline{zx} \cup \overline{xz'}$ is either an arc or a simple triod. If $\overline{zx} \cup \overline{xz'}$ is an arc, then $\overline{xz'} \subset \mathcal{T}(\{z'\}) \cap \overline{zz'}$ and $\overline{zx} \subset \mathcal{T}(\{z\}) \cap \overline{zz'}$, contrary to our hypothesis. If $\overline{zx} \cup \overline{xz'}$ is a simple triod, with vertex t, then note that

$t \notin \{z, z'\}$, $\overline{tz'} \subset \mathcal{T}(\{z'\}) \cap \overline{zz'}$ and $\overline{tz} \subset \mathcal{T}(\{z\}) \cap \overline{zz'}$, contrary to our hypothesis. Therefore, our claim is proven. Hence, the set $Z = \{z \in X \mid x \in \mathcal{T}(\{z\})\}$ lies in an arc having x as one of its end points. Let $p(x)$ be the other end point (different form x) of the minimal arc containing Z. Observe that Z is closed and $x \in \mathcal{T}(\{p(x)\})$. Let $\{x_n\}_{n=1}^{\infty}$ be a sequence of points of X that converges to x. Then,

$$\overline{p(x)x} \subset \liminf \overline{p(x)x_n} \subset \limsup \overline{p(x)x_n}.$$

Let $q \in \limsup \overline{p(x)x_n}$. For each subcontinuum K of X containing x in its interior, we have that $p(x) \in K$ and $\overline{p(x)x_n} \subset K$ for almost all $n \in \mathbb{N}$. Hence, $\limsup \overline{p(x)x_n} \subset K$. Thus, $q \in K$ and $x \in \mathcal{T}(\{q\})$. Then $q \in Z \subset \overline{p(x)x}$. As a consequence of this, $\limsup \overline{p(x)x_n} \subset \overline{p(x)x}$. Therefore, the sequence of arcs $\{\overline{p(x)x_n}\}_{n=1}^{\infty}$ converges to the arc $\overline{p(x)x}$. Hence, X is pointwise smooth. □

Note that in the proof of Theorem 7.3.15, if X satisfies the statement of that theorem, then for each $x \in X$, we may choose the point $p(x)$ in such a way that $x \in \mathcal{T}(\{p(x)\})$. Hence, we have the following.

7.3.16 Corollary. *A dendroid X is pointwise smooth if and only if for each $x \in X$, there exists a point $p(x)$ in X such that for each sequence $\{x_n\}_{n=1}^{\infty}$ converging to x, the sequence of arcs $\{\overline{p(x)x_n}\}_{n=1}^{\infty}$ converges to the arc $\overline{p(x)x}$ and $x \in \mathcal{T}(\{p(x)\})$.*

7.4 Local Connectedness

We present several characterizations of local connectedness of continua. We show the characterizations for products, cones, suspensions and hyperspaces of metric continua. We end with a characterization of local connectedness for unicoherent continua.

The next result is contained in Theorem 5.3.6, and we present a different proof here.

7.4.1 Theorem. *Let X and Y be continua. If $Z = X \times Y$ is a continuum for which $\mathcal{T}_Z$ is continuous, then Z is locally connected. In particular, X and Y are locally connected.*

Proof. Let A be a nonempty closed subset of Z. By Lemma 1.6.9, there exists a net $\{K_\lambda\}_{\lambda\in\Lambda}$ of finite subsets of X converging to A. Since $\mathcal{T}_Z$ is continuous, $\{\mathcal{T}_Z(K_\lambda)\}_{\lambda\in\Lambda}$ converges to $\mathcal{T}_Z(A)$ ([97, Proposition 3.38] and [104, Theorem 4]). By Theorem 4.1.8, $\mathcal{T}_Z(K_\lambda) = K_\lambda$ for each $\lambda \in \Lambda$. Hence, $\mathcal{T}_Z(A) = A$. Since A is an arbitrary nonempty closed subset of Z, by Theorem 2.1.37, Z is locally connected. Note that X and Y are locally connected by Theorem 2.1.52. □

As a consequence of Theorems 7.4.1 and 2.1.52, we obtain the following.

7.4.2 Corollary. *If X is a metric continuum such that $\mathcal{T}_{K(X)}$ is continuous for $K(X)$, then $K(X)$ is locally connected. In particular, X is locally connected.*

Proof. Let A be a nonempty closed subset of X. By Lemma 1.6.9, there exists a sequence $\{M_n\}_{n=1}^{\infty}$ of finite subsets of $K(X)$ converging to A. Since $\mathcal{T}_{K(X)}$ is continuous, $\{\mathcal{T}_{K(X)}(M_n)\}_{n=1}^{\infty}$ converges to $\mathcal{T}_{K(X)}(A)$. By Theorem 4.1.9, $\mathcal{T}_{K(X)}(M_n) = M_n$ for each $n \in \mathbb{N}$. Hence, $\mathcal{T}_{K(X)}(A) = A$. Since A is an arbitrary nonempty closed subset of $K(X)$, by Theorem 2.1.37, $K(X)$ is locally connected. Observe that X is locally connected by Theorem 2.1.52. □

A similar proof to the one given for Theorem 7.4.2 gives the following.

7.4.3 Corollary. *If X is a metric continuum such that $\mathcal{T}_{\Sigma(X)}$ is continuous for $\Sigma(X)$, then $\Sigma(X)$ is locally connected. In particular, X is locally connected.*

7.4.4 Corollary. *Let X be a metric continuum, and let $n \geq 2$ be an integer. If $\mathcal{T}_{\mathcal{F}_n(X)}$ is continuous for $\mathcal{F}_n(X)$, then X is locally connected.*

Proof. By [80, Theorem 4], $\mathcal{F}_n(X)$ is aposyndetic. Hence, by Corollary 5.1.17, $\mathcal{F}_n(X)$ is locally connected. Therefore, X is locally connected by Lemma 1.6.10. □

7.4.5 Theorem. *Let X be a metric continuum. If*

$$\mathcal{L} \in \{2^X, \mathcal{C}_n(X)\ (n \in \mathbb{N})\}$$

and $\mathcal{T}_{\mathcal{L}}$ is continuous for $\mathcal{L}$, then X is locally connected.

Proof. Suppose $\mathcal{T}_{\mathcal{L}}$ is continuous for $\mathcal{L}$. Since $\mathcal{L}$ is aposyndetic ([50, Theorem 1] for 2^X and [92, Corollary 6.3.3] for $\mathcal{C}_n(X)$), by Corollary 5.1.17, $\mathcal{L}$ is locally connected. Hence, X is locally connected ([105, (1.92)] for 2^X and [92, Theorem 6.1.6] for $\mathcal{C}_n(X)$). □

To present our next application of local connectedness for metric continua, we need some connectivity properties of $\mathcal{T}$.

7.4.6 Lemma. *Let X be a metric continuum. If S is a nonempty totally disconnected closed subset of X such that there exists a point $p \in \mathcal{T}(S) \setminus S$ and such that for each proper subset S' of S, $p \in X \setminus \mathcal{T}(S')$, then $\mathcal{T}(S)$ is connected.*

Proof. Let S be a totally disconnected closed subset of X satisfying the properties stated. Let S_0 be a nonempty subset of S, which is both open and closed in S. Since $p \in X \setminus \mathcal{T}(S \setminus S_0)$, there exists a subcontinuum W of X such that $p \in \operatorname{Int}(W) \subset W \subset X \setminus (S \setminus S_0)$. Let $\{U_n\}_{n=1}^{\infty}$ and $\{V_n\}_{n=1}^{\infty}$ be decreasing sequences of open subsets of X such that for each $n \in \mathbb{N}$, $S \setminus S_0 \subset U_n$, $S_0 \subset V_n$, $U_1 \cap \operatorname{Cl}(V_1) = U_1 \cap W = \operatorname{Cl}(V_1) \cap \{p\} = \emptyset$, $S \setminus S_0 = \bigcap_{n=1}^{\infty} U_n$ and $S_0 = \bigcap_{n=1}^{\infty} V_n$.

For each $n \in \mathbb{N}$, let C_n be the component of $X \setminus U_n$ such that $W \subset C_n$. Since p does not belong to the interior of the component of $C_n \setminus V_n$ in which it lies, there exists a sequence $\{D_n^j\}_{j=1}^{\infty}$ of distinct components of $C_n \setminus V_n$ such that $p \in D_n = \lim_{j \to \infty} D_n^j$. Since, by Theorem 1.4.36, $D_n^j \cap \operatorname{Cl}(V_n) \neq \emptyset$, for each $n \in \mathbb{N}$, D_n is a continuum, $D_n \subset C_n \setminus V_n$, $p \in D_n$ and $D_n \cap \text{Cl}(V_n) \neq \emptyset$. Let $D = \lim_{n\to\infty} D_n$. Then D is a continuum containing p and $D \cap S_0 \neq \emptyset$.

We assert that $D \subset \mathcal{T}(S)$. Suppose this is not true. Then there exist $q \in D \setminus \mathcal{T}(S)$ and a subcontinuum W' of X such that $q \in \text{Int}(W') \subset W' \subset X \setminus S$. It follows that there exists $N_1 \in \mathbb{N}$ such that for $n > N_1$, $W' \subset X \setminus (U_n \cup V_n)$. Since $q \in \text{Int}(W') \cap D$, there exists $N_2 \in \mathbb{N}$ such that for $n > N_2$, $(\text{Int}(W')) \cap D_n \neq \emptyset$. Let $m > \max\{N_1, N_2\}$. Since $(\text{Int}(W')) \cap D_m \neq \emptyset$, there exists $N_3 \in \mathbb{N}$ such that for $j > N_3$, $(\text{Int}(W')) \cap D_m^j \neq \emptyset$. Let $j > \max\{N_1, N_2, N_3\}$. Then $D_m^j \subset C_m$ and $W' \cap C_m \neq \emptyset$. Thus, $W' \subset C_m \setminus V_m$, a contradiction, since $j > \max\{N_1, N_2, N_3\}$ and $(\text{Int}(W')) \cap D_m^j \neq \emptyset$. Therefore, $D \subset \mathcal{T}(S)$.

Now, let $s \in S$. Then there exists a sequence $\{S_n\}_{n=1}^\infty$ of subsets of S such that for each $n \in \mathbb{N}$, S_n is open and closed in S, and $\{s\} = \bigcap_{n=1}^\infty S_n$. By the previous construction, for each $n \in \mathbb{N}$, there exists a subcontinuum A_n of $\mathcal{T}(S)$ such that $p \in A_n$ and $A_n \cap S_n \neq \emptyset$. Since $\mathcal{C}_1(X)$ is compact (Theorem 1.6.7), without loss of generality, we assume that the sequence $\{A_n\}_{n=1}^\infty$ converges. Let $A_s = \lim_{n\to\infty} A_n$. Then A_s is a subcontinuum of $\mathcal{T}(S)$ and $\{p, s\} \subset A_s$.

Let $A = \bigcup_{s\in S} A_s$. Then A is connected (for every $s \in S$, $p \in A_s$) and $S \subset A \subset \mathcal{T}(S)$. For each $x \in \mathcal{T}(S)$, let C_x be the component of $\mathcal{T}(S)$ such that $x \in C_x$. By Corollary 2.1.20, $C_x \cap S \neq \emptyset$. Thus, $C_x \cap A \neq \emptyset$ for every $x \in \mathcal{T}(S)$, and it follows that

$$\mathcal{T}(S) = A \cup \left(\bigcup_{x\in\mathcal{T}(S)} C_x \right)$$

is connected. □

The following theorem shows that we may remove the "totally disconnected" hypothesis in Lemma 7.4.6.

7.4.7 Theorem. *Let X be a metric continuum. Suppose A is a nonempty closed subset of X, and there exists a point $p \in \mathcal{T}_X(A) \setminus A$ such that for each proper closed subset A' of A, $p \in X \setminus \mathcal{T}_X(A')$. Then $\mathcal{T}_X(A)$ is connected.*

Proof. If $\mathcal{T}_X(A) = X$, there is nothing to do. Then suppose $\mathcal{T}_X(A)$ is a proper closed subset of X. Let $\{A_\lambda\}_{\lambda\in\Lambda}$ be the family of components of A. Then $\mathcal{G} = \{A_\lambda\}_{\lambda\in\Lambda} \cup \{\{x\} \mid x \in X \setminus A\}$ is an upper semicontinuous decomposition of X, and $X/\mathcal{G}$ is a metric continuum [92, Theorem 1.7.3]. Let $q\colon X \twoheadrightarrow X/\mathcal{G}$ be the quotient map. Note that q is monotone.

Since A is a closed subset of X, $q(A) = q\left(\bigcup_{\lambda\in\Lambda} A_\lambda\right) = \bigcup_{\lambda\in\Lambda} q(A_\lambda)$ is a closed totally disconnected subset of $X/\mathcal{G}$. Thus, $\mathcal{T}_{X/\mathcal{G}}q(A)$ is connected, Lemma 7.4.6. Since q is monotone, by Theorem 2.1.51 (b), we have that $\mathcal{T}_X(A) = \mathcal{T}_X q^{-1}q(A) \subset q^{-1}\mathcal{T}_{X/\mathcal{G}}q(A)$. By Theorem 2.1.51 (c), we have $q^{-1}\mathcal{T}_{X/\mathcal{G}}q(A) = q^{-1}q\mathcal{T}_X q^{-1}q(A) = q^{-1}q\mathcal{T}_X(A)$. Since $A \subset \mathcal{T}(A)$, $q^{-1}q\mathcal{T}_X(A) = \mathcal{T}_X(A)$. Hence, $q^{-1}\mathcal{T}_{X/\mathcal{G}}q(A) = \mathcal{T}_X(A)$. Therefore, $\mathcal{T}_X(A)$ is connected. □

The next theorem is an improvement made by David P. Bellamy. His proof enables us to put all the results remaining in this section in the class of continua, instead of in the class of metric continua, as they were originally stated.

7.4.8 Theorem. *Let X be a continuum. Suppose A is a nonempty closed subset of X, and there exists $x \in \mathcal{T}(A) \setminus A$. Then there exists a closed subset D of X such that*

(a) $D \subset A$,
(b) $x \in \mathcal{T}(D)$,
(c) if E is a nonempty closed subset of X and $E \subsetneq D$, then $x \in X \setminus \mathcal{T}(E)$ and
(d) $\mathcal{T}(D)$ *is a continuum.*

Proof. Let X, A and x be as given. Let

$$\mathfrak{A} = \{B \subseteq A \mid B \text{ is closed and } x \in \mathcal{T}(B)\}.$$

Then $\mathfrak{A}$ is partially ordered by set inclusion ($\subseteq$). Thus, by the Hausdorff Maximal Principle [70, **24**, p. 32], $\mathfrak{A}$ contains a maximal linearly ordered subset, $\mathcal{M}$. Since $\mathcal{M}$ is linearly ordered by set inclusion, $\mathcal{T}(\bigcap \mathcal{M}) = \bigcap\{\mathcal{T}(M) \mid M \in \mathcal{M}\}$, Lemma 2.2.9. Let $D = \bigcap \mathcal{M}$. Note that $D \in \mathcal{M}$. Hence, $x \in \mathcal{T}(D)$. To prove that $\mathcal{T}(D)$ is a continuum, suppose that $\mathcal{T}(D) = P \cup Q$, where P and Q are disjoint closed subsets of X. Then x belongs to either P or Q. Assume that $x \in P$. Thus, by Lemma 2.1.46, $P = \mathcal{T}(D \cap P)$, so $x \in \mathcal{T}(D \cap P)$. Since $D \cap P$ is a proper closed subset of D, and D is the smallest element of $\mathcal{M}$, $D \cap P$ cannot belong to $\mathcal{M}$. This is a contradiction to the maximality of $\mathcal{M}$, since $D \cap P$ satisfies both properties required for membership in $\mathfrak{A}$ and is a subset of every member of $\mathcal{M}$. Therefore, $\mathcal{T}(D)$ is a continuum.

Now, suppose E is an arbitrary proper closed subset of D. If $x \in \mathcal{T}(E)$, then if $E \in \mathfrak{A}$, $\mathcal{M} \cup \{E\}$ cannot properly contain $\mathcal{M}$. Hence, $E \notin \mathfrak{A}$. This is again a contradiction. Thus, $x \in X \setminus \mathcal{T}(E)$. □

In [106, 5.12] it is shown that if X is a metric continuum and x is a point at which X is not connected im kleinen, then there exists a subcontinuum K of X such that $x \in K$ and X is not connected im kleinen at any of the points of K. In the following theorem, we show that something similar happens with the images of $\mathcal{T}$.

7.4.9 Theorem. *Let X be a continuum. Suppose A is a nonempty closed subset of X such that there exists $x \in \mathcal{T}(A) \setminus A$. Then there exists a nondegenerate subcontinuum K of X such that $x \in K \subset \mathcal{T}(A)$.*

Proof. Let A be a nonempty closed subset of X, and let $x \in \mathcal{T}(A) \setminus A$. By Theorem 7.4.8, there exists a nonempty closed subset D of X satisfying conditions (a), (b), (c) and (d). Let V be an open subset of X such that $x \in V \subset \mathrm{Cl}(V) \subset X \setminus A$. Let H be the component of $V \cap \mathcal{T}(D)$ containing x, and let $K = \mathrm{Cl}(H)$. Since $V \cap \mathcal{T}(D)$ is open in $\mathcal{T}(D)$, by Theorem 1.4.36, K intersects $\mathrm{Bd}(V)$. Thus, K is a nondegenerate subcontinuum of X contained in $\mathcal{T}(A)$. □

The next theorem characterizes local connectedness in unicoherent continua.

7.4.10 Theorem. *Let X be a unicoherent continuum. Then X is locally connected if and only if for each nonempty closed subset C of X which separates X between two points x and y, there exists a component E of C which separates X between x and y.*

Proof. First, suppose that X is unicoherent and locally connected. Let C be a nonempty closed subset of X which separates X between x and y. Let A be the component of $X \setminus C$ containing x, and let B be the component of $X \setminus \mathrm{Cl}(A)$ containing y. Since X is locally connected, by Lemma 1.4.17, A and B are open subsets of X. Also, $X \setminus (\mathrm{Cl}(A) \cap \mathrm{Cl}(B)) \subset (X \setminus \mathrm{Cl}(B)) \cup B$, $x \in X \setminus \mathrm{Cl}(B)$ and $y \in B$. Thus, $\mathrm{Cl}(A) \cap \mathrm{Cl}(B)$ is a closed subset of X which separates X between x and y.

Let $\{S_\lambda\}_{\lambda \in \Lambda}$ be the family of components of $X \setminus \mathrm{Cl}(A)$ such that $S_\lambda \cap B = \emptyset$ for each $\lambda \in \Lambda$. Then, by Theorem 1.4.36, $\mathrm{Cl}(A) \cup \left(\bigcup_{\lambda \in \Lambda} \mathrm{Cl}(S_\lambda)\right)$ is connected. Observe that $X = \mathrm{Cl}(A) \cup \mathrm{Cl}\left(\bigcup_{\lambda \in \Lambda} \mathrm{Cl}(S_\lambda)\right) \cup \mathrm{Cl}(B)$. Since X is unicoherent,

$$\mathrm{Cl}(A) \cap \mathrm{Cl}(B) = \left(\mathrm{Cl}(A) \cup \mathrm{Cl}\left(\bigcup_{\lambda \in \Lambda} \mathrm{Cl}(S_\lambda)\right)\right) \cap \mathrm{Cl}(B)$$

is a continuum. Let E be the component of C that contains $\mathrm{Cl}(A) \cap \mathrm{Cl}(B)$. If x and y belong to the same component of $X \setminus E \subset X \setminus (\mathrm{Cl}(A) \cap \mathrm{Cl}(B))$, then x and y are in the same component of $X \setminus (\mathrm{Cl}(A) \cap \mathrm{Cl}(B))$. Since this is not true, E separates X between x and y.

Now, suppose X is a unicoherent continuum satisfying the conditions stated, and suppose X is not locally connected, hence, not connected im kleinen at some point $p \in X$. Then, by Theorem 2.1.30, there exists a nonempty closed subset A of X such that $p \in \mathcal{T}(A) \setminus A$. By Theorem 7.4.8, there exists a nonempty closed subset B of X satisfying conditions (a), (b), (c) and (d). Let $x \in B$, and let U_0 be an open subset of X such that $p \in U_0 \subset \mathrm{Cl}(U_0) \subset X \setminus B$. Then $\mathrm{Bd}(U_0)$ is a closed set that separates X between x and p. Hence, there exists a component N of $\mathrm{Bd}(U_0)$ which separates X between x and p.

Let U and V be disjoint open sets such that $X \setminus N = U \cup V$, $p \in U$ and $x \in V$. Then, by Lemma 1.4.32, $H = N \cup U$ and $K = N \cup V$ are continua, and $X = H \cup K$. Since $x \in B \cap K$, we have that $B \cap K \neq \emptyset$. If $B \cap H = \emptyset$, then $p \in X \setminus \mathcal{T}(B)$, which is contrary to (b) of Theorem 7.4.8. Thus, $B \cap H \neq \emptyset$. Since $B \cap K$ and $B \cap H$ are two nonempty closed proper subsets of B, it follows from (c) of Theorem 7.4.8 that there exist two subcontinua L_1 and L_2 of X such that $p \in \mathrm{Int}(L_1) \subset L_1 \subset X \setminus (B \cap H)$ and $p \in \mathrm{Int}(L_2) \subset L_2 \subset X \setminus (B \cap K)$. Again, if $L_1 \cap (B \cap K) = \emptyset$ or $L_2 \cap (B \cap H) = \emptyset$, it follows that $p \in X \setminus \mathcal{T}(B)$. Thus, assume that $L_1 \cap (B \cap K) \neq \emptyset$ and $L_2 \cap (B \cap H) \neq \emptyset$. Let $X_1 = L_1 \cup K$ and $X_2 = L_2 \cup H$. Then X_1 and X_2 are continua, and $X = X_1 \cup X_2$. Now, since X is unicoherent, $X_1 \cap X_2$ is a continuum. Note that $p \in \mathrm{Int}(X_1 \cap X_2) \subset (X_1 \cap X_2) \subset X \setminus B$. Hence,

$p \in X \setminus \mathcal{T}(B)$, which is contrary to condition (b) of Theorem 7.4.8. Therefore, X is locally connected. □

7.5 Shore Sets

We present sufficient conditions for a closed subset of a dendroid to have its image under the set function $\mathcal{T}$ be a shore set in the dendroid.

7.5.1 Definition. Let X be a metric continuum, and let A be a nonempty subset of X. Then A is a *shore set* in X if for every $\varepsilon > 0$, there exists a subcontinuum W of X such that $W \cap A = \emptyset$ and $\mathcal{H}(W, X) < \varepsilon$, where $\mathcal{H}$ is the Hausdorff metric. If $A = \{p\}$, then p is called a *shore point* of X.

7.5.2 Definition. Let X be a metric continuum, and let p be a point of X. Then p is a *centre point* of X if there exist two points a and b of X such that for each $\varepsilon > 0$, there exists a subcontinuum W of X with $p \in W$ and $\text{diam}(W) < \varepsilon$, and there exist two open subsets U and V of X, containing a and b, respectively, such that every arc in X from a point of U to a point of V intersects W.

7.5.3 Notation. Let X be a dendroid, and let p and x be two points of X. Then $Q_p(x) = \{z \in X \mid x \in \overline{zp}\}$.

7.5.4 Notation. Let X be a dendroid. Then $C_e(X) = \{z \in X \mid z \text{ is a centre of } X\}$.

7.5.5 Proposition. *If X is a $\mathcal{T}$-additive aposyndetic metric continuum and A is a closed shore set of X, then $\mathcal{T}(A)$ is a shore set of X.*

Proof. Since X is aposyndetic and $\mathcal{T}$-additive, by Theorem 2.2.14, X is locally connected. Hence, $\mathcal{T}(A) = A$, Theorem 2.1.37. Therefore, $\mathcal{T}(A)$ is a shore set of X. □

7.5.6 Corollary. *If X is an aposyndetic dendroid and A is a closed shore set of X, then $\mathcal{T}(A)$ is a shore set of X.*

Proof. Note that, by Theorem 2.2.12, X is $\mathcal{T}$-additive. Now the corollary follows from Proposition 7.5.5. □

7.5.7 Theorem. *Let X be a dendroid, let p be a centre of X and let A be a nonempty closed subset of X. If $\mathcal{T}(A) \subset Q_p(y)$, for some $y \in \mathcal{T}(A)$ and $\mathcal{T}(A) \cap C_e(X) = \emptyset$, then $\mathcal{T}(A)$ is a shore set of X.*

Proof. Let p, A and y be as in the statement of the theorem. Since $y \in \mathcal{T}(A)$, and $\mathcal{T}(A) \cap C_e(X) = \emptyset$, by [107, Lemma 1], we have that $\text{Int}(Q_p(y)) = \emptyset$. Note that $X \setminus Q_p(y)$ is arcwise connected. Thus, by [76, Lemma 2.4], $Q_p(y)$ is a shore set of X. Hence, since $\mathcal{T}(A) \subset Q_p(y)$, $\mathcal{T}(A)$ is a shore set of X. □

The next result is [76, Theorem 2.9].

7.5.8 Theorem. *Let X be a dendroid, and let A be a nonempty closed subset of X with only finitely many components such that $A \cap C_e(X) = \emptyset$. Then A is a shore set of X.*

As a consequence of Theorem 7.5.8, we have the following three corollaries.

7.5.9 Corollary. *Let X be a dendroid, and let A be a nonempty subset of X such that $\mathcal{T}(A)$ only has a finite number of components and $\mathcal{T}(A) \cap C_e(X) = \emptyset$. Then $\mathcal{T}(A)$ is a shore set of X.*

7.5.10 Corollary. *Let X be a dendroid, and let A be a nonempty subset of X such that $\mathcal{T}(A)$ is connected and $\mathcal{T}(A) \cap C_e(X) = \emptyset$. Then $\mathcal{T}(A)$ is a shore set of X.*

7.5.11 Corollary. *Let X be a dendroid, and let A be subcontinuum of X such that A is a shore set of X. If $\mathcal{T}(A) \cap C_e(X) = \emptyset$, then $\mathcal{T}(A)$ is a shore set of X.*

7.6 Generalized Inverse Limits

We show that the generalized inverse limit of homogeneous metric continua using the set function $\mathcal{T}|_{\mathcal{F}_1(X)}$ as an upper semicontinuous bonding function is an infinite-dimensional homogeneous metric continuum. We present three examples.

7.6.1 Notation. Given a one-dimensional metric continuum X, $\widehat{X}$ denotes the continuous curves of pseudo-arcs associated to X constructed by Wayne Lewis [78], and $\widehat{\mathcal{G}} = \{P_{\hat{x}} \mid \hat{x} \in \widehat{X}\}$ denotes the decomposition of $\widehat{X}$ into pseudo-arcs.

As a consequence of [78, Theorem 4], we have the following.

7.6.2 Lemma. *Let X be a one-dimensional metric continuum. If $h\colon \widehat{X} \to \widehat{X}$ is a homeomorphism, then $h(P_{\hat{x}}) = P_{h(\hat{x})}$.*

A proof of the next result, due to Wayne Lewis, may be found in [78, Corollary 6].

7.6.3 Lemma. *If X is a one-dimensional homogeneous metric continuum, then $\widehat{X}$ is a homogeneous metric continuum.*

7.6.4 Theorem. *If X is a one-dimensional homogeneous metric continuum, then $\varprojlim f_{\widehat{\mathcal{G}}}$ (Notation 7.6.1) is an infinite-dimensional decomposable homogeneous metric continuum.*

Proof. By Theorem 1.4.76, $\varprojlim f_{\widehat{\mathcal{G}}}$ is an infinite-dimensional decomposable metric continuum. Let $(\hat{x}_n)_{n=1}^{\omega}$ and $(\hat{y}_n)_{n=1}^{\omega}$ be two elements of $\varprojlim f_{\widehat{\mathcal{G}}}$. Since, by Lemma 7.6.3, $\widehat{X}$ is homogeneous, for each $n \in \mathbb{N}$, there exists a homeomorphism $h_n\colon \widehat{X} \to \widehat{X}$ such that $h_n(\hat{x}_n) = \hat{y}_n$. Let $h\colon \varprojlim f_{\widehat{\mathcal{G}}} \to \varprojlim f_{\widehat{\mathcal{G}}}$ be given by $h\left((\hat{z}_n)_{n=1}^{\omega}\right) = \left(h_n(\hat{z}_n)\right)_{n=1}^{\omega}$. Note that, by Lemma 7.6.2, h is well defined. Clearly, h is a homeomorphism. Therefore, $\varprojlim f_{\widehat{\mathcal{G}}}$ is homogeneous. □

With a similar argument to the one given in Theorem 7.6.4 and using the metric version of Jones' Aposyndetic Decomposition Theorem (Theorem 3.3.8), we have the following.

7.6.5 Theorem. *If X is a decomposable homogeneous nonaposyndetic metric continuum and $\mathcal{G} = \{\mathcal{T}(\{x\}) \mid x \in X\}$, then $\varprojlim(f_{\mathcal{T}})_{\mathcal{G}}$, where $f_{\mathcal{T}}\colon X \to \mathcal{C}_1(X)$ is given by $f_{\mathcal{T}}(x) = \mathcal{T}(\{x\})$, is an infinite-dimensional decomposable homogeneous continuum.*

We finish this section with three examples.

7.6.6 Notation. If X is a metric continuum, define $f_{\mathcal{T}}\colon X \to \mathcal{C}_1(X)$ by $f_{\mathcal{T}}(x) = \mathcal{T}(\{x\})$. Note that, by Theorem 2.1.27, $f_{\mathcal{T}}$ is well defined. Also, by Theorem 5.1.1, $f_{\mathcal{T}}$ is upper semicontinuous. Hence, by [65, Theorem 4.7], $\varprojlim f_{\mathcal{T}}$ is a metric continuum.

7.6.7 Example. Let X be a metric compactification of $[0,1)$ with an arc as a remainder. Observe that $\mathcal{G} = \{f_{\mathcal{T}}(x) \mid x \in X\}$ is an upper semicontinuous decomposition of X and $f_{\mathcal{T}}(x) = \{x\}$ for all $x \in [0,1)$. Then $\varprojlim f_{\mathcal{T}}$ is homeomorphic to $[0,1]^{\omega} \cup R$, where $\mathrm{Cl}_{\varprojlim f_{\mathcal{T}}}(R)$ is homeomorphic to X. In fact, $\mathrm{Cl}_{\varprojlim f_{\mathcal{T}}}(R) = \Delta^{\omega}_{[0,1]^{\omega}} \cup R$.

Similarly, we obtain the following two examples:

7.6.8 Example. Let X be a metric compactification of $[0,1)$ with a simple closed curve, $\mathcal{S}^1$, as a remainder. Note that $\mathcal{G} = \{f_{\mathcal{T}}(x) \mid x \in X\}$ is an upper semicontinuous decomposition of X and $f_{\mathcal{T}}(x) = \{x\}$ for all $x \in [0,1)$. Then $\varprojlim f_{\mathcal{T}}$ is homeomorphic to $(\mathcal{S}^1)^{\omega} \cup R$, where $\mathrm{Cl}_{\varprojlim f_{\mathcal{T}}}(R)$ is homeomorphic to X. In fact, $\mathrm{Cl}_{\varprojlim f_{\mathcal{T}}}(R) = \Delta^{\omega}_{(\mathcal{S}^1)^{\omega}} \cup R$.

7.6.9 Example. Let X be a metric compactification of $(0,1)$ with an arc as a remainder. Note that $\mathcal{G} = \{f_{\mathcal{T}}(x) \mid x \in X\}$ is an upper semicontinuous decomposition of X and $f_{\mathcal{T}}(x) = \{x\}$ for all $x \in (0,1)$. Then $\varprojlim f_{\mathcal{T}}$ is homeomorphic to $[0,1]^{\omega} \cup R$, where $\mathrm{Cl}_{\varprojlim f_{\mathcal{T}}}(R)$ is homeomorphic to X. In fact, $\mathrm{Cl}_{\varprojlim f_{\mathcal{T}}}(R) = \Delta^{\omega}_{[0,1]^{\omega}} \cup R$.

7.7 $\mathcal{T}$ and $\mathcal{K}$

We present relationships between the set functions $\mathcal{T}$ and $\mathcal{K}$ for weakly irreducible, $\mathcal{T}$-symmetric, $\mathcal{K}$-symmetric, $\mathcal{T}$-additive and $\mathcal{K}$-additive continua.

First, note the next property of the set function $\mathcal{K}$:

7.7.1 Theorem. *Let X be a continuum. Then,*

$$\mathcal{K}(A) = \bigcup\{\mathcal{K}(\{a\}) \mid a \in A\}$$

for each $A \in \mathcal{C}(X)$.

Proof. Since $\bigcup\{\mathcal{K}(\{a\}) \mid a \in A\} \subseteq \mathcal{K}(A)$, we only need to show that $\mathcal{K}(A) \subseteq \bigcup\{\mathcal{K}(\{a\}) \mid a \in A\}$. Let $z \in X \setminus \bigcup\{\mathcal{K}(\{a\}) \mid a \in A\}$. Then $z \in \bigcap\{X \setminus \mathcal{K}(\{a\}) \mid a \in A\}$. Thus, for each $a \in A$, there exists a subcontinuum M_a of X such that $a \in \operatorname{Int}(M_a)$ and $z \notin M_a$. Then $\{\operatorname{Int}(M_a) \mid a \in A\}$ is an open cover of A. Since A is compact, there exist $a_1, \dots, a_n$ in A such that $A \subseteq \bigcup_{j=1}^{n} \operatorname{Int}(M_{a_j})$. Let $M = \bigcup_{j=1}^{n} M_{a_j}$. Then M is a subcontinuum of X such that $A \subseteq \operatorname{Int}(M)$ and $z \notin M$. Hence, $z \in X \setminus \mathcal{K}(A)$. Therefore, $\mathcal{K}(A) = \bigcup\{\mathcal{K}(\{a\}) \mid a \in A\}$. □

7.7.2 Theorem. *Let X be a continuum. If A is a subcontinuum of X, then $\mathcal{K}(A) = \{x \in X \mid \mathcal{T}(\{x\}) \cap A \neq \emptyset\}$.*

Proof. Let x be a point of X such that $\mathcal{T}(\{x\}) \cap A \neq \emptyset$. Then there exists a point z in A such that $z \in \mathcal{T}(\{x\})$. Hence, for each subcontinuum K of X that contains z in its interior, we have that $x \in K$. Thus, for each subcontinuum L of X such that $A \subset \operatorname{Int}(L)$, we have that $x \in L$. Therefore, $x \in \mathcal{K}(A)$.

Now, let x be a point of X such that $\mathcal{T}(\{x\}) \cap A = \emptyset$. Then for each point a of A, there exists a subcontinuum W_a of X such that $a \in \operatorname{Int}(W_a) \subset W_a \subset X \setminus \{x\}$. Hence, $\{\operatorname{Int}(W_a) \mid a \in A\}$ is an open cover of X. Since A is compact, there exist $a_1, \dots, a_n$ in A such that $A \subset \bigcup_{j=1}^{n} \operatorname{Int}(W_{a_j}) \subset \bigcup_{j=1}^{n} W_{a_j} \subset X \setminus \{x\}$. Hence, since A is connected, $\bigcup_{j=1}^{n} W_{a_j}$ is a continuum containing A in its interior and missing x. Therefore, $x \in X \setminus \mathcal{K}(A)$. □

7.7.3 Lemma. *Let X be a continuum. Then $p \in \mathcal{K}(\{q\})$ if and only if $q \in \mathcal{T}(\{p\})$ for every pair of points p and q of X.*

Proof. If $p \in X \setminus \mathcal{K}(\{q\})$, then there exists a subcontinuum W of X such that $q \in \operatorname{Int}(W)$ and $p \in X \setminus W$. This implies that $q \in X \setminus \mathcal{T}(\{p\})$.

If $q \in X \setminus \mathcal{T}(\{p\})$, then there exists a subcontinuum W of X such that $q \in \operatorname{Int}(W) \subset W \subset X \setminus \{p\}$. Hence, $p \in X \setminus \mathcal{K}(\{q\})$. □

7.7.4 Corollary. *If X is a point $\mathcal{T}$-symmetric continuum, then $\mathcal{T}(\{x\}) = \mathcal{K}(\{x\})$, for all $x \in X$.*

Proof. Let x be a point of X, and let $p \in \mathcal{T}(\{x\})$. Since X is point $\mathcal{T}$-symmetric, $x \in \mathcal{T}(\{p\})$. Hence, by Lemma 7.7.3, $p \in \mathcal{K}(\{x\})$.

Now, let $p \in \mathcal{K}(\{x\})$. Then, by Lemma 7.7.3, $x \in \mathcal{T}(\{p\})$. Since X is point $\mathcal{T}$-symmetric, $p \in \mathcal{T}(\{x\})$. Therefore, $\mathcal{T}(\{x\}) = \mathcal{K}(\{x\})$. □

As a consequence of Theorem 7.7.1 and Corollary 7.7.4, we have the following.

7.7.5 Corollary. *If X is a point $\mathcal{T}$-symmetric continuum, then $\mathcal{K}(W)$ is a continuum for each subcontinuum W of X.*

Proof. Let W be a subcontinuum of X. By Theorem 7.7.1, $\mathcal{K}(W) = \bigcup\{\mathcal{K}(\{w\} \mid w \in W\}$. By Corollary 7.7.4, for every $w \in W$, $\mathcal{K}(\{w\}) = \mathcal{T}(\{w\})$. Hence, $\mathcal{K}(W) = \bigcup\{\mathcal{T}(\{w\}) \mid w \in W\}$. Since each $\mathcal{T}(\{w\})$ is a continuum (Theorem 2.1.27), we obtain that $\mathcal{K}(W)$ is a continuum. □

7.7.6 Theorem. *If X is a point $\mathcal{T}$-symmetric continuum and W is a subcontinuum of X, then $\mathcal{K}(W) \subset \mathcal{T}(W)$.*

Proof. Let X and W be as stated. By Theorem 7.7.1, $\mathcal{K}(W) = \bigcup\{\mathcal{K}(\{w\}) \mid w \in W\}$. By Corollary 7.7.4, $\mathcal{K}(W) = \bigcup\{\mathcal{T}(\{w\}) \mid w \in W\}$. Note that, by Proposition 2.1.7, for each $w \in W$, $\mathcal{T}(\{w\}) \subset \mathcal{T}(W)$. Therefore, $\mathcal{K}(W) \subset \mathcal{T}(W)$. □

The following example is due to David P. Bellamy, and it shows that [9, Theorem 3.22] is not true.

7.7.7 Example. Let X and Y be cones over the Cantor set, Example 2.1.15, such that the intersection of X and Y is a single point p, which is in the base of each, and let $Z = X \cup Y$. Let ν_X and ν_Y be the vertexes of X and Y, respectively. Consider the arcs $\overline{\nu_X p}$ and $\overline{\nu_Y p}$. Note that $\mathcal{T}_Z(\overline{\nu_X p}) = X$. Hence, $\mathrm{Int}(\mathcal{T}_Z(\overline{\nu_X p})) \neq \emptyset$ and $\mathcal{K}_Z(\overline{\nu_X p}) = (\overline{\nu_X p}) \cup (\overline{\nu_Y p})$. Thus, $\mathcal{K}(\overline{\nu_X p}) \not\subset \mathcal{T}_Z(\overline{\nu_X p})$.

7.7.8 Theorem. *Let X be a hereditarily decomposable metric continuum. Then the set $\{x \in X \mid \mathrm{Int}(\mathcal{T}(\{x\})) = \emptyset\}$ is a dense G_δ subset of X.*

Proof. Let $\{x_n\}_{n=1}^{\infty}$ be a countable dense subset of X. Since X is hereditarily decomposable, by [116, Theorem 9], we have that $\mathrm{Int}(\mathcal{K}(\{x\})) = \emptyset$ for all $x \in X$. In particular, $\mathrm{Int}(\mathcal{K}(\{x_n\})) = \emptyset$, for every $n \in \mathbb{N}$. Then $X \setminus \bigcup_{n=1}^{\infty} \mathcal{K}(\{x_n\})$ is a dense G_δ subset of X. Let $x \in X \setminus \bigcup_{n=1}^{\infty} \mathcal{K}(\{x_n\})$. Then $x \in X \setminus \mathcal{K}(\{x_n\})$ for each $n \in \mathbb{N}$. Hence, $x_n \in X \setminus \mathcal{T}(\{x\})$ for all $n \in \mathbb{N}$ (Lemma 7.7.3). Therefore, $\mathrm{Int}(\mathcal{T}(\{x\})) = \emptyset$. □

7.7.9 Corollary. *Let X be a continuum. Then $\mathcal{T}(\{p\}) = \{p\}$, for each $p \in X$, if and only if $\mathcal{K}(\{p\}) = \{p\}$, for every $p \in X$.*

Proof. Suppose $\mathcal{K}(\{p\}) = \{p\}$, for every $p \in X$. Let q be a point of X. For each $p \in X \setminus \{q\}$, $\mathcal{K}(\{p\}) = \{p\}$. Then there exists a subcontinuum W_p of X such that $p \in \mathrm{Int}(W_p)$ and $q \notin W_p$. Hence, $p \notin \mathcal{T}(\{q\})$. Therefore, $\mathcal{T}(\{q\}) = \{q\}$.

Now, assume that $\mathcal{T}(\{p\}) = \{p\}$ for each $p \in X$. Let q be a point in X. For each $p \in X \setminus \{q\}$, $\mathcal{T}(\{p\}) = \{p\}$. Then there exists a subcontinuum W_p of X such that $q \in \mathrm{Int}(W_p) \subset W_p \subset X \setminus \{p\}$. Thus, $p \notin \mathcal{K}(\{q\})$. Therefore, $\mathcal{K}(\{q\}) = \{q\}$. □

7.7.10 Theorem. *If X is a $\mathcal{T}$-symmetric continuum, then $\mathcal{T}(A) \subset \mathcal{K}(A)$, for all nonempty closed subsets A of X.*

Proof. Let A be a nonempty closed subset of X, and let $x \in \mathcal{T}(A)$. Then $\{x\} \cap \mathcal{T}(A) \neq \emptyset$ and, since X is $\mathcal{T}$-symmetric, $A \cap \mathcal{T}(\{x\}) \neq \emptyset$. Hence, there exists $a \in A$

such that $a \in \mathcal{T}(\{x\})$. Thus, if W is a subcontinuum of X such that $A \subset \text{Int}(W)$, we have that $a \in \text{Int}(W)$ and $x \in W$. Therefore, $x \in \mathcal{K}(A)$. □

As a consequence of Theorems 2.2.2 and 7.7.10, we have the following.

7.7.11 Corollary. *Let X be a weakly irreducible continuum. Then for each nonempty closed subset A of X, $\mathcal{T}(A) \subseteq \mathcal{K}(A)$.*

Since each irreducible continuum is weakly irreducible (Theorem 1.4.43), we have the next corollary.

7.7.12 Corollary. *Let X be an irreducible continuum. Then for every nonempty closed subset A of X, $\mathcal{T}(A) \subseteq \mathcal{K}(A)$.*

7.7.13 Definition. A continuum X is *$\mathcal{K}$-symmetric* provided that for each pair of nonempty closed subsets A and B of X, we have that $A \cap \mathcal{K}(B) = \emptyset$ if and only if $\mathcal{K}(A) \cap B = \emptyset$. The continuum X is *$\mathcal{K}$-additive* if for each family Λ of nonempty closed subsets of X whose union is closed in X, $\mathcal{K}\left(\bigcup\{L \mid L \in \Lambda\}\right) = \bigcup\{\mathcal{K}(L) \mid L \in \Lambda\}$.

7.7.14 Theorem. *If X is a $\mathcal{K}$-symmetric continuum, then $\mathcal{K}(A) \subseteq \mathcal{T}(A)$, for each nonempty closed subset A of X.*

Proof. Let A be a nonempty closed subset of X, and let $x \in \mathcal{K}(A)$. Then $\{x\} \cap \mathcal{K}(A) \neq \emptyset$ and, since X is $\mathcal{K}$-symmetric, $A \cap \mathcal{K}(\{x\}) \neq \emptyset$. Hence, there exists $a \in A$ such that $a \in \mathcal{K}(\{x\})$. Thus, if W is a subcontinuum of X such that $x \in \text{Int}(W)$, we have that $a \in W$. As a consequence of this, $x \in \mathcal{T}(\{a\}) \subset \mathcal{T}(A)$ (Proposition 2.1.7). □

7.7.15 Corollary. *If X is a $\mathcal{K}$-symmetric, $\mathcal{T}$-symmetric continuum, then $\mathcal{T}(A) = \mathcal{K}(A)$ for all nonempty closed subsets A of X.*

As a consequence of Theorems 2.2.2 and Corollary 7.7.15, we obtain the following.

7.7.16 Corollary. *If X is a $\mathcal{K}$-symmetric weakly irreducible continuum, then $\mathcal{T}(A) = \mathcal{K}(A)$ for each nonempty closed subset A of X.*

7.7.17 Theorem. *If X is a $\mathcal{K}$-symmetric and $\mathcal{T}$-additive continuum, then $\mathcal{T}(A) = \mathcal{K}(A)$ for every nonempty closed subset A of X.*

Proof. Let A be a nonempty closed subset of X. Since X is $\mathcal{K}$-symmetric, by Theorem 7.7.14, $\mathcal{K}(A) \subset \mathcal{T}(A)$. Let $x \in \mathcal{T}(A) = \bigcup\{\mathcal{T}(\{a\}) \mid a \in A\}$ (Corollary 2.2.13). Then there exists $a \in A$ such that $x \in \mathcal{T}(\{a\})$. By Lemma 7.7.3, $a \in \mathcal{K}(\{x\})$. Since X is $\mathcal{K}$-symmetric, $x \in \mathcal{K}(\{a\}) \subset \mathcal{K}(A)$. □

7.7.18 Theorem. *If X is a $\mathcal{T}$-symmetric and $\mathcal{K}$-additive continuum, then $\mathcal{T}(A) = \mathcal{K}(A)$ for all nonempty closed subsets A of X.*

Proof. Let A be a nonempty closed subset of X. Since X is $\mathcal{T}$-symmetric, by Theorem 7.7.10, $\mathcal{T}(A) \subset \mathcal{K}(A)$. Let $x \in \mathcal{K}(A) = \bigcup\{\mathcal{K}(\{a\}) \mid a \in A\}$ (this equality follows directly from the definition of $\mathcal{K}$-additivity). Then there exists

$a \in A$ such that $x \in \mathcal{K}(\{a\})$. By Lemma 7.7.3, $a \in \mathcal{T}(\{x\})$. Since X is $\mathcal{T}$-symmetric, $x \in \mathcal{T}(\{a\}) \subset \mathcal{T}(A)$. □

The next theorem shows that the reverse inclusion to the one given in Corollary 7.7.11 is true for subcontinua.

7.7.19 Theorem. *Let X be a weakly irreducible continuum. Then $\mathcal{T}(A) = \mathcal{K}(A)$ for each subcontinuum A of X.*

Proof. Let X be a weakly irreducible continuum, and let A be a subcontinuum of X. Let $x \in X \setminus \mathcal{T}(A)$. Then there exists a subcontinuum W of X such that $x \in \mathrm{Int}(W) \subseteq W \subseteq X \setminus A$. Since X is weakly irreducible and W is a subcontinuum of X, then $X \setminus W$ only has finitely many components. Let M be the component of $X \setminus W$ containing A. Since M is connected and open in X, $\mathrm{Cl}(M)$ is a subcontinuum of X such that $A \subseteq M \subseteq \mathrm{Cl}(M)$ and $x \notin \mathrm{Cl}(M)$. Thus, $x \in X \setminus \mathcal{K}(A)$. This implies that $\mathcal{K}(A) \subseteq \mathcal{T}(A)$. The other inclusion follows from Corollary 7.7.11. □

7.7.20 Corollary. *If X is an irreducible continuum, then $\mathcal{T}(A) = \mathcal{K}(A)$ for every subcontinuum A of X.*

7.7.21 Definition. Let X be a continuum, and let A be a subset of X. We say that a point $z \in X$ is a *weak cut point of X that separates A* provided that $z \in M$ for every subcontinuum M of X such that $A \subseteq M$.

The proof of the following lemma follows from Definition 7.7.21.

7.7.22 Lemma. *Let X be a continuum, and let A be a nonempty closed subset of X. If $W_C(A)$ is the set of all weak cut points of X that separate A, then $W_C(A) \subseteq \mathcal{K}(A)$.*

7.7.23 Lemma. *Let X be an irreducible continuum, and let A be a nonempty closed subset of X. If W is a subcontinuum of X such that $\mathcal{T}(A) \subseteq \mathrm{Int}(W)$, then $\mathcal{K}(A) \subseteq \mathrm{Int}(W)$.*

Proof. Let X be an irreducible continuum between p and q. Let A be a nonempty closed subset of X, and let W be a subcontinuum of X such that $\mathcal{T}(A) \subseteq \mathrm{Int}(W)$. Suppose $\mathcal{K}(A) \nsubseteq \mathrm{Int}(W)$, and let $y \in \mathcal{K}(A) \setminus \mathrm{Int}(W)$. Since $\mathcal{T}(A) \subseteq \mathrm{Int}(W)$, $y \in X \setminus \mathcal{T}(A)$. Then there exists a subcontinuum M of X such that $y \in \mathrm{Int}(M) \subseteq M \subseteq X \setminus A$. Since X is irreducible, $X \setminus M$ has at most two components (Theorem 1.4.40). If A is contained in one of these components, say C_1, then $\mathrm{Cl}(C_1)$ is a subcontinuum of X such that $A \subseteq \mathrm{Int}(\mathrm{Cl}(C_1)) \subseteq \mathrm{Cl}(C_1) \subseteq X \setminus \{y\}$, but this contradicts the fact that $y \in \mathcal{K}(A)$. Hence, $X \setminus M = C_1 \cup C_2$ where $A \cap C_j \neq \emptyset$ for $j \in \{1, 2\}$, and we may assume that $p \in C_1$ and $q \in C_2$. Then every point of $\mathrm{Int}(M)$ is a weak cut point of X that separates A. If there exists a point $z \in \mathrm{Int}(M)$, which is not a weak cut point of X that separates A, there exists a subcontinuum R of X such that $A \subseteq R$ and $z \notin R$. But this implies that $\mathrm{Cl}(C_1) \cup R \cup \mathrm{Cl}(C_2)$ is a proper subcontinuum of X containing p and q, which is a contradiction. Hence, by Lemma 7.7.22, $\mathrm{Int}(M) \subseteq W$. Thus, $y \in \mathrm{Int}(W)$, which contradicts our assumption. Therefore, $\mathcal{K}(A) \subset \mathrm{Int}(W)$. □

7.7.24 Theorem. *If X is an irreducible metric continuum, then $\mathcal{T}(\mathcal{K}(A)) = \mathcal{K}(\mathcal{T}(A))$ for each nonempty closed subset A of X.*

Proof. Let A be a nonempty closed subset of X. First, we show that $\mathcal{K}(\mathcal{T}(A)) \subseteq \mathcal{T}(\mathcal{K}(A))$. By Corollary 7.7.12, we have that $\mathcal{T}(A) \subseteq \mathcal{K}(A)$. Hence, $\mathcal{K}(\mathcal{T}(A)) \subseteq \mathcal{K}(\mathcal{K}(A))$. By [43, Theorem 3.13], $\mathcal{K}(A)$ is connected and, by Corollary 7.7.20, $\mathcal{K}(\mathcal{K}(A)) = \mathcal{T}(\mathcal{K}(A))$. Therefore, $\mathcal{K}(\mathcal{T}(A)) \subseteq \mathcal{T}(\mathcal{K}(A))$.

For the other inclusion, let $x \in X \setminus \mathcal{K}(\mathcal{T}(A))$. Then there exists a subcontinuum W of X such that $\mathcal{T}(A) \subseteq \mathrm{Int}(W)$ and $x \notin W$. By Lemma 7.7.23, $\mathcal{K}(A) \subseteq \mathrm{Int}(W)$. Since X is irreducible, $X \setminus W$ has at most two components (Theorem 1.4.40). Let C be the component of $X \setminus W$ containing x. Then $\mathrm{Cl}(C)$ is a subcontinuum of X such that $x \in \mathrm{Int}(\mathrm{Cl}(C)) \subseteq \mathrm{Cl}(C) \subseteq X \setminus \mathcal{K}(A)$. Thus, $x \in X \setminus \mathcal{T}(\mathcal{K}(A))$. Hence, $\mathcal{T}(\mathcal{K}(A)) \subset \mathcal{K}(\mathcal{T}(A))$. Therefore, $\mathcal{K}(\mathcal{T}(A)) = \mathcal{T}(\mathcal{K}(A))$. □

References for Chapter 7

Section 7.1: [38, 45, 80, 85, 88, 92, 93, 105, 119, 126].
Section 7.2: [7, 22].
Section 7.3: [26–28, 42, 47].
Section 7.4: [12, 50, 51, 70, 92, 97, 104–106, 128].
Section 7.5: [76, 107].
Section 7.6: [48, 65, 78].
Section 7.7: [9, 28, 43, 53, 103, 116].

Chapter 8
Questions

8.1 Questions About the Set Function $\mathcal{T}$

We include the questions on the set function $\mathcal{T}$ from [92]. We write new comments and present new questions. We include the author and the reference in which the question is asked at the end of each question.

8.1.1 Question. If the set function $\mathcal{T}$ is continuous for the continuum X, is it true that X is $\mathcal{T}$-additive? (D. P. Bellamy [25])

Comment: A bushel of *Extra Fancy Stayman Winesap apples* for the solution. (D. P. Bellamy)

Comment: Let us observe that, since for continua X for which $\mathcal{T}$ is continuous, being $\mathcal{T}$-additive is equivalent to being point $\mathcal{T}$-symmetric (Corollary 5.1.16), by Theorem 3.2.9, Question 8.1.1 is equivalent to Question 8.1.2. (S. Macías)

Comment: A positive answer is given by J. Camargo and C. Uzcátegui in [21] for the metric case. Professor Bellamy decided that the work done in [21] was worth obtaining the bushel of *Extra Fancy Stayman Winesap apples*. A positive answer for the general case is in [90], see Theorem 5.1.10. (S. Macías)

8.1.2 Question. If $\mathcal{T}$ is continuous for the continuum X, is it true that the collection $\{\mathcal{T}(\{p\}) \mid p \in X\}$ is a continuous decomposition of X such that the quotient space is locally connected? (D. P. Bellamy [25])

Comment: A positive answer is given by J. Camargo and C. Uzcátegui in [21] for the metric case, a positive answer for the general case is in [90], see Theorem 3.2.9. (S. Macías)

8.1.3 Question. If X and Y are indecomposable continua, is $\mathcal{T}$ idempotent on $X \times Y$? Even for only closed sets in $X \times Y$? (D. P. Bellamy [25])

S. Macías, *Set Function $\mathcal{T}$*, Developments in Mathematics 67,
https://doi.org/10.1007/978-3-030-65081-0_8

Comment: A negative answer to the first question is presented in [86], see Corollary 2.3.23. (S. Macías)

8.1.4 Question. If $\mathcal{T}$ is continuous for X and X is a decomposable continuum, is it true that for each $p \in X$, $\text{Int}(\mathcal{T}(\{p\})) = \emptyset$? (D. P. Bellamy [25])

Comment: As a consequence of Theorem 3.2.9, this question has a positive answer. (S. Macías)

8.1.5 Question. Let X be a continuum. If $X/\mathcal{T}$ denotes the finest decomposition space of X which shrinks each $\mathcal{T}(\{p\})$ to a point, is $X/\mathcal{T}$ locally connected? (D. P. Bellamy [77])

Comment: The answer is affirmative if it is assumed that X is $\mathcal{T}$-additive.

8.1.6 Question. If X is an indecomposable continuum and W is a subcontinuum of $X \times X$ with nonempty interior, is $\mathcal{T}(W) = X \times X$? (F. B. Jones [77])

Comment: C. L. Hagopian [56] has shown that this question has an affirmative answer if X is a metric chainable indecomposable continuum. (S. Macías)

8.1.7 Question. If X is an atriodic continuum (or X contains no uncountable collection of pairwise disjoint triods) and X does not have a weak cut point, then is there a continuum $W \subset X$ such that $\text{Int}(W) \neq \emptyset$ and $\mathcal{T}(W) \neq X$? (H. Cook [77])

Comment: Let X be a continuum, and let x be a point of X. We say that x is a *weak cut point* of X provided that there exist two points y and z in X such that if W is a subcontinuum of X and $\{y, z\} \subset W$, then $x \in W$.

8.1.8 Question. If $\mathcal{T}$ is continuous for the continuum X and $f\colon X \twoheadrightarrow Z$ is a continuous and monotone surjection, is $\mathcal{T}$ continuous for Z also? (D. P. Bellamy [77])

Comment: A negative answer is given by A. Illanes and R. Leonel in [64]. (S. Macías)

8.1.9 Question. If X is a strictly point $\mathcal{T}$-asymmetric dendroid, then is X smooth? (D. P. Bellamy [77])

Comment: A continuum X is *strictly point $\mathcal{T}$-asymmetric* if for any two distinct points p and q of X with $p \in \mathcal{T}(\{q\})$, we have that $q \notin \mathcal{T}(\{p\})$.

Comment: L. Fernández has given a negative answer [42], see Example 7.3.6. (S. Macías)

8.1.10 Question. Let X be a continuum. Suppose the restriction of $\mathcal{T}$ to the hyperspace of subcontinua of X is continuous. Does this imply that $\mathcal{T}$ is continuous for X? (D. P. Bellamy [6])

Comment: A positive partial answer to this question is presented in Theorem 2.1.38. Another positive partial answer for the product of two metric continua is given in [19], see Theorem 5.3.6. (S. Macías)

8.1.11 Question. Do open maps preserve $\mathcal{T}$-additivity? $\mathcal{T}$-symmetry? (D. P. Bellamy [77])

8.1.12 Question. Is there a metric θ-continuum X, that is not a metric θ_1-continuum, such that $\text{Int}(\mathcal{T}(\{x\})) = \emptyset$ and $\mathcal{T}^2(\{x\}) = X$ for each x in X? (E. E. Grace and E. J. Vought [54])

8.1.13 Question. Are Theorem 2.3.27 and Corollary 2.3.28 true for continua with span zero? (S. Macías [86])

Comment: A continuum X has *span zero* provided that for each subcontinuum Z of $X \times X$ such that $\pi_1(Z) = \pi_2(Z)$, where $\pi_2, \pi_1 \colon X \times X \twoheadrightarrow X$ are the projection maps, we have that $Z \cap \Delta_{X^2} \neq \emptyset$.

8.1.14 Question. Does there exist a characterization of the idempotency of $\mathcal{T}$ on closed sets for continua? (S. Macías [87])

8.1.15 Question. Is Theorem 4.1.9 true when K is closed and $\dim(K) = 0$? (S. Macías [92])

8.1.16 Question. Is Theorem 4.1.10 true when K is closed and $\dim(K) = 0$? (S. Macías [92])

8.1.17 Question. If X is a continuum, then is $\mathfrak{T}(X)$ an F_σ subset of 2^X? (D. P. Bellamy, L. Fernández and S. Macías [9])

8.1.18 Question. Let X and Y be continua, and let $f \colon X \twoheadrightarrow Y$ be a monotone map. When is it true that $|\mathfrak{T}(Y)| = |\mathfrak{T}(X)|$? (D. P. Bellamy, L. Fernández and S. Macías [9])

Comment: A partial positive answer is known for atomic maps [89], see Theorem 4.1.32 (S. Macías).

8.1.19 Question. Is the converse of Theorem 4.2.3 true for a continuum X for which $\mathcal{T}$ is idempotent? (D. P. Bellamy, L. Fernández and S. Macías [9])

Comment: H. Abobaker and W. J. Charatonik gave a negative answer to this question [1], see Example 4.2.6. (S. Macías)

8.1.20 Question. Let X be a metric continuum. Does there exist a countable ordinal α such that $\mathcal{T}^\infty = \mathcal{T}^\alpha$? (D. P. Bellamy, L. Fernández and S. Macías [9])

8.1.21 Question. For what type of continuum X are the equations of Theorems 2.4.14 and 2.4.15 true for all subsets K and L of X of a certain type? (H. S. Davis, D. P. Stadtlander and P. M. Swingle [32])

8.1.22 Question. For what kind of continuum X are the equations of Theorems 2.4.14 and 2.4.15 true for some subsets K and L of X of a certain type? (H. S. Davis, D. P. Stadtlander and P. M. Swingle [32])

8.1.23 Question. Let X be a continuum and let A be a subset of X, if there exists $n \in \mathbb{N}$ such that $\mathcal{T}^n(A) = \mathcal{T}^{n-1}(A)$, then does there exists $m \in \{1, \ldots, n\}$ such that $\mathcal{K}^m(A) = \mathcal{K}^{m-1}(A)$? (S. Gorka [53])

8.1.24 Question. If X is an ω-indecomposable continuum, then does it follow that there exists only one collection of indecomposable subcontinua of X satisfying Proposition 6.2.12? (J. Camargo, S. Macías and C. Uzcátegui [20])

8.1.25 Question. If X is an ω-indecomposable continuum, then does there exist a family of indecomposable continua $\{M_k\}_{k=1}^{\infty}$ such that there exists a composant C_k of M_k, for each $k \in \mathbb{N}$, where $C_k \cap \mathrm{Cl}_X\left(\bigcup_{j \neq k} M_j\right) = \emptyset$? (J. Camargo, S. Macías and C. Uzcátegui [20])

8.1.26 Question. What necessary or sufficient conditions must a continuum X have in order that $\mathcal{T}(2^X)$ be connected?, or $\mathcal{T}(2^X)$ be compact? or $\mathcal{T}(2^X)$ be a continuum? (J. Camargo, S. Macías and C. Uzcátegui [20])

8.1.27 Question. Let X be a continuum for which there exists $m \in \mathbb{N}$ such that $\mathcal{T}^m(A) = \mathcal{T}^{m-1}(A)$ for each $A \in \mathcal{C}_\infty(X) = \bigcup_{n=1}^{\infty} \mathcal{C}_n(X)$. Then does it follow that $\mathcal{T}^m(B) = \mathcal{T}^{m-1}(B)$ for every $B \in 2^X$? (J. Camargo, S. Macías and M. Ruiz [19])

8.1.28 Question. What kind of continua have the property that the idempotency on continua implies the idempotency on closed sets? (J. Camargo, S. Macías and M. Ruiz [19])

Comment: A partial answer is given by J. Camargo, S. Macías and M. Ruiz in [19], see Theorem 2.3.37. (S. Macías)

8.1.29 Question. Let $Z = X \times Y$ where X and Y are continua. If $\mathcal{T}_Z(\mathcal{C}_1(Z))$ is compact, then does it follow that Z is locally connected? (J. Camargo, S. Macías and M. Ruiz [19])

8.1.30 Question. Given a continuum X, is $\mathcal{T}(2^X)$ a Borel set of 2^X? (J. Camargo, S. Macías and M. Ruiz [19])

Comment: The class of *Borel sets* of a continuum is the σ-algebra generated by the open subsets of X. A partial positive answer is given in Corollary 6.1.5.

8.1.31 Question. Is it true that if A is a shore set of a dendroid, then $\mathrm{Int}(\mathcal{T}(A)) = \emptyset$? (R. Leonel [76])

8.1.32 Question. For a closed subset A of a metric continuum X, when is $\mathcal{T}(A)$ a shore set of X? (R. Leonel [76])

8.1.33 Question. For a shore set A of a metric continuum X, when is $\mathcal{T}(A)$ a shore set of X? (R. Leonel [76])

References

1. H. Abobaker, W. J. Charatonik, $\mathcal{T}$ closed Sets, Topology Appl., 222 (2017), 274–277.
2. L. V. Ahlfors, *Conformal Invariants, Topics in Geometric Function Theory*, Series in Higher Mathematics, McGraw-Hill Book Co., New York, 1973.
3. D. P. Bellamy, Continua for Which the Set Function $\mathcal{T}$ is Continuous, Trans. Amer. Math. Soc., 1511 (1970), 581–587.
4. D. P. Bellamy, Set Functions and Continuous Maps, in *General Topology and Modern Analysis*, (L. F. McAuley and M. M. Rao, eds.), Academic Press, (1981), 31–38.
5. D. P. Bellamy, Some Topics in Modern Continua Theory, in *Continua Decompositions Manifolds*, (R H Bing, W. T. Eaton and M. P. Starbird, eds.), University of Texas Press, (1983), 1–26.
6. D. P. Bellamy, Questions In and Out of Context, talk given in the VI Joint AMS-SMM Meeting, Houston, TX, May 13–15, 2004 (LaTeX edition by Jonnthan Hatch.)
7. D. P. Bellamy and J. J. Charatonik, The Set Function $\mathcal{T}$ and Contractibility of Continua, Bull. Acad. Polon. Sci. Sér. Sci. Math. Astronom. Phys., 25 (1977), 47–49.
8. D. P. Bellamy and H. S. Davis, Continuum Neighborhoods and Filterbases, Proc. Amer. Math. Soc., 27 (1971), 371–374.
9. D. P. Bellamy, L. Fernández and S. Macías, On $\mathcal{T}$-closed Sets, Topology Appl., 195 (2015), 209–225. (Special issue honoring the memory of Professor Mary Ellen Rudin.)
10. D. P. Bellamy and C. L. Hagopian, Mapping continua onto their cones, Colloquium Math., 41 (1979), 53–56.
11. D. P. Bellamy and L. Lum, The Cyclic Connectivity of Homogeneous Arcwise Connected Continua, Trans. Amer. Math. Soc., 266 (1981), 389–396.
12. D. E. Bennet, A Characterization of Locally Connectedness By Means of the Set Function $\mathcal{T}$, Fund. Math., 86 (1974), 137–141.
13. R H Bing, Concerning Hereditarily Indecomposable Continua, Pacific J. Math., 1 (1951), 43–51.
14. R H Bing and F. B. Jones, Another Homogeneous Plane Continuum, Trans. Amer. Math. Soc., 90 (1959), 171–192.
15. K. Borsuk and S. Ulam, On Symmetric Products of Topological Spaces, Bull. Amer. Math. Soc., 37 (1931), 875–882.
16. C. E. Burgess, Continua and Their Complementary Domains in the Plane, II, Duke Math. J., 19 (1952), 223–230.
17. C. E. Burgess, Continua Which are the Sum of a Finite Number of Indecomposable Continua, Proc. Amer. Math. Soc., 4 (1953), 234–239.

S. Macías, *Set Function* $\mathcal{T}$, Developments in Mathematics 67,
https://doi.org/10.1007/978-3-030-65081-0

18. C. E. Burgess, Separation Properties and n-indecomposable Continua, Duke Math. J., 23 (1956), 595–599.
19. J. Camargo, S. Macías and M, Ruiz, Some Aspects Related to the Jones' Set Function $\mathcal{T}$, Topology Appl., 266 (2019), 106835.
20. J. Camargo, S. Macías, C. Uzcátegui, On the Images of Jones' Set Function $\mathcal{T}$, Colloquium Math., 153 (2018), 1–19.
21. J. Camargo, C. Uzcátegui, Continuity of the Jones' Set Function $\mathcal{T}$, Proc. Am. Math. Soc., 145 (2017), 893–899.
22. J. J. Charatonik, The Set Function $\mathcal{T}$ and Homotopies, Colloquium Math., 39 (1978), 271–274.
23. W. J. Charatonik, A Homogeneous Continuum Without the Property of Kelley, Topology Appl., 96 (1999), 209–216.
24. C. O. Christenson and W. L. Voxman, *Aspects of Topology*, Monographs and Textbooks in Pure and Applied Math., Vol. 39, Marcel Dekker, New York, Basel, 1977.
25. H. Cook, W. T. Ingram, and A. Lelek, *A List of Problems Known as Houston Problem Book*, in *Continua with The Houston Problem Book*, H. Cook, W. T. Ingram, K. T. Kuperberg, A. Lelek, and P. Minc, editors. Lecture Notes in Pure and Applied Mathematics, Vol. 170, Marcel Dekker, New York, Basel, Hong Kong (1995), 365–398.
26. S. T. Czuba, The Set Function $\mathcal{T}$ and R-continuum. Bull. Acad. Polon. Sci. Ser. Sci. Math., 27 (1979), 303–308.
27. S. T. Czuba, The Concept of Pointwise Smooth Dendroids, Uspekhi Mat. Nauk., 34 (1979), 215–217.
28. S. T. Czuba, Some Other Characterizations of Pointwise Smooth Dendroids, Comment. Math. Prace Mat., 24 (1984), 195–200.
29. H. S. Davis, A Note on Connectedness im Kleinen, Proc. Amer. Math. Soc., 19 (1968), 1237–1241.
30. H. S. Davis, Relationships Between Continuum Neighborhoods in Inverse Limit Spaces and Separations in Inverse Limit Sequences, Proc. Amer. Math. Soc., 64 (1977), 149–153.
31. H. S. Davis and P. H. Doyle, Invertible Continua, Portugal. Math., 26 (1967), 487–491.
32. H. S. Davis, D. P. Stadtlander and P. M. Swingle, Properties of the Set Functions $\mathcal{T}^n$, Portugal. Math., 21 (1962), 113–133.
33. H. S. Davis, D. P. Stadtlander and P. M. Swingle, Semigroups, Continua and the Set Functions $\mathcal{T}^n$, Duke Math. J., 29 (1962), 265–280.
34. R. F. Dickman, L. R. Rubin and P. M. Swingle, Characterization of n-spheres by an Excluded Middle Membrane Principle, Michigan Math. J., 11 (1964), 53–59.
35. C. H. Dowker, Mapping Theorems for Non-compact Spaces, Amer. J. Math., 69 (1947), 200–242.
36. J. Dugundji, *Topology,* Allyn and Bacon, Inc., Boston, London, Sydney, Toronto, 1966.
37. E. Dyer and M. E. Hamstrom, Completely Regular Mappings, Fund. Math., 45 (1958), 103–118.
38. C. Eberhart and S. B. Nadler, Jr., Irreducible Whitney Levels, Houston J. Math., 6 (1980), 355–363.
39. A. Emeryk and Z. Horbanowicz, On Atomic Mappings, Colloquium Math., 27 (1973), 49–55.
40. R. Engelking, *General Topology*, Sigma Series in Pure Mathematics, Vol. 6, Heldermann, Berlin, 1989.
41. R. Fenn, What is the Geometry of a Surface?, The Amer. Math. Monthly, 90 (1983), 87–98.
42. L. Fernández, On Strictly Point $\mathcal{T}$-asymmetric Continua, Topology Proc., 35 (2010), 91–96.
43. L. Fernández and S. Macías, The Set Functions $\mathcal{T}$ and $\mathcal{K}$ and Irreducible Continua, Colloquium Math., 121 (2010), 79–91.
44. R. W. FitzGerald, The Cartesian Product of Non-degenerate Compact Continua is n-point Aposyndetic, *Topology Conference* (Arizona State Univ., Tempe, Ariz., 1967), Arizona State University, Tempe, Ariz., (1968), 324–326.
45. R. W. FitzGerald, Connected Sets With a Finite Disconnection Property, in *Studies in Topology* (N. M. Stavrakas and K. R. Allen, Eds.), Academic Press, (1974), 139–173.

46. R. W. FitzGerald and P. M. Swingle, Core Decompositions of Continua, Fund. Math., 61 (1967), 33–50.
47. J. B. Fugate, G. R. Gordh, Jr. and L. Lum, Arc-smooth Continua, Trans. Amer. Math. Soc., 265 (1981), 545–561.
48. C. Good and S. Macías, Symmetric Products of Generalized Metric Spaces, Topology Appl., 206 (2016), 93–114.
49. C. Good and S. Macías, What is Topological About Topological Dynamics? Discrete Contin. Cyn. Syst., 38 (2018), 1007–1031.
50. J. T. Goodykoontz, Jr., Aposyndetic Properties of Hyperspaces, Pacific J. Math., 47 (1973), 91–98.
51. J. T. Goodykoontz, Some Functions on Hyperspaces of Hereditarily Unicoherent Continua, Fund. Math., 95 (1977), 1–10.
52. G. R. Gordh, Jr. and C. B. Hughes, On Freely Decomposable Mappings of Continua, Glasnik Math., 14 (34) (1979), 137–146.
53. S. Gorka, *Several Set Functions and Continuous Maps*, Ph. D. Dissertation, University of Delaware, 1997.
54. E. E. Grace and E. J. Vought, Monotone Decompositions of θ_n-continua, Trans. Amer. Math. Soc., 263 (1981), 261–270.
55. S. Greenwood and S. Macías, A Note on Inverse Limits with Upper Semicontinuous Functions, preprint.
56. C. L. Hagopian, Mutual Aposyndesis, Proc. Amer. Math. Soc., 23 (1969), 615–622.
57. C. L. Hagopian, A Cut Point Theorem for Plane Continua, Duke Math. J., 38 (1971), 509–512.
58. C. L. Hagopian, Schlais' Theorem Extends to λ Connected Plane Continua, Proc. Amer. Math. Soc., 40 (1973), 265–267.
59. C. L. Hagopian, Characterizations of λ Connected Plane Continua, Pacific J. Math., 49 (1973), 371–375.
60. J. G. Hocking and G. S. Young, *Topology*, Dover Publications, Inc., New York, 1988.
61. S. T. Hu, *Theory of Retracts*, Wayne State University Press, Detroit, 1965.
62. R. P. Hunter, On the semigroup structure of continua, Trans. Amer. Math. Soc., 93 (1959), 356–368.
63. W. Hurewicz and H. Wallman, *Dimension theory*, Princeton Univ. Press, Princeton, NJ, 1948.
64. A. Illanes and R. Leonel, Continuity of Jones' Function is not Preserved Under Monotone Mappings, Topology Appl., 231 (2017), 136–158.
65. W. T. Ingram and W. S. Mahavier, Inverse limits of upper semicontinuous set valued functions, Houston J. Math., 32 (2006), 119–130.
66. F. B. Jones, Concerning the Boundary of a Complementary Domain of a Continuous Curve, Bull. Amer. Math. Soc., 45 (1939), 428–435.
67. F. B. Jones, Aposyndetic Continua and Certain Boundary Problems, Amer. J. Math., 53 (1941), 545–553.
68. F. B. Jones, Concerning Nonaposyndetic Continua, Amer. J. Math., 70 (1948), 403–413.
69. A. Kechris, *Classical Descriptive Set Theory*, Graduate Texts in Mathematics 156, Springer-Verlag, 1995.
70. J. L. Kelley, *General Topology*, Graduate Texts in Mathematics 27, Springer-Verlag, 1991.
71. R. J. Koch, A Note on Weak Cutpoints in Clans, Duke Math. J., XXIV (1957), 611–616.
72. R. J. Koch and A. D. Wallace, Admissibility of Semigroups Structures on Continua, Trans. Amer. Math. Soc., 88 (1958), 277–287.
73. J. Krasinkiewicz, On Two Theorems of Dyer, Colloquium Math., 50 (1986), 201–208.
74. K. Kuratowski, *Topology*, Vol I, Academic Press, New York, N. Y., 1966.
75. K. Kuratowski, *Topology*, Vol. II, Academic Press, New York, N. Y., 1968.
76. R. Leonel, Shore and Center Points of a Continuum, Topology Proc., 46 (2015), 205–212.
77. W. Lewis, Continuum Theory Problems, Topology Proc., 8 (1983), 361–394.
78. W. Lewis, Continuous Curves of Pseudo-arcs, Houston J. Math., 11 (1985), 91–99.
79. S. Macías, On Symmetric Products of Continua, Topology Appl., 92 (1999), 173–182.

80. S. Macías, Aposyndetic Properties of Symmetric Products of Continua, Topology Proc., 22 (1997), 281–296.
81. S. Macías, Conncectedness of the Hyperspace of Closed Subsets With at Most n Components, Q. & A. in General Topology, 19 (2001), 133–138.
82. S. Macías, A Class of One-dimensional Nonlocally Connected Continua for Which the Set Function $\mathcal{T}$ is Continuous, Houston J. Math., 32 (2006), 161–165.
83. S. Macías, Homogeneous Continua for Which the Set Function $\mathcal{T}$ is Continuous, Topology Appl., 153 (2006), 3397–3401.
84. S. Macías, A Decomposition Theorem for a Class of Continua for Which the Set Function $\mathcal{T}$ is Continuous, Colloquium Math., 109 (2007), 163–170.
85. S. Macías, On Continuously Irreducible Continua, Topology Appl., 156 (2009), 2357–2363.
86. S. Macías, On the Idempotency of the Set Function $\mathcal{T}$, Houston J. Math., 37 (2011), 1297–1305.
87. S. Macías, A Note on The Set Functions $\mathcal{T}$ and $\mathcal{K}$, Topology Proc., 46 (2015), 55–65.
88. S. Macías, On Continuously Type A' θ-continua, J. P. Journal of Geometry and Topology, 18 (2015), 1–14.
89. S. Macías, Atomic Maps and $\mathcal{T}$-closed Sets, Topology Proc., 50 (2017), 97–100.
90. S. Macías, On Jones' set function $\mathcal{T}$ and the property of Kelley for Hausdorff continua, Topology Appl., 226 (2017), 51–65.
91. S. Macías, Hausdorff Continua and the Uniform Property of Effros, Topology Appl., 230 (2017), 338–352.
92. S. Macías, *Topics on Continua, 2nd Edition*, Springer-Cham, 2018.
93. S. Macías y S. B. Nadler, Jr., Z-sets in Hyperspaces, Q. & A. in General Topology, 19 (2001), 227–241.
94. S. Macías and S. B. Nadler, Jr., On Hereditarily Decomposable Homogeneous Continua, Topology Proc., 34 (2009), 131–145.
95. W. Makuchowski, On Local Connectedness in Hyperspaces, Bull. Pol. Acad. Sci., 47 (1999), 119–126.
96. A. Martínez Rodríguez, *Propiedades de la función $\mathcal{T}$ de Jones y otras funciones tipo conjunto*, Master's Thesis, Facultad de Ciencias, U. A. E. Mex., (2018) (Spanish)
97. A. McCluskey and B. McMaster, *Undergraduate Topology*, Oxford University Press, 2014.
98. E. Michael, Topologies on Spaces of Subsets, Trans. Amer. Math. Soc., 71 (1951), 152–182.
99. J. van Mill, *Infinite-Dimensional Topology*, North Holland, Amsterdam, 1989.
100. L. Mohler and L. G. Oversteegen, On the Structure of Tranches in Continuously Irreducible Continua, Colloquium Math., 54 (1987), 23–28.
101. M. A. Molina, *Algunos Aspectos Sobre la Función $\mathcal{T}$ de Jones*, Bachelor's Thesis, Facultad de Ciencias, U. N. A. M., 1998. (Spanish)
102. R. L. Moore, *Foundations of Point Set Theory*, Rev. Ed., Amer. Math. Soc. Colloq. Publ., Vol. 13, Amer. Math. Soc., Providence, Rhode Island, 1962.
103. W. T. Moreland, Jr. *Some Properties of Four Set Valued Set Functions*, Master's Thesis, University of Delaware, 1970.
104. S. Mrówka, On the Convergence of Nets of Sets, Fund. Math., 45 (1958), 237–246.
105. S. B. Nadler, Jr., *Hyperspaces of Sets*, Monographs and Textbooks in Pure and Applied Math., Vol. 49, Marcel Dekker, New York, Basel, 1978. Reprinted in: Aportaciones Matemáticas de la Sociedad Matemática Mexicana, Serie Textos # 33, 2006.
106. S. B. Nadler, Jr., *Continuum Theory: An Introduction,* Monographs and Textbooks in Pure and Applied Math., Vol. 158, Marcel Dekker, New York, Basel, Hong Kong, 1992.
107. V. C. Nall, Centers and Shore Points of a Dendroid, Topology Appl., 154 (2007), 2167–2172.
108. A. Pełczyński, A Remark on Spaces 2^X for Zero-dimensional X, Bull. Pol. Acad. Sci. Math. Astr. Phys., 13 (1965), 85–89.
109. J. R. Prajs, A Homogeneous Arcwise Connected Non-locally-connected Curve, Amer. J. Math., 124 (2002), 649–675.
110. J. R. Prajs, Mutually Aposyndetic Decomposition of Homogeneous Continua, Canad. J. Math., 62 (2010), 182–201.

111. T. M. Reagan, *Quotient Uniformities*, Master Thesis, Naval Posrgraduate School, 1970.
112. J. T. Rogers, Jr., Orbits of Higher–dimensional Hereditarily Indecomposable Continua, Proc. Amer. Math. Soc., 95 (1985), 483–486.
113. J. T. Rogers, Jr., Hyperbolic Ends And Continua, Michigan Math. J., 34 (1987), 337–347.
114. J. T. Rogers, Jr., Decompositions of Continua Over the Hyperbolic Plane, Trans. Amer. Math. Soc., 310 (1988), 277–291.
115. J. T. Rogers, Jr., Higher Dimensional Aposyndetic Decompositions, Proc. Amer. Math. Soc., 131 (2003), 3285–3288.
116. H. E. Schlais, Non-aposyndetic and Non-hereditary Decomposability, Pac. J. Math., 45 (1973), 643–652.
117. P. Scott, The Geometries of 3-manifolds, Bull. London Math. Soc., 15 (1983), 401–487.
118. D. Stadtlander, Further Properties of the Set Function $\mathcal{T}$, J. Natur. Sci. and Math., 7 (1967), 91–94.
119. E. S. Thomas, Jr., Monotone Decompositions of Irreducible Continua, Dissertationes Math. (Rozprawy Mat.), 50 (1966), 1–74.
120. E. L. VandenBoss, *Set Functions and Local Connectivity*, Ph. D. Dissertation, Michigan State University 1970.
121. K. Villarreal, Fibered Products of Homogeneous Continua, Trans. Amer. Math. Soc., 338 (1993), 933–939.
122. E. J. Vought, Monotone Decompositions into Trees of Hausdorff Continua Irreducible About a Finite Subset, Pacific J. Math., 54 (1974), 253–261.
123. E. J. Vought, Monotone Decompositions of Hausdorff Continua, Proc. Amer. Math. Soc., 56 (1976), 371–376.
124. E. J. Vought, ω-connected Continua and Jones' $\mathcal{K}$ Function, Proc. Amer. Math. Soc., 91 (1984), 633–636.
125. A. D. Wallace, The Position of C-sets in Semigroups, Proc. Amer. Math. Soc., 6 (1955), 639–642.
126. L. E. Ward, Extending Whitney Maps, Pacific J. Math., 93 (1981), 464–469.
127. G. T. Whyburn, Semi-locally-connected Sets, Amer. J. Math., 61 (1939), 733–749.
128. G. T. Whyburn, *Analytic Topology,* Amer. Math. Soc. Colloq. Publ., Vol. 28, Amer. Math. Soc., Providence, R. I., 1942.
129. S. Willard, *General Topology*, Addison-Wesley Publishing Co., 1970.

Index

S. Macías, *Set Function $\mathcal{T}$*, Developments in Mathematics 67,
https://doi.org/10.1007/978-3-030-65081-0

www.ingramcontent.com/pod-product-compliance
Ingram Content Group UK Ltd.
Pitfield, Milton Keynes, MK11 3LW, UK
UKHW021533300726
14060UKWH00011B/393

9783030650803